*f*P

Abundance

The Future Is Better Than You Think

PETER H. DIAMANDIS
AND STEVEN KOTLER

Free Press
New York London Toronto Sydney New Delhi

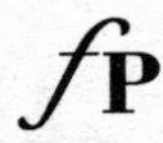

Free Press
A Division of Simon & Schuster, Inc.
1230 Avenue of the Americas
New York, NY 10020

First Free Press hardcover edition February 2012

Designed by Maura Fadden Rosenthal/Mspace

Manufactured in the United States of America

15 17 19 20 18 16

Library of Congress Cataloging-in-Publication Data

Diamandis, Peter H.
Abundance : the future is better than you think /
Peter H. Diamandis and Steven Kotler. — 1st ed.
p. cm.
1. Technological innovations—Forecasting. 2. Technological forecasting.
3. Technology—Social aspects. I. Kotler, Steven, 1967– II. Title.
T173.8.D536 2012
303.48'3—dc23 2011039926

ISBN 978-1-4516-1421-3
ISBN 978-1-4516-1684-2 (ebook)

Peter's Dedication

During the writing of this book, my wife, Kristen, gave birth to our two sons, Jet James Diamandis and Daxton Harry Diamandis. It is to her, and to them, that I dedicate this book. May Dax and Jet live in a world of true Abundance.

Steven's Dedication

When I was younger it was a quintet of men who taught me the importance of dreaming big: Daniel Kamionkowski, Joshua Lauber, Steve Peppercorn, Howard Shack, and Michael Wharton. When I was older it was a trio of women who taught me how hard one has to fight to make those dreams a reality: my wife, Joy Nicholson; Dr. Kathleen Ramsey; and Dr. Patricia Wright. It's to all of you I dedicate this book.

Contents

PART FOUR: THE FORCES OF ABUNDANCE

PART FIVE: PEAK OF THE PYRAMID

PART SIX: STEERING FASTER

A Note from the Authors

A Historical Perspective

These are turbulent times. A quick glance at the headlines is enough to set anybody on edge and—with the endless media stream that has lately become our lives—it's hard to get away from those headlines. Worse, evolution shaped the human brain to be acutely aware of all potential dangers. As will be explored in later chapters, this dire combination has a profound impact on human perception: It literally shuts off our ability to take in good news.

This creates something of a challenge for us, as *Abundance* is a tale of good news. At its core, this book examines the hard facts, the science and engineering, the social trends and economic forces that are rapidly transforming our world. But we are not so naïve as to think that there won't be bumps along the way. Some of those will be big bumps: economic meltdowns, natural disasters, terrorist attacks. During these times, the concept of abundance will seem far-off, alien, even nonsensical, but a quick look at history shows that progress continues through the good times and the bad.

The twentieth century, for example, witnessed both incredible advancement and unspeakable tragedy. The 1918 influenza epidemic killed fifty million people, World War II killed another sixty million. There were tsunamis, hurricanes, earthquakes, fires, floods, even plagues of locust. Despite such unrest, this period also saw infant mortality decrease by 90 percent, maternal mortality decrease by 99 percent, and, overall, human lifespan increase by more than 100 percent. In the past two decades, the United States has experienced tremendous economic upheaval. Yet today, even the poorest Americans have access to a telephone, television, and a flush toilet—three luxuries that even the wealthiest couldn't imagine at the turn of the last century. In fact, as will soon be clear, using almost any metric currently available, quality of life has improved more in the past century than ever before. So while there are likely to be plenty of rude, heartbreaking interruptions along the way, as this book will demonstrate, global living standards will continue to improve regardless of the horrors that dominate the headlines.

Why You Should Care

This is a book about improving global living standards and the standards that need the most help are those found in the developing world. This raises a second question. For those of us living in the developed world, why should we care? After all, there are plenty of important issues facing us here at home. Both US unemployment rates and foreclosure rates are soaring, so humanitarian reasons aside, should we really waste our time working toward an age of global abundance?

The short answer is yes. Our days of isolation are behind us. In today's world, what happens "over there" impacts "over here." Pandemics do not respect borders, terrorist organizations operate on a global scale, and overpopulation is everybody's problem. What's the best way to solve these issues? Raise global standards of living. Research shows that the wealthier, more educated, and healthier a nation, the less violence and civil unrest among its populace, and the less likely that unrest will spread across its borders. As such, stable governments are better prepared to stop an infectious disease outbreak before it becomes a global pandemic. And, as a bonus, there is a direct correlation between quality of life and population growth rates—as quality increases, birth rates decrease. The point is this: In today's hyperlinked world, solving problems anywhere, solves problems everywhere.

Moreover, the greatest tool we have for tackling our grand challenges is the human mind. The information and communications revolution now underway is rapidly spreading across the planet. Over the next eight years, three billion new individuals will be coming online, joining the global conversation, and contributing to the global economy. Their ideas—ideas we've never before had access to—will result in new discoveries, products, and inventions that will benefit us all.

A Collaboration of Two Minds

Peter and Steven first met in 2000, when Steven wrote an article on the X PRIZE for *GQ* magazine. Peter enjoyed Steven's writing style and approached him about a book collaboration on the concept of abundance. Peter had come to this organizing principal through his creation of the X PRIZE Foundation and Singularity University and his work on innovation and exponential technologies. Steven had been considering similar ideas and brought his unique perspective and expertise on neuroscience, psychology, technology, education, energy, and the environment to this book. This effort is a true partnership, as the ideas and the writing in *Abundance* were shared equally between Peter and Steven.

Peter H. Diamandis
Santa Monica, California

Steven Kotler
Chimayo, New Mexico

PART ONE
PERSPECTIVE

CHAPTER ONE
OUR GRANDEST CHALLENGE

The Lesson of Aluminum

Gaius Plinius Cecilius Secundus, known as Pliny the Elder, was born in Italy in the year AD 23. He was a naval and army commander in the early Roman Empire, later an author, naturalist, and natural philosopher, best known for his *Naturalis Historia,* a thirty-seven-volume encyclopedia describing, well, everything there was to describe. His opus includes a book on cosmology, another on farming, a third on magic. It took him four volumes to cover world geography, nine for flora and fauna, and another nine for medicine. In one of his later volumes, *Earth,* book XXXV, Pliny tells the story of a goldsmith who brought an unusual dinner plate to the court of Emperor Tiberius.

The plate was a stunner, made from a new metal, very light, shiny, almost as bright as silver. The goldsmith claimed he'd extracted it from plain clay, using a secret technique, the formula known only to himself and the gods. Tiberius, though, was a little concerned. The emperor was one of Rome's great generals, a warmonger who conquered most of what is now Europe and amassed a fortune of gold and silver along the way. He was also a financial expert who knew the value of his treasure would seriously decline if people suddenly had access to a shiny new metal rarer than gold. "Therefore," recounts Pliny, "instead of giving the goldsmith the regard expected, he ordered him to be beheaded."

This shiny new metal was aluminum, and that beheading marked its loss to the world for nearly two millennia. It next reappeared during the early

1800s but was still rare enough to be considered the most valuable metal in the world. Napoléon III himself threw a banquet for the king of Siam where the honored guests were given aluminum utensils, while the others had to make do with gold.

Aluminum's rarity comes down to chemistry. Technically, behind oxygen and silicon, it's the third most abundant element in the Earth's crust, making up 8.3 percent of the weight of the world. Today it's cheap, ubiquitous, and used with a throwaway mind-set, but—as Napoléon's banquet demonstrates—this wasn't always the case. Because of aluminum's high affinity for oxygen, it never appears in nature as a pure metal. Instead it's found tightly bound as oxides and silicates in a claylike material called bauxite.

While bauxite is 52 percent aluminum, separating out the pure metal ore was a complex and difficult task. But between 1825 and 1845, Hans Christian Oersted and Frederick Wohler discovered that heating anhydrous aluminum chloride with potassium amalgam and then distilling away the mercury left a residue of pure aluminum. In 1854 Henri Sainte-Claire Deville created the first commercial process for extraction, driving down the price by 90 percent. Yet the metal was still costly and in short supply.

It was the creation of a new breakthrough technology known as electrolysis, discovered independently and almost simultaneously in 1886 by American chemist Charles Martin Hall and Frenchman Paul Héroult, that changed everything. The Hall-Héroult process, as it is now known, uses electricity to liberate aluminum from bauxite. Suddenly everyone on the planet had access to ridiculous amounts of cheap, light, pliable metal.

Save the beheading, there's nothing too unusual in this story. History's littered with tales of once-rare resources made plentiful by innovation. The reason is pretty straightforward: scarcity is often contextual. Imagine a giant orange tree packed with fruit. If I pluck all the oranges from the lower branches, I am effectively out of accessible fruit. From my limited perspective, oranges are now scarce. But once someone invents a piece of technology called a ladder, I've suddenly got new reach. Problem solved. Technology is a resource-liberating mechanism. It can make the once scarce the now abundant.

To expand on this a bit, let's take a look at the planned city of Masdar, now under construction by the Abu Dhabi Future Energy Company. Located on the edge of Abu Dhabi, out past the oil refinery and the airport, Masdar will soon house 50,000 residents, while another 40,000 work there. They will do so without producing any waste or releasing any carbon. No cars will be allowed within the city's perimeter and no fossil fuels will be consumed inside its walls. Abu Dhabi is the fourth-largest OPEC producer, with 10 percent of known oil reserves. *Fortune* magazine once called it the wealthiest city in the world. All of which makes it interesting that they're willing to spend $20 billion of that wealth building the world's first post-petroleum city.

In February 2009 I traveled to Abu Dhabi to find out just how interesting. Soon after arriving, I left my hotel, hopped in a cab, and took a ride out to the Masdar construction site. It was a journey back in time. I was staying at the Emirates Palace, which is both one of the most expensive hotels ever built and one of the few places I know of where someone (someone, that is, with a budget much different from mine) can rent a gold-plated suite for $11,500 a night. Until the discovery of oil in 1960, Abu Dhabi had been a community of nomadic herders and pearl divers. As my taxi drove past the "Welcome to the future home of Masdar" sign, I saw evidence of this. I was hoping the world's first post-petroleum city might look something like a *Star Trek* set. What I found was a few construction trailers parked in a barren plot of desert.

During my visit, I had the chance to meet Jay Witherspoon, the technical director for the whole project. Witherspoon explained the challenges they were facing and the reasons for those challenges. Masdar, he said, was being built on a conceptual foundation known as One Planet Living (OPL). To understand OPL, Witherspoon explained, I first had to understand three facts. Fact one: Currently humanity uses 30 percent more of our planet's natural resources than we can replace. Fact two: If everyone on this planet wanted to live with the lifestyle of the average European, we would need three planets' worth of resources to pull it off. Fact three: If everyone on this planet wished to live like an average North American, then we'd need five planets to pull it off. OPL, then, is a global initiative meant to combat these shortages.

The OPL initiative, created by BioRegional Development and the World Wildlife Fund, is really a set of ten core principles. They stretch from preserving indigenous cultures to the development of cradle-to-cradle sustainable materials, but really they're all about learning to share. Masdar is one of the most expensive construction projects in history. The entire city is being built for a post-petroleum future where oil shortages and water war are a significant threat. But this is where the lesson of aluminum becomes relevant.

Even in a world without oil, Masdar is still bathed in sunlight. A lot of sunlight. The amount of solar energy that hits our atmosphere has been well established at 174 petawatts (1.740 × 10^17 watts), plus or minus 3.5 percent. Out of this total solar flux, approximately half reaches the Earth's surface. Since humanity currently consumes about 16 terawatts annually (going by 2008 numbers), there's over five thousand times more solar energy falling on the planet's surface than we use in a year. Once again, it's not an issue of scarcity, it's an issue of accessibility.

Moreover, as far as water wars are concerned, Masdar sits on the Persian Gulf—which is a mighty aqueous body. The Earth itself is a water planet, covered 70 percent by oceans. But these oceans, like the Persian Gulf, are far too salty for consumption or crop production. In fact, 97.3 percent of all water on this planet is salt water. What if, though, in the same way that electrolysis easily transformed bauxite into aluminum, a new technology could desalinate just a minute fraction of our oceans? How thirsty is Masdar then?

The point is this: When seen through the lens of technology, few resources are truly scarce; they're mainly inaccessible. Yet the threat of scarcity still dominates our worldview.

The Limits to Growth

Scarcity has been an issue since life first emerged on this planet, but its contemporary incarnation—what many call the "scarcity model"—dates to the late eighteenth century, when British scholar Thomas Robert Malthus realized that while food production expands linearly, population grows exponentially. Because of this, Malthus was certain there was going to come a point in time when we would exceed our capacity to feed ourselves. As he put it, "The power of population is indefinitely greater than the power of the Earth to produce subsistence for man."

In the years since, plenty of thinkers have echoed this concern. By the early 1960s something of a consensus had been reached. In 1966 Dr. Martin Luther King Jr. pointed out: "Unlike the plagues of the dark ages or contemporary diseases, which we do not understand, the modern plague of overpopulation is soluble by means we have discovered and with resources we possess." Two years later, Stanford University biologist Dr. Paul R. Ehrlich sounded an even louder alarm with the publication of *The Population Bomb.* But it was the downstream result of a small meeting held in 1968 that really alerted the world to the depth of the crisis.

That year, Scottish scientist Alexander King and Italian industrialist Aurelio Peccei gathered together a multidisciplinary group of top international thinkers at a small villa in Rome. The Club of Rome, as this group was soon known, had come together to discuss the problems of short-term thinking in a long-term world.

In 1972 they published the results of that discussion. *The Limits to Growth* became an instant classic, selling twelve million copies in thirty languages, and scaring almost everyone who read it. Using a model developed by the founder of system dynamics, Jay Forrester, the club compared worldwide population growth rates to global resource consumption rates. The science behind this model is complicated, the message was not. Quite simply: we are running out of resources, and we are running out of time.

It's been over four decades since that report came out. While many of their more dire predictions have failed to materialize, for the most part, the years haven't softened the assessment. Today we are still finding proof of its veracity most places we look. One in four mammals now faces extinction, while 90 percent of the large fish are already gone. Our aquifers are starting to dry up, our soil growing too salty for crop production. We're running out of oil, running low on uranium. Even phosphorus—one of the principal ingredients in fertilizer—is in short supply. In the time it takes to read this sentence, one child will die of hunger. By the time you've made it through this paragraph, another will be dead from thirst (or from drinking dirty water to quench that thirst).

And this, the experts say, is just the warm-up round.

There are now more than seven billion people on the planet. If trends don't reverse, by 2050, we'll be closer to ten billion. Scientists who study the carrying capacity of the Earth—the measure of how many people can live here sustainably—have fluctuated massively in their estimations. Wild-eyed optimists believe it's close to two billion. Dour pessimists think it might be three hundred million. But if you agree with even the most uplifting of these predictions—as Dr. Nina Fedoroff, science and technology advisor to the US secretary of state, recently told reporters—only one conclusion can be drawn: "We need to decrease the growth rate of the global population; the planet cannot support many more people."

Some things, though, are easier said than done.

The most infamous example of top-down population control was the Nazis' eugenics program, but there have been a few other nightmares as well. India performed tubal ligations and vasectomies on thousands of people during the middle 1970s. Some were paid for their sacrifice; others were simply forced into the procedure. The results drove the ruling party out of power and created a controversy that still rages today. China, meanwhile, has spent thirty years under a one-child-per-family policy (while it's often discussed as a blanket program, this policy actually extends to only about 36 percent of the population). According to the government, the results have been 300 million fewer people. According to Amnesty International, the results have been an increase in bribery, corruption, suicide rates, abortion rates, forced sterilization procedures, and persistent rumors of infanticide. (A male child is preferable, so rumors hold that newborn girls are being murdered.) Either way, as our species has sadly discovered, top-down population control is barbaric, both in theory and in practice.

This seems to leave only one remaining option. If you can't shed people, you have to stretch the resources those people use. And stretch them dramatically. How to do this has been a matter of much debate, but these days the principles of OPL have been put forth as the only viable option. This option bothered me, but not because I wasn't committed to the idea of greater efficiency. Seriously—use less, gain more—who would be opposed to efficiency? Rather, the source of my concern was that efficiency was being forwarded as the only option available. But everything I was doing with my life told me there were additional paths worth pursuing.

The organization I run, the X PRIZE Foundation, is a nonprofit dedicated to bringing about radical breakthroughs for the benefit of humanity through the design and operation of large incentive-prize competitions. One month before traveling to Masdar, I'd chaired our annual "Visioneering" board meeting, where maverick inventors like Dean Kamen and Craig Venter, brilliant technology entrepreneurs such as Larry Page and Elon Musk, and international business giants like Ratan Tata and Anousheh

Ansari were debating how to drive radical breakthroughs in energy, life sciences, education, and global development. These are all people who have created world-changing industries where none had existed before. Most of them accomplished this feat by solving problems that had long been considered unsolvable. Taken together, they are a group whose track record showed that one of the better responses to the threat of scarcity is not to try to slice our pie thinner—rather it's to figure out how to make more pies.

The Possibility of Abundance

Of course, the make-more-pies approach is nothing new, but there are a few key differences this time around. These differences will comprise the bulk of this book, but the short version is that for the first time in history, our capabilities have begun to catch up to our ambitions. Humanity is now entering a period of radical transformation in which technology has the potential to significantly raise the basic standards of living for every man, woman, and child on the planet. Within a generation, we will be able to provide goods and services, once reserved for the wealthy few, to any and all who need them. Or desire them. Abundance for all is actually within our grasp.

In this modern age of cynicism, many of us bridle in the face of such proclamation, but elements of this transformation are already underway. Over the past twenty years, wireless technologies and the Internet have become ubiquitous, affordable, and available to almost everyone. Africa has skipped a technological generation, by-passing the landlines that stripe our Western skies for the wireless way. Mobile phone penetration is growing exponentially, from 2 percent in 2000, to 28 percent in 2009, to an expected 70 percent in 2013. Already folks with no education and little to eat have gained access to cellular connectivity unheard of just thirty years ago. Right now a Masai warrior with a cell phone has better mobile phone capabilities than the president of the United States did twenty-five years ago. And if he's on a smart phone with access to Google, then he has better access to information than the president did just fifteen years ago. By the end of 2013, the vast majority of humanity will be caught in this same World Wide Web of instantaneous, low-cost communications and information. In other words, we are now living in a world of information and communication abundance.

In a similar fashion, the advancement of new, transformational technologies—computational systems, networks and sensors, artificial intelligence, robotics, biotechnology, bioinformatics, 3-D printing, nanotechnology, human-machine interfaces, and biomedical engineering—will soon enable the vast majority of humanity to experience what only the affluent have access to today. Even better, these technologies aren't the only change agents in play.

There are three additional forces at work, each augmented by the power of exponentially growing technologies, each with significant, abundance-producing potential. A Do-It-Yourself (DIY) revolution has been brewing for the past fifty years, but lately it's begun to bubble over. In today's world, the purview of backyard tinkerers has extended far beyond custom cars and homebrew computers, and now reaches into once-esoteric fields like genetics and robotics. What's more, these days, small groups of motivated DIY-ers can accomplish what was once the sole province of large corporations and governments. The aerospace giants felt it was impossible, but Burt Rutan flew into space. Craig Venter tied the mighty US government in the race to sequence the human genome. The newfound power of these maverick innovators is the first of our three forces.

The second force is money—a lot of money—being spent in a very particular way. The high-tech revolution created an entirely new breed of wealthy technophilanthropists who are using their fortunes to solve global, abundance-related challenges. Bill Gates is crusading against malaria; Mark Zuckerberg is working to reinvent education; while Pierre and Pam Omidyar are focused on bringing electricity to the developing world. And this list goes on and on. Taken together, our second driver is a technophilanthropic force unrivaled in history.

Lastly, there are the very poorest of the poor, the so-called bottom billion, who are finally plugging into the global economy and are poised to become what I call "the rising billion." The creation of a global transportation network was the initial step down this path, but it's the combination of the Internet, microfinance, and wireless communication technology that's transforming the poorest of the poor into an emerging market force. Acting alone, each of these three forces has enormous potential. But acting together, amplified by exponentially growing technologies, the once-unimaginable becomes the now actually possible.

So what *is* possible?

Imagine a world of nine billion people with clean water, nutritious food, affordable housing, personalized education, top-tier medical care, and nonpolluting, ubiquitous energy. Building this better world is humanity's grandest challenge. What follows is the story of how we can rise to meet it.

CHAPTER TWO
BUILDING THE PYRAMID

The Trouble with Definitions

Abundance is a radical vision and before we can start striving for it, we must first start by defining it. In trying to map this territory, some economists take a bottom-up approach and begin with poverty, but this can be tricky. The US government defines poverty using two different metrics: "absolute poverty" and "relative poverty." Absolute poverty measures the number of people living under a certain income threshold. Relative poverty is a keeping up with the Jones' measure, comparing an individual's income with the average income for an entire economy. But the difficulty with both terms is that abundance is a global vision and neither hold up well when spread beyond borders.

For example, in 2008 the World Bank revised its international poverty line—an absolute poverty metric—from the longstanding "those living on less than a $1 a day" to "those living on less than $1.25 a day." By that figure, someone who works six days a week for fifty-two weeks earns $390 for their year. But that same year, the US government claimed the 39.1 million individuals found in the forty-eight contiguous states (Alaska and Hawaii had slightly different numbers) who earned $10,400 also lived in absolute poverty. Clearly, there's a pretty big gap between these totals. How to rectify that disparity—as would have to be done if your interest was setting a uniform target for the global reduction of poverty—is a problem for the absolute poverty measure.

A problem with the relative poverty measure is that it doesn't matter how much you earn in relation to your neighbors if that money can't buy what you need. The easy availability of goods and services is another critical factor in determining quality of life, but that availability varies tremendously according to one's geography. Today most poverty-stricken Americans have a television, telephone, electricity, running water, and indoor plumbing. Most Africans do not. If you transferred the goods and services enjoyed by those who live in California's version of poverty to the average Somalian living on less than a $1.25 a day, that Somalian is suddenly fabulously rich. And this makes any relative poverty measure less than useful for setting global standards.

Furthermore, both of these terms grow even shakier when played out over time. Today Americans living below the poverty line are not just light-years ahead of most Africans; they're light-years ahead of the wealthiest Americans from just a century ago. Today 99 percent of Americans living below the poverty line have electricity, water, flushing toilets, and a refrigerator; 95 percent have a television; 88 percent have a telephone; 71 percent have a car; and 70 percent even have air-conditioning. This may not seem like much, but one hundred years ago men like Henry Ford and Cornelius Vanderbilt were among the richest on the planet, but they enjoyed few of these luxuries.

A Practical Definition

Perhaps a better way to edge toward a definition of abundance is to start with what I am not talking about. I am not talking about Trump Towers, Mercedes-Benz, and Gucci. Abundance is not about providing everyone on this planet with a life of luxury—rather it's about providing all with a life of possibility. To be able to live such a life requires having the basics covered and then some. It also means stanching some fairly ridiculous bleeding. Feeding the hungry, providing access to clean water, ending indoor air pollution, and wiping out malaria—four entirely preventable conditions that kill, respectively, seven, three, three, and two people per minute worldwide is a must. But ultimately, abundance is about creating a world of possibility: a world where everyone's days are spent dreaming and doing, not scrapping and scraping.

Certainly, the above ideas are still too fuzzy, but they're a decent place to start. In trying to solidify this target, I look at levels of need loosely related to American psychologist Abraham Maslow's now-famous pyramid. From 1937 to 1951, Maslow was an up-and-comer on staff at Brooklyn College, being mentored by anthropologist Ruth Benedict and Gestalt psychologist Max Wertheimer. Back then, most of psychology was focused on fixing pathological problems rather than celebrating psychological possibilities, but Maslow had other ideas. He thought both Benedict and Wertheimer such "wonderful human beings" that he began studying their behavior, trying to figure out what it was they were doing right.

Over time, he began studying the behavior of other exemplars of ultimate human performance. Albert Einstein, Eleanor Roosevelt, and Frederick Douglass each came under his scrutiny. Maslow was looking for common traits and common circumstances, wanting to explain why these folks could attain such unbelievable heights, while so many others continued to flounder.

To illustrate his thinking, Maslow created his "Hierarchy of Human Needs," a theory arranged like a pyramid. In his pyramid, there are five levels of human needs—with the top tier of the pyramid belonging to "self-actualization" or a human being's need to reach their full potential. According to Maslow, the needs at each level have to be satisfied before a person can progress to the next. For this reason, physical needs like air, water, food, warmth, sex, and sleep are at the pyramid's base, followed closely by safety needs like protection, security, law, order, and stability. His middle tier is occupied by love and belongingness: family, relationships, affection, and work; and above that is esteem: achievement, status, responsibility, and reputation. At the very top are his "self-actualized needs," which are about personal growth and fulfillment—though they really constitute one's devotion to a higher purpose and a willingness to serve society.

My pyramid of abundance, while a little more compressed than Maslow's, follows a similar scheme for similar reasons. There are three levels, with the bottom belonging to food, water, shelter, and other basic survival concerns; the middle is devoted to catalysts for further growth like abundant energy, ample educational opportunities, and access to ubiquitous communications and information; while the highest tier is reserved for freedom and health, two core prerequisites enabling an individual to contribute to society.

Let's take a closer look.

The Base of the Pyramid

At the base of my pyramid, creating global abundance means taking care of simple physiological needs: providing sufficient water, food, and shelter. Having three to five liters of clean drinking water per person per day and 2,000 calories or more of balanced and nutritious food gives everyone on the planet the necessary water and food requirements for optimal health. Making sure that everyone receives a full complement of vitamins and minerals, either through one's food or in the form of a supplement, is also critical. For example, simply by providing populations with the requisite amount of Vitamin A removes the leading cause of preventable blindness in children from the global health equation. On top of these things, an additional twenty-five liters of water is necessary for bathing, cooking, and cleaning, and, considering that 837 million people now live in slums—and the United Nations predicts that this number will rise to 2 billion by 2050—a durable shelter that protects against the elements and further provides adequate reading light, ventilation, and sanitation, is also a must.

Of course, in the developed world, this may not sound like much, but it's a game-changer most everywhere else—and not just for the obvious reasons. The unobvious reasons begin with Thomas Friedman's Flat World. On this small planet, our grand challenges are not isolated concerns. Rather, they are stacked up like rows of dominoes. If we topple one domino, by meeting one challenge, plenty of others will follow suit. The results are a feedback loop of positive gain. Even better, the reverberations of this cascade stretch far beyond borders—which means that providing for basic physiological needs in developing countries also improves quality of life in the developed ones as well.

This is such an important point that before we return to the abundance pyramid, it's worth diving deeper into the upside of one of these goals: providing everyone on the planet with clean water.

The Upside of Water

Currently a billion people lack access to safe drinking water, and 2.6 billion lack access to basic sanitation. As a result, half of the world's hospitalizations are due to people drinking water contaminated with infectious agents, toxic chemicals and radiological hazards. According to the World Health Organization (WHO), just one of those infectious agents—the bacteria that cause diarrhea—accounts for 4.1 percent of the global disease burden, killing 1.8 million children a year. Right now more folks have access to a cell phone than a toilet. In fact, the ancient Romans had better water quality than half the people alive today.

So what happens if we solve this one problem? According to calculations done by Peter Gleick at the Pacific Institute, an estimated 135 million people will die before 2020 because they lack safe drinking water and proper sanitation. First and foremost, access to clean water means saving these lives. But it also means sub-Saharan Africa no longer loses the 5 percent of its gross domestic product (GDP) that's currently wasted on the health spending, productivity losses and labor diversions all associated with dirty water. Furthermore, because dehydration also lowers one's ability to absorb nutrients, providing clean water helps those suffering from hunger and malnutrition. As a bonus, an entire litany of diseases and disease vectors gets wiped off the planet, as do a number of environmental concerns (fewer trees will be chopped down to boil water; fewer fossil fuels will be burned to purify water). And this is merely the beginning.

One of the advantages we now possess in addressing the world's woes is information. We have a lot of it, especially about population growth and its various drivers and effects. For example, couple what we know about the planet's carrying capacity with what we know about population growth rates and no surprise that so many feel we are heading for disaster. So dire does this threat appear that one of the frequent criticisms leveled at the concept of abundance is that by solving problems like dirty water, the result, however high-minded in intent, will only serve to boost global population and worsen our situation.

On a certain level, this is absolutely correct. If the 884 million currently facing water shortages suddenly get enough to drink, this will certainly keep a great many of them alive for a good while longer. A population spike will result. But there are sound evolutionary reasons why it won't last.

Homo sapiens has been on the planet for roughly 150,000 years, yet until 1900, there was only one country in the world with an infant mortality rate below 10 percent. Since children take care of their parents later in life, in places where a lot of children die, by having a large family, parents are ensuring themselves a more comfortable old age. The good news is the inverse is also true. As Microsoft cofounder Bill Gates pointed out in his recent talk on the subject: "The key thing you can do to reduce population growth is actually improve health. . . . [T]here is a perfect correlation, as you improve health, within half a generation, the population growth rate goes down."

And the reason Gates knows this is because he's seen a plethora of population data that has been gathered over the last forty years. Morocco, for example, is now a young nation. Over half the population is under the age of twenty-five; almost one-third is under fifteen. Having this many kids around is a fairly recent historical development, but not for lack of trying. Back in 1971, when child-mortality rates were high and average life-expectancy rates were low, Moroccan women had an average of 7.8 children. But after making great strides in improving water, sanitation, health care, and women's rights, these days, Morocco's baby boom is winding down. The average number of births per woman is now 2.7, while the population growth rate has dipped below 1.6 percent—and all because people are living longer, healthier, freer lives.

John Oldfield, managing director of the WASH Advocacy Initiative, which is dedicated to solving global water challenges, explains it this way: "The best way to control population is through increasing child survival, educating girls, and making knowledge about and availability of birth control ubiquitous. By far the most important of these is increasing child survival. In communities where childhood death rates hover near one-third, most parents opt to significantly overshoot their desired family size. They will have replacement births, insurance births, lottery births—and the population soars. It's counterintuitive, but eradicating smallpox and vaccine-preventable disease and stopping diarrheal diseases and malaria are the best family planning programs yet devised. More disease, especially affecting the poor, will raise infant and child mortality which, in turn, will raise the birth rate. With fewer childhood deaths, you get lower fertility rates—it's really that straightforward."

By solving our water worries, we're also alleviating world hunger, relieving poverty, lowering the global disease burden, slowing rampant population growth, and preserving the biosphere. Children will no longer be yanked out of school to gather water and the firewood needed to boil water, so education levels will begin to rise. Since women also waste hours a day running these same errands, providing clean water also betters everything from quality of family life to quantity of family income (because mom now has time to get a job). But the best news is that water is merely one example of this interdependent phenomenon. The solutions to all of our grand challenges are similarly stacked and toppling any of these dominoes sets off a positive chain reaction—which is yet another reason why abundance for all is closer than many suspect.

The Pursuit of Catallaxy

Once our basic survival needs are fulfilled, the next level up the abundance pyramid is energy, education, and information/communication. Why this particular trio of advantages? Because these three pay double dividends. In the short term, they raise standards of living. In the long run, they pave the way for two of the greatest abundance assets in history: specialization and exchange. Energy provides the means to do work; education allows workers to specialize; information/communication abundance not only furthers specialization (through expanding educational opportunities), it allows specialists to exchange specialties, thus creating what economist Friedrich Hayek called catallaxy: the ever-expanding possibility generated by the division of labor. In his excellent book *The Rational Optimist: How Prosperity Evolves,* Matt Ridley elaborates: "'If I sew you a hide tunic today, you can sew me one tomorrow' brings limited rewards and diminishing returns. '[But] . . . I make the clothes, you catch the food' brings increasing returns. Indeed, it has the beautiful property that it does not even need to be fair. For barter to work, two individuals do not need to offer things of equal value. Trade is often unequal but it still benefits both sides."

Out of this trilogy, energy is clearly the biggest game changer. So how much energy does it take to change the game? Let's start in Nigeria. In Africa's most populous country, the average household has five people living in a single room. Under these conditions, four lights should provide ample illumination. Typically, a 60-watt incandescent bulb is enough to read by—and that's the figure we'll use for our calculation—but today that same luminosity can be provided by a 15-watt fluorescent and in the future with even less energy by using even more efficient LED (light-emitting diode) technology. Let's add to the list an efficient, 16-cubic-foot refrigerator that runs on 150 watts and keeps critical foods and drugs from perishing; a two-burner cookstove at 1,200 watts; two electric fans for ventilation at 100 watts each; a couple of laptop computers at 45 watts each, and—since we're splurging—an LCD TV, DVD player, and radio for 100 watts (although laptops will eventually displace these needs). Include another 35 watts for charging five cell phones, and we get a total of 1.73 kilowatts peak load. If we assume average usage for these items, we end up with a target minimum of 8.7 kilowatt-hours per household per day. While that's about a quarter of the power consumed in an average US household (an average household of 2.6 people consumes 16.4 KWh per day, or 6.32 KWh per person per day, excluding the gas and oil used for heating), it's a radical improvement for Nigeria.

It's also a radical improvement a lot of other places as well. For example, the two-burner electric cookstove is a simple device, but it would bring magnificent change to the 3.5 billion people who now cook food and get light and heat by burning biomass: wood, dung, and crop residue. According to a 2002 WHO report, 36 percent of acute upper respiratory infections, 22 percent of chronic obstructive pulmonary disease, and 1.5 percent of all cancers are all caused by indoor air pollution resulting from this practice. Thus an electric cookstove relieves 4 percent of the global disease burden.

Ever better—and just like water—the electric cookstove is another example of an interconnected solution. A 2007 UN report found that 90 percent of all wood removals in Africa are used for energy. Thus providing the power to run a cookstove will also help preserve endangered forests and the entire litany of ecosystem services those forests provide. Ecosystem services are things like crop pollination, carbon sequestration, climate regulation, water purification, air purification, nutrient dispersal, nutrient recycling, waste processing, flood control, pest control, disease control, and so forth, that the environment provides for us free of charge. This is a big deal for two reasons. The first is that the value of the ecosystems services our environment now provides (for free) has been calculated at $36 trillion a year—a figure roughly equal to the entire annual global economy. The second reason is that—as the $200 million experiment that was Biosphere 2 so clearly proved—none of these are services we can yet provide for ourselves.

But the cookstove's advantages are not only ecological. Freed from the burden of fuel gathering, women and children can get jobs and educations and, since all of these factors further lower child mortality and enhance women's rights, a concurrent reduction in population growth will occur. What's more, if a cookstove alone can bring this much positive change, consider the upside of the proposed 8.7 kilowatt-hours of power running a much fuller compliment of appliances.

Reading, Writing, and Ready

Another profound change would be education, specifically teaching every child on the planet the basics of literacy, mathematics, life skills, and critical thinking. Here, too, this may seem too thin an offering, but most experts feel this proposed quartet of grade school basics is the foundation for self-improvement, which is obviously abundance's backbone. Moreover, self-improvement doesn't mean what it used to. Since the advent of the Internet, these basics are the background needed to understand a significant portion of online materials, thus providing the fundamentals necessary to access what is clearly the greatest self-improvement tool in history.

This emphasis on personal growth and personal responsibility is key because we are in the midst of an education revolution. As experts like Sir Ken Robinson—who was knighted for his contribution to education—have said repeatedly, these days, antiquated classrooms are the least of our worries. "Suddenly degrees aren't worth anything," says Robinson. "When I was a student, if you had a degree, you had a job. If you didn't have a job, it was because you didn't want one."

The problem is both that there are many places in the world without any education infrastructure and, in those places where it does exist, they rely on a pedagogical framework that is seriously outdated. Most of today's educational systems are built upon the same learning hierarchy: math and science at the top, humanities in the middle, art on the bottom. The reason for this is because these systems were developed in the nineteenth century, in the midst of the industrial revolution, when this hierarchy provided the best foundation for success. This is no longer the case. In a rapidly changing technological culture and an ever-growing information-based economy, creative ideas are the ultimate resource. Yet our current educational system does little to nourish this resource.

Moreover, our current system is built around fact-based learning, but the Internet makes almost every fact desirable instantly available. This means we're training our children in skills they rarely need, while ignoring those they absolutely do. Teaching kids how to nourish their creativity and curiosity, while still providing a sound foundation in critical thinking, literacy and math, is the best way to prepare them for a future of increasingly rapid technological change.

Even better is the technological change that's coming. Unlike the one-size-fits-all framework that is our current educational system—because tomorrow's version is arriving via personal computers (or personal computing devices like a smart phone)—it's decentralized, personalized, and extremely interactive. *Decentralized* means learning cannot easily be curtailed by autocratic governments and is considerably more immune to socioeconomic upheaval. *Personalized* means that it can be tailored to an individual's needs and preferred learning style. These are both significant improvements, but many feel that its interactivity that could bring the biggest gains. As Nicholas Negroponte, founder of the Massachusetts Institute of Technology (MIT) Media Lab and the organization One Laptop Per Child (OLPC)—whose goal is to put a laptop in the hands of every school-age child in the world—explains: "Epistemologists from John Dewey to Paulo Freire to Seymour Papert agree that you learn through doing. This suggests that if you want more learning, you want more doing. Thus OLPC puts an emphasis on software tools for exploring and expressing, rather than instruction. Love is a better master than duty. Using the laptop as the agency for engaging children in constructing knowledge based upon their personal interests and providing them tools for sharing and critiquing these constructions will lead them to become learners and teachers."

Turning on the Data Tap

The final item at this level of our pyramid is information and communication abundance. The topic has already been touched upon, but the impact of these improvements cannot be overstated. In Kenya, a job placement service known as KAZI 560 uses mobile phones to connect potential workers with potential employers. In its first seven years, some 60,000 Kenyans have found employment via this network. In Zambia, farmers without bank accounts now rely on mobile phones to buy seeds and fertilizer, boosting their profits by almost 20 percent. In Niger, in 2005, cell phones served as a de facto national food distribution system, and effectively warded off a famine. In 2007, business executive Isis Nyong'o (then with MTV, now with Google) told the BBC that the impact of the mobile phone in Africa has "had about the same effect as a democratic change of leadership."

Perhaps more important, cell phones produced this change nearly organically. The technology did not have to be "sold" in any traditional sense. Instead, cell phones spread virally, and nearly unstoppably. To borrow Malcolm Gladwell's phrase, the idea tipped. Once people understood the technology and once the technology became vaguely affordable (vaguely, that is, because cell phones in the third world are often micro-financed), their rate of growth became exponential—just look at Nigeria.

In 2001, 134 million Nigerians were sharing 500,000 landlines. That same year, the government began encouraging market competition in wireless communications and the market responded. By 2007, Nigeria had 30 million cellular subscribers. This obviously produced a big boost in the local economy, but it's important to remember that it wasn't just Nigerians who benefited. When Nokia's profits hit $1 billion in 2009, the company said that market penetration in Africa was largely responsible. In 2010, when the Finnish multinational sold its billionth handset, it came as no surprise that the sale took place in Nigeria.

The Peak of the Pyramid

Abundance is an all-inclusive idea. It means everyone. It means the individual must matter, and matter like never before. In light of this, my abundance pyramid culminates with a pair of concepts that strengthen the individual's ability to matter: health and freedom. We'll start with health.

If the individual matters, then the individual's well-being matters; thus preserving good health and providing good health care are core components of an abundant world. And one thing is most certain: the creation of this world starts by stopping the needless deaths of millions resulting from ailments either entirely preventable or already easy to treat.

Acute respiratory infections are one of the leading causes of serious illnesses worldwide, accounting for about two million deaths each year and ranking first among causes of disability-adjusted life-years lost in developing countries. The populations most at risk are the young, the elderly, and the immunocompromised. Why is this the case? Because these infections typically go undiagnosed. Pneumonia, a disease we've been able to treat for almost a century, still accounts for 19 percent of all deaths in children under five. More perplexing, the drugs to treat pneumonia are generic. They're cheap and ubiquitous. This means that the problem is mostly one of diagnosis and/or distribution.

These days, to perform a blood test, you need access to sterile equipment and trained personnel. Clearly, it doesn't take much to take a blood sample, but after being gathered, it has to be sent to appropriate labs and then everyone must wait days, sometimes weeks, for the results. Not only are the tests prohibitively expensive, but in the developing world, where public transportation can be nonexistent, it's hard enough for most people just to get to the doctor in the first place, let alone return weeks later to learn the results and obtain treatment.

A technology now under development, known as Lab-on-a-Chip (LOC), has the potential to solve these problems. Packaged into a portable, cell-phone-sized device, LOC will allow doctors, nurses, and even patients themselves to take a sample of bodily fluid (such as urine, sputum, or a single drop of blood) and run dozens, if not hundreds, of diagnostics on the spot and in a matter of minutes. "It's a game-changing technology," says John T. McDevitt, a Rice University professor of bioengineering and chemistry and an early pioneer in the field. "In the developing world, it will bring reliable health care to billions who don't currently have it. In the developed world, like here in the US—where medical costs go up another 8 percent every year and 16.5 percent of the economy goes to health care—if personalized medical technologies like the lab-on-a-chip aren't brought to bear on the situation, we're going to bankrupt the country."

Another upside to LOC technologies is their ability to gather data. Because these chips are online, the information they collect—like, say, an outbreak of swine flu—can be immediately uploaded to a cloud, where it can be analyzed for deeper patterns. "For the first time," says McDevitt, "we'll have access to large quantities of global medical data. This will be crucial in halting the spread of new, emerging diseases and pandemics."

Moreover, LOCs are but one such technology currently in development. According to a 2010 report by PricewaterhouseCoopers, the field of personalized medicine—an industry that really didn't exist before 2001 (as the sequencing of the human genome is often cited as its start date)—is growing at a rate of 15 percent a year. By 2015, the global market for personalized medicine is projected to reach $452 billion. All of which is to say, we will soon have the means, methods, and motivation to value an individual's well-being like never before.

Freedom

The final element in our pyramid of abundance is freedom. This may seem a tall order, but it's a critical one. In his 1999 book *Development as Freedom,* the Nobel Prize–winning economist Amartya Sen pointed out that political liberty moves in lockstep with sustainable development. Since abundance, by definition, is a sustainable goal, then a certain level of freedom is the prerequisite for reaching that goal. Luckily, a certain level of freedom also emerges organically in response to certain new technologies—especially those of the communication and information variety.

This idea is not new. In his 1962 book *The Structural Transformation of the Public Sphere: An Inquiry into a Category of Bourgeois Society,* social philosopher Jurgen Habermas argues that empowering people with tools for open expression puts increasing pressure on undemocratic leaders while concurrently expanding the rights of the public. But even a thinker as bright as Habermas could not have predicted what Jared Cohen discovered in June 2009.

Cohen is a young Gen-Y, Internet-savvy, Harvard graduate who joined President Barack Obama's State Department for the chance to work under Secretary of State Hillary Clinton. It was Cohen who, in the midst of the June 2009 postelection protests in Iran, reached out to Twitter founder Jack Dorsey and urged the company to reschedule its planned website maintenance so that Iranians could keep tweeting. Given that all other forms of communication had been blocked or shut down, Twitter became the Iranian pipeline to the outside world.

The importance of this pipeline has been the subject of much debate. The Webby Awards, the leading international awards honoring online excellence, put the so-called Twitter Revolution on its list of top ten Internet moments of the decade (alongside the 2008 presidential campaign and the Google IPO), while others have pointed out that tweets don't stop bullets.

But either way, the revolution certainly proved that information technologies are extremely potent change agents. "By using new media to extend horizontal linkages and press the current regime," wrote political analyst Patrick Quirk in *Foreign Policy Focus,* "this generation has reinforced the foundation of a potentially robust force for democratic change."

Nor is this change merely an Iranian phenomena. A 2009 report by the Swedish International Development Cooperation Agency examined the impact of information and communications technologies (ICTs) for advancing democracy and empowerment in Kenya, Tanzania, and Uganda and found: "Access to and the strategic use of ICTs have been shown to have the potential to help bring about economic development, poverty reduction and democratization—including freedom of speech, the free flow of information and the promotion of human rights."

The Bigger Challenge

So there you have it: a first look at our hard targets. As far as a time frame for reaching these targets, everything outlined in the preceding pages (and much more to be discussed later) should be achievable within twenty-five years, with noticeable change possible within the next decade. Of course, now that we've defined our targets and our timetable, there's another problem to resolve: the fact that all this seems a little too far-fetched.

An end to most of what ails us by 2035? Can he really be serious?

And therein lies the focus of the next few chapters. While parts 2, 3, and 5 of this book are devoted to the technologies involved in this change, part 4 examines the three forces that are coming together to make such abundance possible, and part 6 examines ways to accelerate and direct this process. The remainder of part 1 is devoted to exploring why many of us, when hearing of the promise of abundance, simply cannot believe in the possibility.

People cry foul for a number of reasons. There are some who believe that the hole of disease, hunger, and war we're currently in appears too deep to climb out of, forget about anything else. For others, the time frame is too short and not enough technological progress will be made in the next few decades to dent these concerns. Then there are those who see our problems worsening: the rich getting richer and the poor falling further behind, while the list of global threats—pandemics, terrorism, escalating regional conflicts—grows unabated. These are all valid concerns, and we'll address each of them in chapters to come. But first it's helpful to understand a little more about the roots of this cynicism, and why it's this reaction—the inability of people to see the positive trends through the sea of bad news—that may be the biggest stumbling block on the road toward abundance.

CHAPTER THREE

SEEING THE FOREST THROUGH THE TREES

Daniel Kahneman

Abundance is a big vision compressed into a small time frame. The next twenty-five years can remake the world, but this won't happen on its own. There are plenty of issues to be faced, not all of them technological in nature. Overcoming the psychological blocks—cynicism, pessimism, and all those other crutches of contemporary thinking—that keep many of us from believing in the possibility of abundance is just as important. To accomplish this, we need to understand the way our brain shapes our beliefs and our beliefs shape our reality. There is perhaps no one better suited to help us look at this issue than the Nobel Prize–winning economist Daniel Kahneman.

Kahneman was born Jewish in Tel Aviv in 1934, but his childhood was spent in Nazi-occupied Paris. One afternoon in 1942, he was playing at a Christian friend's house, lost track of time, and stayed long past the Nazi-imposed six o'clock curfew. After realizing his error, Kahneman turned his sweater inside out to hide the Star of David that Jews were forced to wear on their clothes and set off to slink home. He didn't get far before he bumped into an SS soldier coming toward him on a deserted street. There was nowhere to hide. Certain that the soldier was about to notice the star, Kahneman picked up his pace, but the soldier stopped him anyway. Yet instead of arresting him, as recounted in Kahneman's Nobel autobiography: "[H]e beckoned me over, picked me up, and hugged me . . . He was speaking to me with great emotion, in German. When he put me down, he opened his wallet, showed me a picture of a boy, and gave me some money. I went home more certain than ever that my mother was right: people were endlessly complicated and interesting."

Kahneman never forgot this encounter. His family survived the war and relocated to Israel, where his curiosity about human behavior turned into a degree in psychology. After graduating from Hebrew University in 1954, Kahneman was immediately drafted into the Israeli Defense Forces. Because of his psychology background, the army asked him to help assess candidates for officer training. Kahneman took the job—and the study of human behavior hasn't been the same since.

The Israelis had developed a very compelling test for would-be officers. Candidates were assembled into small groups, dressed in neutral uniforms, and given a difficult task, such as lift a telephone pole off the ground and pass it over a seven-foot wall without the pole touching either the ground or the wall. "Under stress of the event," writes Kahneman, "we felt the soldiers' true nature would reveal itself, and we would be able to tell who would be a good leader and who would not."

But it didn't work as planned. "The trouble was, in fact, we could not tell. Every month or so we had a 'statistics day,' during which we would get feedback from the officer-training school, indicating the accuracy of our ratings of candidates' potential. The story was always the same: our ability to predict performance at the school was negligible. But the next day, there would be another batch of candidates to be taken to the obstacle field, where we would face them with the wall and see their true natures revealed. I was so impressed by the complete lack of connection between the statistical information and the compelling experience of insight that I coined a term for it: "the illusion of validity."

Kahneman describes the illusion of validity as "the sense that you understand somebody and can predict how they will behave," but it's since been expanded to "a tendency for people to view their own beliefs as reality." The Israelis were certain the telephone pole test would reveal a soldier's true character so they kept using it, despite the fact that there was no correlation between test results and later performance. What was generating this illusion and why people are so susceptible to its charms became the focus of Kahneman's future work: a half-century odyssey that would forever change how we think about how we think—including how we think about abundance.

Cognitive Biases

One reason abundance remains hard to accept is because we live in an extraordinarily uncertain world, and decision making in the face of uncertainty is never easy. In a perfectly rational world, when given a choice, we would assess the probability and the utility of all possible outcomes and then combine these two to make our call. But humans rarely have all the facts, we can't possibly know all the outcomes, and—even if we did—we have neither the temporal flexibility nor the neurological capacity to analyze all the data. Rather, our decisions are made based on limited, often unreliable, information, and further hampered by internal limits (the brain's processing power) and external limits (the time constraints under which we have to make our decision). So we have developed a subconscious strategy, a problem-solving aid for just such situations: we rely upon heuristics.

Heuristics are cognitive shortcuts: time-saving, energy-saving rules of thumb that allow us to simplify the decision-making process. They come in all flavors. In the study of visual perception, clarity is a heuristic used to help us judge distances: the more sharply an object is seen, the closer it appears. In the field of social psychology, heuristics show up when we assign probabilities—like evaluating the possibility that a Hollywood actor is a cocaine addict. To answer this question, the first thing the brain does is check its database for known Hollywood drug users. This is known as the availability heuristic—how available are examples for comparison—and our ease of access to this information becomes a significant portion of our foundation for assessment.

Normally, this is not a bad way to go. Heuristics are an evolutionary solution to an ongoing problem: we have limited mental resources. As such, they have a very long and thoroughly time-tested history of helping us—on average—make better decisions. But what Kahneman discovered is that there are certain situations when our reliance on heuristics leads to what he calls "severe and systematic errors."

Take clarity. Most of the time relying on this heuristic works perfectly for judging the gap between A and B; however, when visibility is poor and the contours of objects are blurry, we tend to overestimate distance. The inverse is also true. When visibility is good and objects are crisp, we err in the opposite direction. "Thus," wrote Kahneman and Hebrew University psychologist Amos Tversky in their 1974 paper "Judgment Under Uncertainty: Heuristics and Biases," "the reliance on clarity as an indication of distance leads to a common bias."

Our common biases have since become known as cognitive biases, which are defined as "patterns of deviation in judgment that occur in particular situations." Researchers have now collected a very long list of these biases, and a great many of them have a direct impact on our ability to believe in the possibility of abundance. For example, confirmation bias is a tendency to search for or interpret information in a way that confirms one's preconceptions—but it can often limit our ability to take in new data and change old opinions. This means that if your opposition to abundance is built around "the hole we're in is too deep to climb out of" hypothesis, any information that confirms your suspicions will be remembered, while conflicting data will not even register.

Here's a great example: Sarah Palin's alleged "death panels." In 2009 and 2010, during debates over the Obama administration's proposed health care reform bill, the idea spread like wildfire despite reliable sources decrying its mendacity. The *New York Times* was puzzled: "The stubborn yet false rumor that President Obama's health care proposals would create government-sponsored 'death panels' to decide which patients were worthy of living seemed to arise from nowhere in recent weeks." But "nowhere" was really our confirmation bias. Far-right Republicans already distrusted Obama, so those reliable death panel denials fell on deaf ears.

And confirmation bias is but only one of a litany of biases impacting abundance. The negativity bias—the tendency to give more weight to negative information and experiences than positive ones—sure isn't helping matters. Then there's anchoring: the predilection for relying too heavily on one piece of information when making decisions. "When people believe the world's falling apart," says Kahneman, "it's often an anchoring problem. At the end of the nineteenth century, London was becoming uninhabitable because of the accumulation of horse manure. People were absolutely panicked. Because of anchoring, they couldn't imagine any other possible solutions. No one had any idea the car was coming and soon they'd be worrying about dirty skies, not dirty streets."

Making this situation more difficult is the fact that our cognitive biases often work in tandem. Because of our negativity bias, standing up in today's climate and claiming that the world is getting better makes you appear addled. But we also suffer from the bandwagon effect—the tendency to do or believe things because others do—so even if you suspect there is real cause for optimism, these two biases will team up and make you doubt your own opinion.

In recent years, scientists have begun to notice larger patterns in our biases. One of those is often described as our "psychological immune system." If you believe your own life hopeless, then what's the point of pushing on? To guard against this, we've developed a psychological immune system: a set of biases that keep us ridiculously cocksure. In hundreds of studies, researchers have consistently found that we overestimate our own attractiveness, intelligence, work ethic, chances for success (be it winning the lottery or getting a promotion), chances of avoiding a negative outcome (bankruptcy, getting cancer), impact on external events, impact on other people, and even the superiority of our own peer group (known as the Lake Wobegon Effect after author Garrison Keillor's fictional happyland "where all the children are above average"). But there's a flip side: while we seriously overestimate ourselves, we significantly underestimate the world at large.

Human beings are designed to be local optimists and global pessimists and this is an even bigger problem for abundance. Kahneman and Tversky's collaborator, Cornell University psychologist Thomas Gilovich, believes the issue is twofold. "First, as anchoring shows, there's a direct link between imagination and perception. Second, we're control fiends and are significantly more optimistic about things we believe we can control. If I ask you what you can do to get a better grade in math—well, you can imagine studying harder, partying less, maybe hiring a tutor. You have control here. And because of this, your psychological immune system makes you feel overconfident. But if I ask what you can do to solve world hunger, all you can imagine is hordes of starving children. There's no sense of control, no overconfidence, and those starving children instead become your anchor—and crowd out all other possibilities."

And one of those other possibilities is that we really do have some control over world hunger. As we shall see in future chapters, because of the growth of exponential technologies, small groups are now being empowered to do what only governments once could—including fighting famine. But before we get there, to really understand all the psychological impediments to such progress, we first have to explore how our brain's architectural design and evolutionary history conspire to keep us pessimistic.

If It Bleeds, It Leads

Every second, an avalanche of data pours in through our senses. To process this deluge, the brain is continuously sifting and sorting information, trying to tease apart the critical from the casual. And since nothing is more critical to the brain than survival, the first filter most of this incoming information encounters is the amygdala.

The amygdala is an almond-shaped sliver of the temporal lobe responsible for primal emotions like rage, hate, and fear. It's our early warning system, an organ always on high alert, whose job is to find anything in our environment that could threaten survival. Anxious under normal conditions, once stimulated, the amygdala becomes hypervigilant. Then our focus tightens and our fight-or-flight response turns on. Heart rate speeds up, nerves fire faster, eyes dilate for improved vision, the skin cools as blood moves toward our muscles for faster reaction times. Cognitively, our pattern-recognition system scours our memories, hunting for similar situations (to help ID the threat) and potential solutions (to help neutralize the threat). But so potent is this response that once turned on, it's almost impossible to shut off, and this is a problem in the modern world.

These days, we are saturated with information. We have millions of news outlets competing for our mind share. And how do they compete? By vying for the amygdala's attention. The old newspaper saw "If it bleeds, it leads" works because the first stop that all incoming information encounters is an organ already primed to look for danger. We're feeding a fiend. Pick up the *Washington Post* and compare the number of positive to negative stories. If your experiment goes anything like mine, you'll find that over 90 percent of the articles are pessimistic. Quite simply, good news doesn't catch our attention. Bad news sells because the amygdala is always looking for something to fear.

But this has an immediate impact on our perception. David Eagleman, a neuroscientist at Baylor College of Medicine, explains that even under mundane circumstances, attention is a limited resource. "Imagine you're watching a short film with a single actor cooking an omelet. The camera cuts to a different angle as the actor continues cooking. Surely you would notice if the actor changed into a different person, right? Two-thirds of observers don't." This happens because attention is a seriously limited resource, and once we're focused on one thing, we often don't notice the next. Of course, any fear response only amplifies the effect. What all of this means is that once the amygdala begins hunting bad news, it's mostly going to find bad news.

Compounding this, our early warning system evolved in an era of immediacy, when threats were of the tiger-in-the-bush variety. Things have changed since. Many of today's dangers are probabilistic—the economy might nose-dive, there could be a terrorist attack—and the amygdala can't tell the difference. Worse, the system is also designed not to shut off until the potential danger has vanished completely, but probabilistic dangers never vanish completely. Add in an impossible-to-avoid media continuously scaring us in an attempt to capture market share, and you have a brain convinced that it's living in a state of siege—a state that's especially troubling, as New York University's Dr. Marc Siegel explains in his book *False Alarm: The Truth About the Epidemic of Fear,* because nothing could be further from the truth:

> Statistically, the industrialized world has never been safer. Many of us are living longer and more uneventfully. Nevertheless, we live in worst-case fear scenarios. Over the past century, we Americans have dramatically reduced our risk in virtually every area of life, resulting in life spans 60 percent longer in 2000 than in 1900. Antibiotics have reduced the likelihood of dying from infections . . . Public health measures dictate standards for drinkable water and breathable air. Our garbage is removed quickly. We live in temperature-controlled, disease-controlled lives. And yet, we worry more than ever before. The natural dangers are no longer there, but the response mechanisms are still in place, and now they are turned on much of the time. We implode, turning our adaptive fear mechanism into a maladaptive panicked response.

For abundance, all this carries a triple penalty. First, it's hard to be optimistic, because the brain's filtering architecture is pessimistic by design. Second, good news is drowned out, because it's in the media's best interest to overemphasize the bad. Third, scientists have recently discovered an even bigger cost: it's not just that these survival instincts make us believe that "the hole we're in is too deep to climb out of," but they also limit our desire to climb out of that hole.

A desire to better the world is predicated partially on empathy and compassion. The good news is that we now know that these prosocial behaviors are hardwired into the brain. The bad news is that these behaviors are wired into the slower-moving, recently evolved prefrontal cortex. But the amygdala evolved long ago, in an era of immediacy, when reaction time was critical for survival. When there's a tiger in the bush, there isn't much time to think, so the brain takes a shortcut: it doesn't.

In dangerous situations, the amygdala directs information around the prefrontal cortex. This is why you jump backward when you see a squiggly shape on the ground before you have time to deduce stick, not snake. But because of the difference in neuronal processing speeds, once our primitive survival instincts take over, our newer, prosocial instincts stay sidelined. Compassion, empathy, altruism—even indignation—become nonfactors. Once the media has us on high alert, for example, the chasm between rich and poor looks too big to bridge because the very emotions that would make us want to close that gap are currently locked out of the system.

"It's No Wonder We're Exhausted"

Over the past 150,000 years, Homo sapiens evolved in a world that was "local and linear," but today's environment is "global and exponential." In our ancestor's local environment, most everything that happened in their day happened within a day's walk. In their linear environment, change was excruciatingly slow—life from generation to the next was effectively the same—and what change did arrive always followed a linear progression. To give you a sense of the difference, if I take thirty linear steps (calling one step a meter) from the front door of my Santa Monica home, I end up thirty meters away. However, if I take thirty exponential steps (one, two, four, eight, sixteen, thirty-two, and so on), I end up a billion meters away, or, effectively lapping the globe twenty-six times.

Today's global and exponential world is very different from the one that our brain evolved to comprehend. Consider the sheer scope of data we now encounter. A week's worth of the *New York Times* contains more information than the average seventeenth-century citizen encountered in a lifetime. And the volume is growing exponentially. "From the very beginning of time until the year 2003," says Google Executive Chairman Eric Schmidt, "humankind created five exabytes of digital information. An exabyte is one billion gigabytes—or a 1 with eighteen zeroes after it. Right now, in the year 2010, the human race is generating five exabytes of information every two days. By the year 2013, the number will be five exabytes produced every ten minutes . . . It's no wonder we're exhausted."

The issue, then, is that we are interpreting a global world with a system built for local landscapes. And because we've never seen it before, exponential change makes even less sense. "Five hundred years ago, technologies were not doubling in power and halving in price every eighteen months," writes Kevin Kelly in his book *What Technology Wants.* "Waterwheels were not becoming cheaper every year. A hammer was not easier to use from one decade to the next. Iron was not increasing in strength. The yield of corn seed varied by the season's climate, instead of improving each year. Every 12 months, you could not upgrade your oxen's yoke to anything much better than what you already had."

The disconnect between the local and linear wiring of our brain and the global and exponential reality of our world is creating what I call a "disruptive convergence." Technologies are exploding and conjoining like never before, and our brains can't easily anticipate such rapid transformation. Our current means of governance and its supporting regulatory structures aren't designed for this pace. Look at the financial markets. Over the past decade, billion-dollar companies like Kodak, Blockbuster, and Tower Records collapsed nearly overnight, while new billion-dollar companies appeared out of nowhere. YouTube went from start-up to being acquired by Google for $1.65 billion in eighteen months. Groupon, meanwhile, went from start-up to a valuation of $6 billion in under two years. Historically, value has never been created this quickly.

This presents us with a fundamental psychological problem. Abundance is a global vision built on the backbone of exponential change, but our local and linear brains are blind to the possibility, the opportunities it may present, and the speed at which it will arrive. Instead we fall prey to what's become known as the "hype cycle." We have inflated expectations when a novel technology is first introduced, followed by short-term disappointment when it doesn't live up to the hype. But this is the important part: we also consistently fail to recognize the post-hype, massively transformative nature of exponential technologies—meaning that we literally have a blind spot for the technological possibilities underlying our vision of abundance.

Dunbar's Number

About twenty years ago, Oxford University evolutionary anthropologist Robin Dunbar discovered another problem with our local and linear perspectives. Dunbar was interested in the number of active interpersonal relationships that the human brain could process at one time. After examining global and historical trends, he found that people tend to self-organize in groups of 150. This explains why the US military, through a long period of trial and error, concluded that 150 is the optimal size for a functional fighting unit. Similarly, when Dunbar examined the traffic patterns from social media sites such as Facebook, he found that while people may have thousands of "friends," they actually interact with only 150 of them. Putting it all together, he realized that humans evolved in groups of 150, and this number—now known as Dunbar's number—is the upper limit to how many interpersonal relationships our brains can process.

In contemporary society—where, for example, the nuclear family has replaced the extended family—very few of us actually maintain 150 relationships. But we still have this primitive pattern imprinted on our brain, so we fill those open slots with whomever we have the most daily "contact"—even if that contact comes only from watching that person on television. Gossip, in its earlier forms, contained information that was critical to survival because, in clans of 150, what happened to anyone had a direct impact on everyone. But this backfires today. The reason we care so much about what happens to the likes of Lady Gaga is not because her shenanigans will ever impact our lives; rather because our brain doesn't realize there's a difference between rock stars we *know about* and relatives we *know.*

On its own, this evolutionary artifact makes television even more addictive (perhaps costing us time and energy that could be spent bettering the planet), but Dunbar's number never acts alone. Nor do any of the neurological processes discussed in this chapter. Our brain is a wonderfully integrated system, so these processes work in concert—and the symphony is not always pretty.

Because of amygdala function and media competition, our airwaves are full of prophets of doom. Because of the negativity bias and the authority bias—our tendency to trust authority figures—we're inclined to believe them. And because of our local and linear brains—of which Dunbar's number is but one example—we treat those authority figures as friends, which triggers the in-group bias (a tendency to give preferential treatment to those people we believe in our own group) and makes us trust them even more.

Once we start believing that the apocalypse is coming, the amygdala goes on high alert, filtering out most anything that says otherwise. Whatever information the amygdala doesn't catch, our confirmation bias—which is now biased toward confirming our eminent destruction—certainly does. Taken in total, the result is a population convinced that the end is near and there's not a damn thing to do about it.

This raises a final concern: what's the truth? If our brain plays this much havoc with our ability to perceive reality, then what does reality really look like? It's an important question. If we're heading for disaster, then having these biases could be an asset. But this is where things get even stranger. In the next chapter, we'll see that those facts have already been confirmed. And those facts are startling. Forget "the hole we're in being too deep to get out of." As we shall soon see, there's really not much of a hole.

CHAPTER FOUR
IT'S NOT AS BAD AS YOU THINK

This Moaning Pessimism

In chapter 2, we outlined our hard targets for abundance. This was an introductory look at our finish line, but the destination is not the journey. To fully understand where we want to go, it helps to have an accurate assessment of our exact starting point. If we can strip away our cynicism, what does our world really look like? How much progress has been made and not noticed?

Matt Ridley has spent the past two decades trying to answer these same questions. Ridley is in his early fifties, a tall Englishman with thinning, brown hair and an easy smile. He's an Oxford-trained zoologist but has spent most of his career as a science writer, specializing in the origins and evolution of behavior. Lately, the behavior that has most caught his attention is a strictly human outpouring: our species' predilection for bad news.

"It's incredible," he says, "this moaning pessimism, this knee-jerk, things-are-going-downhill reaction from people living amid luxury and security that their ancestors would have died for. The tendency to see the emptiness of every glass is pervasive. It's almost as if people cling to bad news like a comfort blanket." In trying to make sense of this pessimism, Ridley, like Kahneman, sees a combination of cognitive biases and evolutionary psychology as the core of the problem. He fingers loss aversion—a tendency for people to regret a loss more than a similar gain—as the bias with the most impact on abundance. Loss aversion is often what keeps people stuck in ruts. It's an unwillingness to change bad habits for fear that the change will leave them in a worse place than before. But this bias is not acting alone. "I also think there could be an evolutionary psychology component," he contends. "We might be gloomy because gloomy people managed to avoid getting eaten by lions in the Pleistocene."

Either way, Ridley has come to believe that our divorce from reality is doing more harm than good, and has lately started to fight back. "It's become a habit now for me to challenge such remarks. Whenever somebody says something grumpy about the world, I just try to think of the other side of the argument and—after examining the facts—again and again I find they have it the wrong way round."

This conversion to positive thinking did not happen overnight. As a cub science reporter, Ridley encountered hundreds of environmentalists fervently prophesying a much glummer future. But fifteen years ago, he started noticing that the doom predicted by these experts was still nowhere in evidence.

Acid rain was the first sign that the facts were not matching the fanfare. Once considered our planet's most dire environmental threat, acid rain develops because burning fossil fuels releases sulfur dioxide and nitrogen oxides into the atmosphere, causing an acidic shift in the pH balance of precipitation—hence the name. First noticed by English scientist Robert Angus Smith in 1852, acid rain took another century to blossom from scientific curiosity to presumed catastrophe. But by the late 1970s, the writing was on the wall. In 1982 Canada's minister of the environment, John Roberts, summed up what many were thinking, telling *Time* magazine, "Acid rain is one of the most devastating forms of pollution imaginable, an insidious malaria of the biosphere."

Back then, Ridley agreed with this opinion. But a few decades passed, and he realized that nothing of the sort was happening. "It wasn't just that the trees weren't dying, it was that they never had been dying—not in any unusual numbers and not because of acid rain. Forests that were supposed to have vanished altogether were healthier than ever."

To be sure, human innovation played a huge role in averting this disaster. In America, that hand-wringing produced everything from amendments to the Clean Air Act to the adoption of catalytic converters for automobiles. The results were a reduction in sulfur dioxide emission from 26 million tons in 1980 to 11.4 million tons in 2008, and nitrogen oxides from 27 million tons to 16.3 million tons during the same period. While some experts feel that current SO_2/NO emission rates are still too high, the fact remains that the eco-apocalypse predicted in the 1970s never did arise.

This absence got Ridley curious. He began looking into other dark prophecies and found a similar pattern. "Predictions about population and famine were seriously wrong," he says, "while epidemics were never as bad as they were supposed to be. Age-adjusted cancer rates, for example, are falling, not rising. Furthermore, I noticed that people who pointed these facts out were heavily criticized but not refuted."

All of this led him to another question: If the really negative predictions weren't coming true, what about the veracity of more common assumptions, such as the idea that the world is getting worse? To figure this out, Ridley began examining global trends: economic and technological; longevity and health-care related; and a host of environmental concerns. The result of this inquiry became the backbone of his 2010 *The Rational Optimist,* a book about why optimism rather than pessimism is the sounder philosophical position for accessing our species' chances at a brighter tomorrow. His uplifting argument sits atop an obvious but often overlooked fact: time is a resource. In fact, time has always been our most precious resource, and this has significant consequences for how we access progress.

Saved Time and Saved Lives

Each of us starts with the same twenty-four hours in the day. How we utilize those hours determines the quality of our lives. We go to extraordinary lengths to manage our time, to save time, to make time. In the past, just meeting our basic needs filled most of our hours. In the present, for a huge chunk of the world, not much has changed. A rural peasant woman in modern Malawi spends 35 percent of her time farming food, 33 percent cooking and cleaning, 17 percent fetching clean drinking water, and 5 percent collecting firewood. This leaves only 10 percent of her day for anything else, including finding the gainful employment needed to pull her off this treadmill. Because of all of this, Ridley feels that the best definition of prosperity is simply "saved time." "Forget dollars, cowrie shells, or gold," he says. "The true measure of something's worth is the hours it takes to acquire it."

So how have people managed to save time over the years? Well, we've tried slavery—both human and animal—and that worked okay until we developed a conscience. We also learned to boost muscle power with more elemental forces: fire, wind, and water, then natural gas, oil, and atoms. But at each step on this path, we have not only developed more power, we've also saved more time.

Light is a fabulous example. In England, artificial lighting was twenty thousand times more expensive circa AD 1300 than it is today. But when Ridley extended the equation and examined how the amount of light bought with an hour's work (at an average wage) has changed over the years, there is an even bigger savings:

> Today [light] will cost less than a half a second of your working time if you are on the average wage: half a second of work for an hour of light! Had you been using a kerosene lamp in the 1880s, you would have had to work for 15 minutes to get the same amount of light. A tallow candle in the 1800s: over six hours' work. And to get that much light from a sesame-oil lamp in Babylon in 1750 BC would have cost you more than fifty hours work.

Put another way, if you compare today's cost of lighting with the cost of sesame oil used in 1750 BC, you'll find a 350,000-fold time-saving difference. And this covers only the savings of work-related time. Since those with electricity rarely knock over a lantern and set the barn on fire or suffer the respiratory ailments resulting from breathing in candle smoke, we have furthered gained those hidden hours once lost to poor health and habitat repair.

Transportation follows an even bigger time-saving developmental curve. For millions of years, we went only where our feet could carry us. Six thousand years ago, we domesticated the horse; a vast improvement, to be sure, but equines have nothing on airplanes. In the 1800s, going from Boston to Chicago via stagecoach took two weeks' time and a month's wages. Today it takes two hours and a day's wage. But when it comes to crossing oceans, well, the horse isn't much use, and our early boats weren't exactly models of efficiency. In 1947 Norwegian adventurer Thor Heyerdahl spent 101 days sailing the raft *Kon-Tiki* from Peru to Hawaii. In a 747, it takes fifteen hours—a 100-day savings that has the added bonus of exponentially decreasing one's chances of dying along the way.

And saved time isn't the only unsung quality-of-life improvement to be found. In fact, as Ridley explains, they turn up almost every place we look:

> Some of the billions alive today still live in misery and want even worse than the worst experienced in the Stone Age. Some are worse off than they were a few months or years before. But the vast majority of people are much better fed, much better sheltered, much better entertained, much better protected against disease and much more likely to live to old age than their ancestors have ever been. The availability of almost everything a person could want has been going rapidly upward for two hundred years and erratically upward for ten thousand years before that: years of life span, mouthfuls of clean water, lungfuls of clean air, hours of privacy, means of traveling faster than you can run, ways of communicating farther than you can shout. Even allowing for the hundreds of millions who still live in abject poverty, disease and want, this generation of human beings has access to more calories, watts, lumen-hours, square-feet, gigabytes, megahertz, light-years, nanometers, bushels per acre, miles per gallon, food miles, air miles, and, of course, dollars than any that went before.

What all this means is that if your case against abundance rests upon "the hole we're in is too deep to climb out of" defense, well, you might want to find a different defense. But if this familiar charge against abundance isn't nearly as bad as most suppose, then what about that other common criticism: the ever-widening gap between rich and poor?

This too is not the problem many suspect. Take India. On August 1, 2010, India's National Council of Applied Economic Research estimated that the number of high-income middle-class households in India (46.7 million) now exceeds the number of low-income middle-class households (41 million) for the first time in history. Moreover, the gap between the two sides is also closing rapidly. In 1995 India had 4.5 million middle-class households. By 2009, that had risen to 29.4 million. Even better, the trend is accelerating. According to the World Bank, the number of people living on less than $1 a day has more than halved since the 1950s to below 18 percent of the world's population. Yes, there are still billions living in back-breaking destitution, but at the current rate of decline, Ridley estimates that the number of people in the world living in "absolute poverty" will hit zero by 2035.

Arguably, the number won't actually drop that low, but absolute poverty measures aren't the only metrics to consider. We also need to examine the availability of goods and services, which, as already established, are two categories that seriously impact quality of life. Here too there have been incredible gains. Between 1980 and 2000, the consumption rate—a measure of goods used by a society—grew in the developing world twice as fast as on the rest of the planet. Because population size and population health and longevity are impacted by consumption, these numbers improved as well. Compared to fifty years ago, today the Chinese are ten times as rich, have one-third fewer babies, and live twenty-eight years longer. In that same half-century time span, Nigerians are twice as well off, with 25 percent fewer children and a nine-year boost in life span. All told, according to the United Nations, poverty was reduced more in the past fifty years than in the previous five hundred.

Moreover, it's a pretty safe bet that these rates won't start rising again. "Once the rise in the position of the lower classes gathers speed," economist Friedrich Hayek wrote in his 1960 book, *The Constitution of Liberty*, "catering to the rich ceases to be the main source of great gain and gives place to efforts directed toward the needs of the masses. Those forces which at first make inequality self-accentuating thus later tend to diminish it."

And this is exactly what's happening in Africa today: the lower classes are gathering speed and gaining independence. For example, the spread of the cell phone is enabling microfinance, and microfinance is enabling the spread of the cell phone, and both are creating greater intraclass opportunity (meaning fewer jobs that directly depend on the rich) and greater prosperity for everyone involved.

Beyond economic measures, both political liberty and civil rights have also improved substantially these past few centuries. Slavery, for example, has gone from a common global practice to one outlawed everywhere. A similar change has occurred in the enshrinement of human rights in the world's constitutions and the spread of electoral processes. Admittedly, in far too many places, these rights and these processes are more window dressing than daily experience, but in less than a century, these memes have risen to such prominence that global surveys find democracy the preferred form of government for more than 80 percent of the world's population.

Perhaps the best news is what Harvard evolutionary psychologist Steven Pinker discovered when he began analyzing global patterns of violence. In his essay "A History of Violence: We're Getting Nicer Every Day," he writes:

> Cruelty as entertainment, human sacrifice to indulge superstition, slavery as a labor-saving device, conquest as the mission statement of government, genocide as a means of acquiring real estate, torture and mutilation as routine punishment, the death penalty for misdemeanors and differences of opinion, assassination as the mechanism of political succession, rape as the spoils of war, pogroms as outlets for frustration, homicide as the major form of conflict resolution—all were unexceptionable features of life for most of human history. But, today, they are rare to nonexistent in the West, far less common elsewhere than they used to be, concealed when they do occur, and widely condemned when they are brought to light.

What all this means is that over the last few hundred years, we humans have covered a considerable stretch of ground. We're living longer, wealthier, healthier, safer lives. We have massively increased access to goods, services, transportation, information, education, medicines, means of communication, human rights, democratic institutions, durable shelter, and on and on. But this isn't the whole of the story. Just as important to this discussion as the progress we've made is the reasons we've made such progress.

Cumulative Progress

Humans share knowledge. We trade ideas and exchange information. In *The Rational Optimist,* Ridley likens this process to sex, and his comparison is more than just florid metaphor. Sex is an exchange of genetic information, a cross-pollination that makes biological evolution cumulative. Ideas too follow this trajectory. They meet and mate and mutate. We call this process learning, science, invention—but whatever the term, it's exactly what Isaac Newton meant when he wrote: "If I have seen further, it is only because I am standing on the shoulders of giants."

Exchange is the beginning, not the end of this line. As the process evolves, specialization comes next. If you're the new blacksmith in town, forced to compete with five other already established blacksmiths, there are only two ways to get ahead. One is to work like mad and perfect your skills, becoming the very best blacksmith of the lot. But this is a risky option. You're going to need to be good enough at blacksmithing that the excellence of your craft overpowers the bonds of nepotism, because in a small town, most of your customers are close friends or relatives. Unfortunately, evolution worked very hard to craft these bonds. But develop a new technology—a slightly better horseshoe or a faster shoeing process—and you incentivize people to look beyond their social network.

This process, Ridley feels, creates a further feedback loop of positive gain: "Specialization encouraged innovation, because it encouraged the investment of time in a tool-making tool. That saved time, and prosperity is simply time saved, is proportional to the division of labor. The more human beings diversified as consumers and specialized as producers, and the more they then exchanged, the better off they have been, are and will be."

For a concrete example, let's return to Thor Heyerdahl's boat trip from Peru to Hawaii. Say you wanted to take that same trip today. What you don't have to do is hike into the forest, fell a tree, spend days tending a slow-burning fire to hollow out that tree's core, work for weeks chiseling that core into a seaworthy vessel, take however long it takes dragging a seaworthy vessel to the beach or however long it takes hauling freshwater or hunting meat or finding enough salt to preserve that meat or doing any of the other tasks that would have to precede sailing to Hawaii. Instead, because specialization has already taken care of all those intermediate steps, you go to a website and book a ticket. That's it. The result is a big boost in your quality of life.

Culture is the ability to store, exchange, and improve ideas. This vast cooperative system has always been one of abundance's largest engines. When the good ideas of your grandfather can be improved upon by the good ideas of your grandchildren, then that engine is up and running. The proof is the enormous bounty of cumulative innovation produced by specialization and exchange. "A large proportion of our high standard of living today derives not just from our ability to more cheaply and productively manufacture the commodities of 1800," writes J. Bradford DeLong, an economist at the University of California at Berkeley, "but from our ability to manufacture whole new types of commodities, some of which do a better job of meeting needs that we had back in 1800, and some of which meet needs that were unimagined back in 1800."

We now have millions of time-saving choices that our forebearers could not begin to imagine. My ancestors could not conceive of a salad bar because they could not imagine a global transportation network capable of providing green beans from Oregon, apples from Poland, and cashews from Vietnam together in the same meal.

"This is the diagnostic feature of modern life," writes Ridley, "the very definition of a high standard of living: diverse consumption, simplified production. Make one thing, use lots. The self-sufficient peasant or hunter-gatherer predecessor is in contrast defined by his multiple production and simple consumption. He does not make just one thing, but many: his shelter, his clothing, his entertainment. Because he only consumes what he produces, he cannot consume very much. Not for him the avocado, Tarantino, or Manolo Blahnik. He is his own brand."

But the very best news in all of this is that we have lately become specialized enough that we now trade in an entirely different kind of good. When people say we have an information-based economy, what they really mean is that what we have figured out is how to exchange information. Information is our latest, our brightest, commodity. "In a world of material goods and material exchange, trade is a zero-sum game," says inventor Dean Kamen. "I've got a hunk of gold and you have a watch. If we trade, then I have a watch and you have a hunk of gold. But if you have an idea and I have an idea, and we exchange them, then we both have two ideas. It's nonzero."

The Best Stats You've Ever Seen

Hans Rosling is in his early sixties, with wire-rimmed glasses, a penchant for elbow-patched tweed, and more energy than most. Starting out as a physician in rural Africa, where years were spent on the trail of konzo—an epidemic paralytic disease that he eventually cured—Rosling went on to cofound the Swedish chapter of Doctors Without Borders, become a professor of international health at one of the world's top medical schools, Sweden's Karolinska Institute, and write one of the most ambitious global health textbooks ever (examining the health of all 6.5 billion people on the planet).

The research for this textbook sent Rosling into the bowels of the UN archives, where reams of data about global poverty rates, fertility rates, life expectancy, wealth distribution, wealth accumulation, and so forth had been carefully disguised as rows of numbers on obscure spreadsheets.

Rosling not only plundered these data but also discovered a new way to visualize them, turning some of the world's best kept secrets into a presentation beyond belief.

The first time I caught Rosling's act was the first time that most people caught it: at the 2006 Technology, Entertainment, and Design (TED) conference in Monterey, California. Rosling's TED presentation—now known as "The Best Stats You've Ever Seen"—began with him onstage, a theater-size screen behind him, a giant graph filling the screen. The graph's horizontal axis was devoted to national fertility rates, while the vertical axis showed national life expectancies. Plotted on this graph were circles of different colors and sizes. The colors represented continents; the circles, nations. The size of the circle correlated to the size of that nation's population, while its position on the graph represented a combination of average family size and average life span for a given year. When Rosling started his talk, a large "1962" appeared across the screen.

"In 1962," he said, pointing toward the screen's upper right corner, "there was a group of countries—the industrialized nations—that had small families and long lives." Then, turning his attention to the bottom left corner: "And here are the developing countries, which have large families and relatively short lives."

This brutal visualization of the 1962 difference between the haves and the have-nots was striking, but it didn't last. With a mouse click, the graph began to animate. The date changed—1963, 1964, 1965, 1966—about one year for every second. As time marched forward, the dots began bouncing about the screen, their movement driven by the UN database. Rosling bounced with them. "Can you see here, it's China moving to the left as health is improving. All the green Latin American countries are moving toward smaller families, all the yellow Arabic countries are getting wealthier and living longer lives." The years ticked by, and progress became clearer. By 2000, excluding the African nations hit by civil war and HIV, most countries were congregated in the upper right corner, toward a better world of longer lives and smaller families.

A new graphic came onto the screen. "Now let's look at the world distribution of income." Along the horizontal axis was a log scale of per capita GDP (the average income per person per year); on the vertical left-hand axis was the child survival rate. Once again, the clock began in 1962. At the bottom left sat Sierra Leone, with a child survival rate of barely 70 percent and an average income of $500 a year. Just above it was the largest ball, China, both financially poor and in poor health. Once again, Rosling clicked his mouse, and his graphic soothsayer moved forward through time. China moved up, then to the right. "This is Mao Tse-tung," he said, "bringing health to China. Then he died . . . And Deng Xiaoping brought money to China."

China was just part of the picture. Most of the world followed the same pattern, the end result being a dense aggregation of countries in the upper right corner, with a pixilated tail of smaller dots trailing down and to the left. It was a graphic representation of the gap between rich and poor, but even with that tail, there wasn't much of a gap. In a 2010 updated presentation, Rosling summarized these findings thus: "Despite the disparities today, we have seen two hundred years of enormous progress. That huge historical gap between the West and the rest is now closing. We have become an entirely new, converging world. And I see a clear trend into the future. With aid, trade, green technology, and peace, it's fully possible that everyone can make it to the healthy, wealthy corner."

So what does this all mean? If Rosling is correct that the gap between rich and poor is mostly a memory, and if Ridley is correct that the hole we're in is none too deep, then the only remaining gripe against abundance is that today's rate of technological progress may be too slow to avert the disasters we now face. But what if this were a different kind of visualization problem, one that wasn't as easily solved by Ridley's theories and Rosling's animated graphics? What if this last issue isn't our current rate of progress; what if, as we shall soon see, it's really our linear brain's inability to comprehend our current rate of exponential progress?

PART TWO

EXPONENTIAL TECHNOLOGIES

CHAPTER FIVE

RAY KURZWEIL AND THE GO-FAST BUTTON

Better Than Your Average Haruspex

If you want to know if technology is accelerating fast enough to bring about an age of global abundance, then you need to know how to predict the future. Of course, this is an ancient art. The Romans, for example, employed a haruspex—a man trained to divine fortune through reading the entrails of disemboweled sheep. These days we've gotten a little better at the process. In fact, when it comes to predicting technological trends, we've gotten it almost down to a science. And perhaps no one is better at this science than Ray Kurzweil.

Kurzweil was born in 1948 and didn't start out trying to be a technological prognosticator, though he didn't start out like most. By the age of five, he wanted to be an inventor, but not just any inventor. His parents, both secular Jews, had fled Austria for New York to escape Hitler. He grew up hearing stories about the horrors of the Nazis but also heard other stories. His maternal grandfather loved to talk about his first trip back to postwar Europe and the amazing opportunity he'd been given to handle Leonardo da Vinci's original writings—an experience he always described in reverential terms. From these tales, Kurzweil learned that human ideas were all powerful. Da Vinci's ideas symbolized the power of invention to transcend human limitations. Hitler's ideas showed the power to destroy. "So from an early age," says Kurzweil, "I placed a critical importance on pursuing ideas that embodied the best of our human values."

By age eight, Kurzweil got even more proof he was on the right track.

That year, he discovered the Tom Swift Jr. books. The plots in this series were mostly the same: Swift would uncover a terrible predicament that threatened the fate of the world, then retreat to his basement laboratory for a hard think. Eventually, the cogs would click into place, he would build some whiz-bang solution and emerge the hero. The moral of the story was clear: ideas, coupled with technology, could solve all of the world's problems.

Since then, Kurzweil has made good on his goal. He's invented dozens of wonders: the world's first CCD flatbed scanner, the world's first text-to-speech synthesizer, the world's first reading machine for the blind—and plenty more. In total, he now holds thirty-nine patents, sixty-three additional patent applications, and twelve honorary doctorates; was inducted into the National Inventors Hall of Fame (yes, we actually have an Inventors Hall of Fame, in Akron, Ohio); and received the National Medal of Technology and the prestigious $500,000 Lemelson-MIT Prize, which recognizes "individuals who translate their ideas into inventions and innovations that improve the world in which we live."

But it wasn't just his inventions that have made Ray Kurzweil famous; it's the reason he invented those inventions that may be his bigger contribution—though this may take a little more explaining.

A Curve on a Piece of Paper

In the early 1950s, scientists began to suspect that there might be hidden patterns in technology's rate of change and that by unearthing those patterns, they might be able to predict the future. One of the first official attempts to do just that was a 1953 US Air Force study that tracked the accelerating progress of flight from the Wright brothers forward. By creating that graph and extrapolating into the future, the Air Force came to what was then a shocking conclusion: a trip to the Moon should soon be possible.

In *What Technology Wants,* Kevin Kelly explains further:

> It is important to remember that in 1953 none of the technology for these futuristic journeys existed. No one knew how to go that fast and survive. Even the most optimistic die-hard visionaries did not expect a lunar land-

> ing any sooner than the proverbial "Year 2000." The only voice telling them they could do it was a curve on a piece of paper. But the curve was right. Just not politically correct. In 1957 the USSR launched Sputnik, right on schedule. Then US rockets zipped to the Moon 12 years later. As [Damien] Broderick notes, humans arrived on the Moon "close to a third of century sooner than loony space travel buffs like Arthur C. Clarke had expected it to occur."

About a decade after the Air Force concluded this study, a man named Gordon Moore uncovered what would soon become the most famed of all tech trends. In 1965, while working at Fairchild Semiconductor (and before cofounding Intel), Moore published a paper entitled "Cramming More Components onto Integrated Circuits," wherein he observed that the number of integrated circuit components on a computer chip had doubled every year since the invention of the integrated circuit in 1958. Moore predicted that the trend would continue "for at least ten years." He was right. The trend did continue for ten years, and then ten more, and ten after that. All told, his prediction has stayed accurate for five decades, becoming so durable that it's come to be known as Moore's law, and is now used by the semiconductor industry as a guide for future planning.

Moore's law states that every eighteen months, the number of transistors on an integrated circuit doubles, which essentially means that every eighteen months, computers get twice as fast for the same price. In 1975 Moore altered his formulation to a doubling every two years, but either way, he's still describing a pattern of exponential growth.

As mentioned, exponential growth is just a simple doubling: 1 becomes 2, 2 becomes 4, 4 becomes 8, but, because most exponential curves start out well below 1, early growth is almost always imperceptible. When you double .0001 to .0002 to .0004 to .0008, on a graph all these plot points look like zero. In fact, at this rate, the curve stays below 1 for a total of thirteen doublings. To most people, it just looks like a horizontal line. But only seven doublings later, that same line has skyrocketed above 100. And it's this kind of explosion, from meager to massive and nearly overnight, that makes exponential growth so powerful. But with our local and linear brains, it's also why such growth can be so shocking.

To see this same pattern unfold in technology, let's examine the Osborne Executive Portable, a bleeding edge computer released in 1982. This bad boy weighed in at about twenty-eight pounds and cost a little over $2,500. Now compare this to the first iPhone, released in 2007, which weighed 1/100th as much, at 1/10th of the cost, while sporting 150 times the processing speed and more than 100,000 times the memory. Putting aside the universe of software applications and wireless connectivity that puts the iPhone light-years ahead of early personal computers, if you were to simply measure the difference in terms of "dollars per ounce per calculation," the iPhone has 150,000 times more price performance than Osborne's Executive.

This astounding increase in computer power, speed, and memory, coupled with a concurrent drop in both price and size, is exponential change at work. By the early 1980s, scientists were beginning to suspect that this pattern didn't show up just in transistor size but also in a larger array of information-based technologies—that is, any technology, like a computer, that's used to input, store, process, retrieve, and transmit digital information.

And this is where Kurzweil returns to our story. In the 1980s, he realized that inventions based on today's technologies would be outdated by the time they got to market. To be really successful, he needed to anticipate where technology would be in three to five years and base his designs on that. So Kurzweil became a student of tech trends. He began plotting his own exponential growth curves, trying to discover how pervasive Moore's law really was.

As it turns out—pretty pervasive.

Google on the Brain

Kurzweil found dozens of technologies that followed a pattern of exponential growth: for example, the expansion of telephone lines in the United States, the amount of Internet data traffic in a year, and the bits per dollar of magnetic data storage. Moreover, it wasn't just that information-based technologies were growing exponentially, it was that they did so regardless of what else was going on in the world. Take computer processing speed. Over the past century, its exponential growth has remained constant—despite the rude imposition of a number of world wars, global depressions, and a whole host of other issues.

In his first book, 1988's *The Age of Intelligent Machines,* Kurzweil used his exponential growth charts to make a handful of predictions about the future. Now, certainly inventors and intellectuals are always making predictions, but his turned out to be uncannily accurate: foretelling the demise of the Soviet Union, a computer's winning the world chess championship, the rise of intelligent, computerized weapons in warfare, autonomous cars, and, perhaps most famously, the World Wide Web. In his 1999 follow-up, *The Age of Spiritual Machines: When Computers Exceed Human Intelligence,* Kurzweil extended this prophetic blueprint to the years 2009, 2019, 2029, and 2099. The accuracy of most of these forecasts won't be known for quite a while, but out of 108 predictions made for 2009, 89 have come true outright and another 13 were damn close, giving Kurzweil a soothsaying record unmatched in the history of futurism.

In his next book, *The Singularity Is Near,* Kurzweil and a team of ten researchers spent almost a decade plotting the exponential future of dozens of technologies, while trying to understand the ramifications this much progress had for the human race. The results are staggering and controversial. To explain why, let's return to the future of computing power.

Today's average low-end computer calculates at roughly 10 to the 11th (10^{11}) or a hundred billion calculations per second. Scientists approximate that the level of pattern recognition necessary to tell Grandfather from Grandmother or distinguish the sound of hoofbeats from the sound of falling rain requires the brain to calculate at speeds of roughly 10 to the 16th (10^{16}) cycles per second, or 10 million billion calculations per second. Using these figures as a baseline and projecting forward using Moore's law, the average $1,000 laptop should be computing at the rate of the human brain in fewer than fifteen years. Fast-forward another twenty-three years, and the average $1,000 laptop is performing 100 million billion billion calculations (10^{26}) per second—which would be equivalent to all the brains of the entire human race.

Here's the controversial part: as our faster computers help us design better technologies, humans will begin incorporating these technologies into our bodies: neuroprosthetics to augment cognition; nanobots to repair the ravages of disease; bionic hearts to stave off decrepitude. In Steven Levy's *In the Plex: How Google Thinks, Works, and Shapes Our Lives,* Google cofounder Larry Page describes the future of search in similar terms: "It [Google] will be included in people's brains. When you think about something you don't know much about, you will automatically get the information." Kurzweil celebrates this coming possibility. Others are uneasy with the transition, believing it's the moment we stop being "us" and start becoming "them"—though this may be a little beside the point.

What's important here is the unbelievable pervasiveness of exponentially growing technologies and the staggering potential these technologies have for improving global standards of living. Sure, a long-term future where we have an AI in our brains sounds neat (at least to me), but what about a near-term future where AIs could be used to diagnose diseases, help educate our children, or oversee a smart grid for energy? The possibilities are immense. But how immense?

In 2007 I realized that if we wanted to start strategically employing exponentially growing technology to improve global standards of living, it wasn't enough to know which fields were accelerating exponentially; we also needed to know where they overlapped and how they might work together. A macroscopic overview was required. But in 2007 one wasn't available. No school in the world offered an integrated curriculum focused on exponentially growing technologies. Perhaps it was time for a new type of university, something both appropriate for a future of rapid technological change and one directly focused on solving the world's grand challenges.

Singularity University

Early universities were devoted to religious teachings, the very first of which was a Buddhist school established in the fifth century in India. This practice continued through the Middle Ages, when the Catholic Church was responsible for many of Europe's top universities. The foundations of faith may have changed, but the core methodology did not. Fact-based learning was king. This emphasis on rote memorization lasted for over a millennia, then shifted in the nineteenth century, when the objective went from the regurgitation of knowledge to the encouragement of productive thinking. Give or take a few details, this is about where we are today.

But how well suited are today's academic institutions to addressing the world's grand challenges? The modern graduate degree has become the realm of the ultraspecialized. A typical doctoral thesis focuses on a topic so insanely obscure that few can decipher its title, forget about content. While such extreme narrowness is important to specialization—which, as Ridley pointed out, has a huge upside—it has also created a world where the best universities rarely produce integrative, macroscopic thinkers.

While I was at MIT studying molecular genetics, I always imagined what it would have been like to explain my research to my great-great-great grandfather.

"Grandpa," I would begin, "do you see the dirt over there?"

"Are you a soil expert?" he might ask.

"No. But in the dirt, there is this microscopic life form called a bacterium."

"Oh, you're an expert in that!"

"No," I'd respond. "Inside the bacteria, there's this thing called DNA."

"So you're an expert in DNA?"

"Not quite. Inside the DNA are these segments called genes—and I'm not an expert in those either—but at the beginning of those genes is what's called a promoter sequence . . ."

"Uh-huh . . ."

"Well, I'm an expert in that!"

The world doesn't need another ultraspecialist-generating research university. We've got that covered. Places like MIT, Stanford, and the California Institute of Technology already do a fine job creating supergeniuses who can geek out in their nano-niche. What's needed is a place where people can go to hear of the biggest and boldest ideas, those exponential possibilities that echo Archimedes: "Give me a lever long enough, and a place to stand, and I will move the world."

In 2008 I took this idea forward, partnering with Ray Kurzweil to found Singularity University (SU). I next involved my old friend, Dr. Simon "Pete" Worden, a retired Air Force general with a doctorate in astronomy, who runs the National Aeronautics and Space Administration Ames Research Center in Mountain View, California. NASA Ames is a big think arm of the space agency, its areas of technical focus perfectly aligned with SU's interests. Worden saw the connection, and pretty soon we had a location for our new university.

After much deliberation, eight exponentially growing fields were chosen as the core of SU's curriculum: biotechnology and bioinformatics; computational systems; networks and sensors; artificial intelligence; robotics; digital manufacturing; medicine; and nanomaterials and nanotechnology. Each of these has the potential to affect billions of people, solve grand challenges, and reinvent industries. So important are these eight fields to our potential for abundance that the next chapter is devoted to exploring each in turn. The goal is to provide a deeper look at exponentials' power to raise global standards of living and to introduce some of the colorful characters who are devoting their lives to doing just that. Where to start? Well, there's probably none more colorful than Dr. J. Craig Venter.

CHAPTER SIX

THE SINGULARITY IS NEARER

A Trip Through Tomorrowland

Craig Venter is sixty-five years old, of average height, with a thick frame, a full beard, and a wide smile. His dress is casual; his eyes are not. They are blue and deep set, and when coupled to the slash of gray running through his right eyebrow and the mild arch to his left, he has the appearance of a modern-day wizard—like Gandalf with a solid stock portfolio and a pair of flip-flops.

Today, besides the flip-flops, Venter is also sporting a bright Hawaiian shirt and faded jeans. This is his tour guide attire, as today he's touring me around his namesake: the J. Craig Venter Institute (JCVI for short). Located in San Diego's "biology alley," JCVI's West Coast arm is a modest two-story research facility, housing sixty scientists and one miniature poodle. The poodle's name is Darwin, and he's a few steps ahead of us, now darting through the building's main entrance hall. He stops at the bottom of a flight of stairs, directly beside an architectural model of a four-tiered building. A plaque beside the model reads: "The first carbon-neutral, green laboratory facility." This is JCVI 2.0, Craig's vision for his future institute.

"If I can get it funded," says Venter, "that's what I want to build."

The price tag on this dream runs north of $40 million, but he'll get it funded. Venter is to biology what Steve Jobs was to computers. Genius with repeat success.

In 1990 the US Department of Energy (DOE) and the National Institutes of Health (NIH) jointly launched the Human Genome Project, a fifteen-year program with the goal of sequencing the three billion base pairs making up the human genome. Some thought the project impossible; others predicted that it would take a half century to complete. Everyone agreed it would be expensive. A budget of $10 billion was set aside, but many felt it wasn't enough. They might still be feeling this way too, except that in 2000 Venter decided to get into the race.

It wasn't even much of a race. Building on work that had come before, Venter and his company, Celera, delivered a fully sequenced human genome in less than one year (tying the government's ten-year effort) for just under $100 million (while the government spent $1.5 billion). Commemorating the occasion, President Bill Clinton said, "Today we are learning the language with which God created life."

As an encore, in May 2010 Venter announced his next success: the creation of a synthetic life form. He described it as "the first self-replicating species we've had on the planet whose parent is a computer." In less than ten years, Venter both unlocked the human genome and created the world's first synthetic life form—genius with repeat success.

To pull off this second feat, Venter strung together over a million base pairs, creating the largest piece of manmade genetic code to date. After engineering this code, it was sent to Blue Heron Biotechnology, a company that specializes in synthesizing DNA. (You can literally email Blue Heron a long string of As, Ts, Cs, and Gs—the four letters of the genetic alphabet—and they will return a vial filled with copies of that exact strand of DNA.) Venter then took the Blue Huron strand and inserted it into a host bacterial cell. The host cell "booted up" the synthetic program and began generating proteins specified by the new DNA. As replication proceeded, each new cell carried only the synthetic instructions, a fact that Venter authenticated by embedding a watermark into the sequence. The watermark, a coded sequence of Ts, Cs, Gs, and As, contains instructions for translating DNA code into English letters (with punctuation) and an accompanying coded message. When translated, this message spells the names of the forty-six people who worked on the project; quotations from novelist James Joyce, as well as physicists Richard Feynman and Robert Oppenheimer; and a URL for a website that anyone who deciphers the code can email.

But the real objective was neither secret messages nor synthetic life. This project was merely the first step. Venter's actual goal is the creation of a very specific new kind of synthetic life—the kind that can manufacture ultra-low-cost fuels. Rather than drilling into the Earth to extract oil, Venter is working on a novel algae, whose molecular machinery can take carbon dioxide and water and create oil or any other kind of fuel. Interested in pure octane? Aviation gasoline? Diesel? No problem. Give your designer algae the proper DNA instructions and let biology do the rest.

To further this dream, Venter has also spent the past five years sailing his research yacht, *Sorcerer II,* around the globe, scooping up algae along the way. The algae is then run through a DNA sequencing machine. Using this technique, Venter has built a library of over forty million different genes, which he can now call upon for designing his future biofuels.

And these fuels are only one of his goals. Venter wants to use similar methods to design human vaccines within twenty-four hours rather than the two to three months currently required. He's thinking about engineering food crops with a fiftyfold production improvement over today's agriculture. Low-cost fuels, high-performing vaccines, and ultrayield agriculture are just three of the reasons that the exponential growth of biotechnology is critical to creating a world of abundance. In the chapters to come, we'll examine this in greater depth, but for now, let's turn to the next category on our list.

Networks and Sensors

It's fall 2009, and Vint Cerf, chief Internet evangelist for Google, is at Singularity University to talk about the future of networks and sensors. In Silicon Valley, where T-shirts and jeans are the normal uniform, Cerf's preference for double-breasted suits and bow ties is unusual. But it's not just his dress that makes him stand out. Nor the fact that he's won the National Medal of Technology, the Turing Award, and the Presidential Medal of Freedom. Rather, what truly sets Cerf apart is that he's one of the people most associated with the design, creation, promotion, guidance, and growth of the Internet.

During his graduate student years, Cerf worked in the networking group that connected the first two nodes of the Advanced Research Projects Agency Network (Arpanet). Next he became a program manager for the Defense Advanced Research Projects Agency (DARPA), funding various groups to develop TCP/IP technology. During the late 1980s, when the

Internet began its transition to a commercial opportunity, Cerf moved to the long-distance telephone company MCI, where he engineered the first commercial email service. He then joined ICANN (Internet Corporation for Assigned Names and Numbers), the key US governance organization for the web, and served as chairman for more than a decade. For all of these reasons, Cerf is considered one of the "fathers of the Internet."

These days, Father is excited about the future of his creation—that is, the future of networks and sensors. A network is any interconnection of signals and information, of which the Internet is the most significant example. A sensor is a device that detects information—temperature, vibration, radiation, and such—that, when hooked up to a network, can also transmit this information. Taken together, the future of networks and sensors is sometimes called the "Internet of things," often imagined as a self-configuring, wireless network of sensors interconnecting, well, all things.

In a recent talk on the subject, Mike Wing, IBM's vice president of strategic communications, describes it this way: "Over the past century but accelerating over the past couple of decades, we have seen the emergence of a kind of global data field. The planet itself—natural systems, human systems, physical objects—has always generated an enormous amount of data, but we weren't able to hear it, to see it, to capture it. Now we can because all of this stuff is now instrumented. And it's all interconnected, so now we can actually have access to it. So, in effect, the planet has grown a central nervous system."

This nervous system is the backbone of the Internet of things. Now imagine its future: trillions of devices—thermometers, cars, light switches, whatever—all connected through a gargantuan network of sensors, each with its own IP addresses, each accessible through the Internet. Suddenly Google can help you find your car keys. Stolen property becomes a thing of the past. When your house is running out of toilet paper or cleaning products or espresso beans, it can automatically reorder supplies. If prosperity is really saved time, then the Internet of things is a big pot of gold.

As powerful as it will be, the impact the Internet of things will have on our personal lives is dwarfed by its business potential. Soon, companies will be able to perfectly match product demand to raw materials orders, streamlining supply chains and minimizing waste to an extraordinary degree. Efficiency goes through the roof. With critical appliances activated only when needed (lights that flick on as someone approaches a building), the energy-saving potential alone would be world changing. And world saving. A few years ago, Cisco teamed up with NASA to put sensors all over the planet to provide real-time information about climate change.

To take the Internet of things to the level predicted—with a projected planetary population of 9 billion and the average person surrounded by 1,000 to 5,000 objects—we'll need 45 thousand billion unique IP addresses (45 × 1012). Unfortunately, today's IP version 4 (IPv4), invented by Cerf and his colleagues in 1977, can provide only about 4 billion addresses (and is likely to run out by 2014). "My only defense," says Cerf, "is that the decision was made at a time when it was uncertain if the Internet would work," later adding that "even a 128-bit address space seemed excessive back then."

Fortunately, Cerf has been leading the charge for the next generation of Internet protocols (creatively called Ipv6), which has enough room for 3.4 × 1038 (340 trillion trillion trillion) unique addresses—roughly 50,000 trillion trillion addresses per person. "Ipv6 enables the Internet of things," he says, "which in turn holds the promise for reinventing almost every industry. How we manufacture, how we control our environment, and how we distribute, use, and recycle resources. When the world around us becomes plugged in and effectively self-aware, it will drive efficiencies like never before. It's a big step toward a world of abundance."

Artificial Intelligence

It's Saturday, July 2010, and Junior is driving me around Stanford University. He's a smooth operator: staying on his side of the road, making elegant turns, stopping at traffic lights, avoiding pedestrians, dogs, and bicyclists. This may not sound like much, but Junior is not your typical driver. Specifically, he's not human. Rather, Junior is an artificial intelligence, an AI, embodied in a 2006 Volkswagen Diesel Passat wagon, to be inexact. To be exact, well, that's a little trickier.

Sure, Junior has all the standard stylings of German engineering, but he also has a Velodyne HD LIDAR system strapped to the roof—which alone costs $80K and generates 1.3 million 3-D data points of information every second. Then there's an omnidirectional HD 6 video camera system; six radar detectors for picking out long-range objects; and one of the most technologically advanced Global Positioning Systems on the planet (worth

$150K). Furthermore, Junior's backseat has two 22-inch monitors and six core Intel Xeons, giving him the processing power of a small supercomputer. And he needs all of this, because Junior is an autonomous vehicle, known in hacker slang as a "robo car."

Junior was built in 2007 at Stanford University by the Stanford Racing Team. He is the second autonomous vehicle built by the team. The first was another VW named Stanley. In 2005 Stanley won DARPA's Grand Challenge, a $2 million incentive prize competition for the fastest autonomous vehicle to complete a 130-mile off-road course. The competition was organized after the 2001 invasion of Afghanistan, to help design robotic vehicles for troop resupply. Junior is the second iteration, designed for DARPA's 2007 follow-up, Urban Challenge (a 60-mile race through a cityscape), in which he placed second.

So successful was the Grand Challenge—and so lucratively tantalizing is the Department of Defense's desire for AI-driven vehicles—that almost every major car company now has an autonomous division. And military applications are only part of the picture. In June 2011 Nevada's governor approved a bill that requires the state to enact regulations that would allow autonomous vehicles to operate on public roads. If the experts have their timing right, that should happen around 2020. Sebastian Thrun, previously the director of the Stanford Artificial Intelligence Laboratory, and now the head of Google's autonomous car lab, feels the benefits will be significant. "There are nearly 50 million auto accidents worldwide each year, with over 1.2 million needless deaths. AI applications such as automatic breaking or lane guidance will keep drivers from injuring themselves when falling asleep at the wheel. This is where artificial intelligence can help save lives every day."

Robocar evangelist Brad Templeton feels that saved lives are just the beginning. "Each year, we spend 50 billion hours and $230 billion in accident costs—or 2 percent to 3 percent of the GDP—because of human driver error. Plus, these vehicles make the adoption of alternative fuel technologies considerably easier. Who cares if the nearest hydrogen filling station is twenty-five miles away, if your car can refuel itself while you sleep?" In the fall of 2011, to further this process along, the X PRIZE Foundation announced its intent to design an annual "human versus machine car race" through a dynamic obstacle course to mark the point in time when autonomous drivers begin outperforming the best human race car drivers in the world.

And autonomous cars are but a small slice of a much larger picture. Diagnosing patients, teaching our children, serving as the backbone for a new energy paradigm—the list of ways that AI will reshape our lives in the years ahead goes on and on. The best proof of this, by the way, is the list of ways that AI has *already* reshaped our lives. Whether it's the lightning-fast response of the Google search engine or the speech recognition used for directory information calls, we are already AI codependent. While some ignore these "weak AI" applications, waiting instead for the "strong AI" of Arthur C. Clarke's HAL 9000 computer from *2001: A Space Odyssey,* it's not like we haven't made progress. "Consider the man-versus-machine chess competition between Garry Kasparov and IBM's Deep Blue," says Kurzweil. "In 1992, when the idea that a computer could play against a world chess champion was first proposed, it was dismissed outright. But the constant doubling of computer power every year enabled the Deep Blue supercomputer to defeat Kasparov only five years later. Today you can buy a championship-level Chess AI for your iPhone for less than ten dollars."

So when will we have true HAL-esque AI? It's hard to say. But IBM recently unveiled two new chip technologies that move us in this direction. The first integrates electrical and optical devices on the same piece of silicon. These chips communicate with light. Electrical signals require electrons, which generate heat, which limits the amount of work a chip can perform and requires a lot of power for cooling. Light has neither limitation. If IBM's estimations are correct, over the next eight years, its new chip design will accelerate supercomputer performance a thousandfold, taking us from our current 2.6 petaflops to an exaflop (that's 10 to the 18th, or a quintillion operations per second)—or one hundred times faster than the human brain.

The second is SyNAPSE, Big Blue's brain-mimicking silicon chip. Each chip has a grid of 256 parallel wires representing dendrites and a perpendicular set of wires for axons. Where these wires intersect are the synapses and one chip has 262,144 of them. In preliminary tests, the chips were able to play a game of Pong, control a virtual car on a racecourse, and identify an image drawn on a screen. These are all tasks that computers have accomplished before, but these new chips don't need specialized programs to complete each task; instead they respond to real-world circumstances and learn from their experience.

Certainly there's no guarantee that these things will be enough to create HAL—strong AI may require more than just a brute force solution—but it's definitely going to rocket us up the abundance pyramid. Just think about what this will mean for the diagnostic potential in personalized medicine; or the educational potential in personalized education. (If you're having trouble imagining these concepts, just hang on for a few chapters, and I'll describe them in detail.) Yet as intriguing as all of this might seem, it's nothing compared to the benefits that AI will provide when combined with our next exponential category: robotics.

Robotics

Scott Hassan is in his midthirties, medium height, with jet-black hair and large almond-shaped eyes. He is a systems programmer, considered one of the best in the business, but his real passion is for building robots. Not industrial car-building machines, or small, cute Roombas, mind you, but real World's Fair, *I, Robot,* help-you-around-the-house type robots.

Certainly we've been striving to create such bots for years. Along the way, we've learned a number of lessons: first, that these robots are a lot harder to build than expected; second, that they're also considerably more expensive. But in both categories, Hassan has an advantage.

In 1996, as a computer science student at Stanford, Hassan met Larry Page and Sergey Brin. The duo was then working on a small side project: the search engine predecessor to Google. Hassan helped with the code, and the Google founders issued him shares. He started eGroups, which was later bought by Yahoo! for $412 million. The bottom line is that unlike other wannabe bot builders, Hassan has the capital needed to dent this field.

Furthermore, he's spent that capital gathering the best and the brightest to his company, Willow Garage (which takes its name from its Willow Road address in Menlo Park). Willow Garage's main project is a personal robot known by the exotic name PR2 (Personal Robot 2). The PR2 has head-mounted stereo cameras and LIDAR, two large arms, two wide shoulders, a broad and rectangular torso, and a four-wheel base. The whole thing looks sort of human, and sort of like R2D2 on steroids. Sure, this might not sound like much, but Hassan's invention is literally a whole new breed of bot.

For decades, robotics progress has been hampered because researchers lacked a stable platform for experimentation. Early computer hackers had the Commodore 64 in common, so innovations could be shared by all. This hasn't been the case with robotics, but that's where the PR2 comes in. Not designed for consumers, Willow Garage's robot is a research and development platform, created specifically so that geeks could go to town. And town is where they have gone. A quick tour of YouTube shows the PR2 opening doors, folding laundry, fetching a beer, playing pool, and cleaning house.

But the bigger breakthrough may be the code that runs the PR2. Instead of making his source code proprietary, Hassan has open-sourced the project. "Proprietary systems slow things down," he says. "We want the best minds around the world working on this problem. Our goal is not to control or own this technology but to accelerate it; put the pedal to the metal to make this happen as soon as possible."

So what's going to happen next, and what does it have to do with a world of abundance? Hassan has a list of beneficial applications, including mechanical nurses taking care of the elderly, and mechanized physicians making health care affordable and accessible. But he is most enthralled by economic possibilities. "In 1950 the global world product was roughly four trillion dollars," he says. "In 2008, fifty-eight years later, it was sixty-one trillion dollars. Where did this fifteenfold increase come from? It came from increased productivity in our factories equipped with automation. About ten years ago, while visiting Japan, I toured a Toyota car manufacturing plant that was able to produce five hundred cars per day with only four hundred employees because of automation. I thought to myself, 'Imagine if you could take this automation and productivity out of the factory and put it into our everyday lives?' I believe this will increase our global economy by orders of magnitude in the decades ahead."

In June 2011 President Obama announced the National Robotics Initiative (NRI), a $70 million multistakeholder effort to "accelerate the development and use of robots in the United States that work beside, or cooperatively, with people." Just like Willow Garage's attempt to create a stable platform for development in the P2P, the NRI is structured around "critical enablers": anchoring technologies that allow manufacturers to standardize processes and products, thus cutting development time and increasing performance. As Helen Greiner, president of the Robotics Technology Consortium, told *PCWorld* magazine: "Investing in robotics is more than just money for research and development, it is a vehicle to transform American lives and revitalize the American economy. Indeed, we are at a critical juncture where we are seeing robotics transition from the laboratory to generate new businesses, create jobs, and confront the important challenges facing our nation."

Digital Manufacturing and Infinite Computing

Carl Bass has been making things for the past thirty-five years: buildings, boats, machines, sculpture, software. He's the CEO of Autodesk, which makes software used by designers, engineers, and artists everywhere. Today he's touring me around his company's demonstration gallery in downtown San Francisco. We pass advanced architectural imaging systems powered by Autodesk's code; screens playing scenes from *Avatar* created with their tools, and ultimately up to a motorcycle and an aircraft engine, both manufactured by a 3-D printer, running—you guessed it—Autodesk software.

3-D printing is the first step toward *Star Trek*'s fabled replicators. Today's machines aren't powered by dilithium crystals, but they can precisely manufacture extremely intricate three-dimensional objects far cheaper and faster than ever before. 3-D printing is the newest form of digital manufacturing (or digital fabrication), a field that has been around for decades. Traditional digital manufacturers utilize computer-controlled routers, lasers, and other cutting tools to precisely shape a piece of metal, wood, or plastic by a subtractive process: slicing and dicing until the desired form is all that's left. Today's 3-D printers do the opposite. They utilize a form of additive manufacturing, where a three-dimensional object is created by laying down successive layers of material.

While early machines were simple and slow, today's versions are quick and nimble and able to print an exceptionally wide range of materials: plastic, glass, steel, even titanium. Industrial designers use 3-D printers to make everything from lamp shades and eyeglasses to custom-fitted prosthetic limbs. Hobbyists are producing functioning robots and flying autonomous aircraft. Biotechnology firms are experimenting with the 3-D printing of organs; while inventor Behrokh Khoshnevis, an engineering professor at the University of Southern California, has developed a large-scale 3-D printer that extrudes concrete for building ultra-low-cost multiroom housing in the developing world. The technology is also poised to leave our world. A Singularity University spin-off, Made in Space, has demonstrated a 3-D printer that works in zero gravity, so astronauts aboard the International Space Station can print spare parts whenever the need arises.

"What gets me most excited," says Bass, "is the idea that every person will soon have access to one of these 3-D printers, just like we have ink-jet printers today. And once that happens, it will change everything. See something on Amazon you like? Instead of placing an order and waiting twenty-four hours for your FedEx package, just hit print and get it in minutes."

3-D printers allow anyone anywhere to create physical items from digital blueprints. Right now the emphasis is on novel geometric shapes; soon we'll be altering the fundamental properties of the materials themselves. "Forget the traditional limitations imposed by conventional manufacturing, in which each part is made of a single material," explains Cornell University associate professor Hod Lipson in an article for *New Scientist.* "We are making materials within materials, and embedding and weaving multiple materials into complex patterns. We can print hard and soft materials in patterns that create bizarre and new structural behaviors."

3-D printing drops manufacturing costs precipitously, as it makes possible an entirely new prototyping process. Previously, invention was a linear game: create something in your head, build it in the real world, see what works, see what fails, start over on the next iteration. This was time consuming, creatively restricting, and prohibitively expensive. 3-D printing changes all of that, enabling "rapid prototyping," so that inventors can literally print dozens of variations on a design with little additional cost and in a fraction of the time previously required for physical prototyping.

And this process will be vastly amplified when coupled to what Carl Bass calls "infinite computing." "For most of my life," he explains, "computing has been treated as a scarce resource. We continue to think about it that way, though it's no longer necessary. My home computer, including electricity, costs less than two-tenths of a penny per CPU core hour. Computing is not only cheap, it's getting cheaper, and we can easily extrapolate this trend to where we come to think of computing as virtually free. In fact, today it's the least expensive resource we can throw at a problem.

"Another dramatic improvement is the scalability now accessible through the cloud. Regardless of the size of the problem, I can deploy hundreds, even thousands of computers to help solve it. While not quite as cheap as computing at home, renting a CPU core hour at Amazon costs less than a nickel."

Perhaps most impressive is the ability of infinite computing to find optimal solutions to complex and abstract questions that were previously unanswerable or too expensive to even consider. Questions such as "How can you design a nuclear power plant able to withstand a Richter 10 earthquake?" or "How can you monitor global disease patterns and detect pandemics in their critical early stages?"—while still not easy—are answerable. Ultimately, though, the most exciting development will be when infinite computing is coupled with 3-D printing. This revolutionary combination thoroughly democratizes design and manufacturing. Suddenly an invention developed in China can be perfected in India, then printed and utilized in Brazil on the same day—giving the developing world a poverty-fighting mechanism unlike anything yet seen.

Medicine

In 2008 the WHO announced that a lack of trained physicians in Africa will threaten the continent's future by the year 2015. In 2006 the Association of American Medical Colleges reported that America's aging baby boomer population will create a massive shortage of 62,900 doctors by 2015, which will rise to 91,500 by 2020. The scarcity of nurses could be even worse. And these are just a few of the reasons why our dream of health care abundance cannot come from traditional wellness professionals.

How do we fill this gap? For starters, we are counting on Lab-on-a-Chip (LOC) technologies. Harvard professor George M. Whitesides, a leader in this emerging field, explains why: "We now have drugs to treat many diseases, from AIDS and malaria to tuberculosis. What we desperately need is accurate, low-cost, easy-to-use, point-of-care diagnostics designed specifically for the sixty percent of the developing world that lives beyond the reach of urban hospitals and medical infrastructures. This is what Lab-on-a-Chip technology can deliver."

Because LOC technology will likely be part of a wireless device, the data it collects for diagnostic purposes can be uploaded to a cloud and analyzed for deeper patterns. "For the first time," says Dr. Anita Goel, a professor at MIT whose company Nanobiosym is working hard to commercialize LOC technology, "we'll have the ability to provide real-time, worldwide disease information that can be uploaded to the cloud and used for detecting and combating the early phase of pandemics."

Now imagine what happens when artificial intelligence gets added to this equation. Sound like a fairy tale? Already, in 2009 the Mayo Clinic used an "artificial neural network" to help physicians rule out the need for invasive procedures by diagnosing patients previously believed to suffer endocarditis, a dangerous heart condition, with 99 percent accuracy. Similar programs have been used to do everything from reading computed tomography (CT) scans to screening for heart murmurs in children. But combining AI, cloud computing, and LOC technology will offer the greatest benefit. Now your cell-phone-sized device can not only analyze blood and sputum but also have a conversation with you about your symptoms, offering a far more accurate diagnosis than was ever before possible and potentially making up for our coming shortage of doctors and nurses. Since patients will be able to use this technology in their own homes, it will also free up time and space in overcrowded emergency rooms. Epidemiologists will have access to incredibly rich data sets, allowing them to make incredibly robust predictions. But the real benefit is that medicine will be transformed from reactive and generic to predictive and personalized.

Nanomaterials and Nanotechnology

Most historians date nanotechnology—that is, the manipulation of matter at the atomic scale—to physicist Richard Feynman's 1959 speech "There's Plenty of Room at the Bottom." But it was K. Eric Drexler's 1986 book, *Engines of Creation: The Coming Era of Nanotechnology,* that really put the idea on the map. The basic notion is simple: build things one atom at a time. What sort of things? Well, for starters, assemblers: little nanomachines that build other nanomachines (or self-replicate). Since these replicators are also programmable, after one has built a billion copies of itself, you can direct those billion to build whatever you want. Even better, because building takes place on an atomic scale, these nanobots, as they are called, can start with whatever materials are on hand—soil, water, air, and so on—pull them apart atom by atom, and use those atoms to construct, well, just about anything you desire.

At first glance, this seems a bit like science fiction, but almost everything we're asking nanobots to do has already been mastered by the simplest life-forms. Duplicate itself a billion times? No problem, the bacteria in your gut will do that in just ten hours. Extract carbon and oxygen out of the air and turn it into a sugar? The scum on top of any pond has been at it for a billion years. And if Kurzweil's exponential charts are even close to accurate, then it won't be long now before our technology surpasses this biology.

Of course, a number of experts feel that once nanotechnology reaches this point, we may lose our ability to properly control it. Drexler himself described a "gray goo" scenario, wherein self-replicating nanobots get free and consume everything in their path. This is not a trivial concern. Nanotechnology is one of a number of exponentially growing fields (also biotechnology, AI, and robotics) with the potential to pose grave dangers. These dangers are not the subject of this book, but it would be a significant oversight not to mention them. Therefore, in our reference section, you'll find a lengthy appendix discussing all of these issues. Please use this as a launch pad for further reading.

While concerns about nanobots and gray goo are decades away (most likely beyond the time line of this book), nanoscience is already giving us incredible returns. Nanocomposites are now considerably stronger than steel and can be created for a fraction of the cost. Single-walled carbon nanotubes exhibit very high electron mobility and are being used to boost power conversion efficiency in solar cells. And Buckminsterfullerenes (C60), or Buckyballs, are soccer-ball-shaped molecules containing sixty carbon atoms with potential uses ranging from superconductor materials to drug delivery systems. All told, as a recent National Science Foundation report on the subject pointed out, "nanotechnology has the potential to enhance human performance, to bring sustainable development for materials, water, energy, and food, to protect against unknown bacteria and viruses, and even to diminish the reasons for breaking the peace [by creating universal abundance]."

Are You Changing the World?

As exciting as these breakthroughs are, there was no place anyone could go to learn about them in a comprehensive manner. It was for this reason I organized the founding conference for Singularity University at the NASA Ames Research Center in September 2008. There were representatives from NASA; academics from Stanford, Berkeley, and other institutions; and industry leaders from Google, Autodesk, Microsoft, Cisco, and Intel. What I remember most clearly from the event was an impromptu speech given by Google's cofounder Larry Page near the end of the first day. Standing before about one hundred attendees, Page made an impassioned speech that this new university must focus on addressing the world's biggest problems. "I now have a very simple metric I use: are you working on something that can change the world? Yes or no? The answer for 99.99999 percent of people is 'no.' I think we need to be training people on how to change the world. Obviously, technologies are the way to do that. That's what we've seen in the past; that's what drives all the change."

And that's what we built. That founding conference gave way to a unique institution. We run graduate studies programs and executive programs and already have over one thousand graduates. Page's challenge has become embedded in the university's DNA. Each year, the graduate students are challenged to develop a company, product, or organization that will positively affect the lives of a billion people within ten years. I call these "ten to the ninth-plus" (or 10^{9+}) companies. While none of these startups has yet to reach its mark (after all, we're only three years in), great progress is being made.

Because of the exponential growth rate of technology, this progress will continue at a rate unlike anything we've ever experienced before. What all this means is that if the hole we're in isn't even a hole, the gap between rich and poor is not much of a gap, and the current rate of technological progress is moving more than fast enough to meet the challenges we now face, then the three most common criticisms against abundance should trouble us no more.

PART THREE
BUILDING THE BASE OF THE PYRAMID

CHAPTER SEVEN

THE TOOLS OF COOPERATION

The Roots of Cooperation

The first two parts of this book explored the promise of abundance and the power of exponentials to further that promise. While there is a breed of techno-utopian who believes that exponentials alone will be enough to bring about this change, that is not the argument being made here. Considering the combinatory power of AI, nanotechnology, and 3-D printing, it does appear that we're heading in that direction, but (most likely) the timeframe required for these developments extends beyond the scope of this book. Here we are interested in the next two to three decades. And to bring about our global vision in that compacted period, exponentials are going to need some help.

But help is on the way. Later in this book, we'll examine the three forces speeding that plow. Certainly all three of these forces—the coming of age of the DIY innovator; a new breed of technophilanthropist; the expanding creative/market power of the rising billion—are augmented by exponential technology. In fact, exponential technology could be viewed as their growth medium, a substrate both anchoring and nurturing the emergence of these forces. Yet exponentially growing technologies are just one part of a larger cooperative process—a process that began a very long time ago.

On our planet, the earliest single-cell life forms were called prokaryotes. No more than a sack of cytoplasm with their DNA free floating in the middle, these cells came into existence roughly three and a half billion years ago. The eukaryotes emerged one and a half billion years later. These cells are more powerful than their prokaryote ancestors because they're more capable and cooperative, employing what we might call biological technology: "devices" such as nuclei, mitochondria, and Golgi apparatus that make the cell more powerful and efficient. There is a tendency to think of these biological technologies as smaller parts in a larger machine—not unlike the engine, chassis, and transmission that combine to form a car—but scientists believe that some of these parts began as separate life forms, individual entities that "decided" to work together toward a greater cause.

This decision is not unusual. We see this same chain of effect in our lives today: new technology creates greater opportunities for specialization, which increases cooperation, which leads to more capability, which generates new technology and starts the whole process over again. We also see it repeated throughout evolution.

One billion years after the emergence of the eukaryotes, the next major technological innovation took place: namely, the creation of multicellular life. In this phase, cells began to specialize, and those specialized cells learned to cooperate in an extraordinary fashion. The results were some very capable life forms. One cell type handled locomotion, while another developed the ability to sense chemical gradients. Pretty soon life forms with individuated tissues and organs began to emerge, among them our own species—whose ten trillion cells and seventy-six organs bespeak a level of complexity almost too great to consider.

"[H]ow do ten trillion cells organize themselves into a human being," asks Canadian wellness professional Paul Ingraham, "often with scarcely a single foul up for several decades? How do ten trillion cells even stand up? Even a simple thing of rising to a height of five or six feet is a fairly impressive trick for a bunch of cells that are, individually, no taller than a coffee stain."

The answer, of course, is this same chain of effect: technology (bones, muscles, neurons) leading toward specialization (the femur, biceps, and femoral nerve) leading toward cooperation (all those parts and many more leading to our bipedal verticality) leading toward greater complexity (every novel possibility that sprung from our upright stance). But the story doesn't end here. In the words of Robert Wright, author of *Nonzero: The Logic of Human Destiny*, "Next humans started a completely second kind of evolution: cultural evolution (the evolutions of ideas, memes, and technologies). Amazingly, that evolution has sustained the trajectory that biological evolution had established towards greater complexity and cooperation."

Nowhere has this causal chain been more evident than in the twentieth century, where, as we shall soon see, cultural evolution yielded the most powerful tools for cooperation the world has ever seen.

From Horses to Hercules

In 1861 William Russell, one of the biggest investors in the Pony Express, decided to use the previous year's presidential election for promotional purposes. His goal was to deliver Abraham Lincoln's inaugural address from the eastern end of the telegraph line, located in Fort Kearny, Nebraska, to the western end of the telegraph line, in Fort Churchill, Nevada, as fast as possible. To pull this off, he spent a small fortune, hired hundreds of extra men, and positioned fresh relay horses every ten miles. As a result, California read Lincoln's words a blistering seventeen days and seven hours after he spoke them.

By comparison, in 2008 the entire country learned that Barack Obama had become the forty-fourth president of the United States the instant he was declared the winner. When Obama gave his inaugural address, his words traveled from Washington, DC, to Sacramento, California, 14,939,040 seconds faster than Lincoln's speech. But his words also hit Ulan Bator, Mongolia, and Karachi, Pakistan, less than a second later. In fact, barring some combination of precognition and global telepathy, this is just about the very fastest such information could possibly travel.

Such rapid progress becomes even more impressive when you consider that our species has been sending messages to one another for 150,000 years. While smoke signals were innovative, and air mail even more so, in the last century, we've gotten so good at this game that no matter the distances involved, and with little more than a smart phone and a Twitter account, anyone's words can reach everyone's screen in an instant. This can happen without additional expenses, extra employees, or a moment of preplanning. It can happen whenever we please and why-ever we please. With an upgrade to a webcam and a laptop, it can happen live and in color. Heck, with the right equipment, it can even happen in 3-D.

This is yet another example of the self-amplifying, positive feedback loop that has been the hallmark of life for billions of years. From the mitochondria-enabled eukaryote to the mobile-phone-enabled Masai warrior, improved technology enables increasing specialization that leads to more opportunities for cooperation. It's a self-amplifying mechanism. In the same way that Moore's law is the result of faster computers being used to design the next generation of faster computers, the tools of cooperation always beget the next generation of tools of cooperation. Obama's speech went instantly global because, during the twentieth century, this same positive feedback loop reached an apex of sorts, producing the two most powerful cooperative tools the world has ever seen.

The first of these tools was the transportation revolution that brought us from beasts of burden to planes, trains, and automobiles in less than two hundred years. In that time, we built highways and skyways and, to borrow Thomas Friedman's phrase, "flattened the world." When famine struck the Sudan, Americans didn't hear about it years later. They got real-time reports and immediately decided to lend a hand. And because that hand could be lent via a C-130 Hercules transport plane rather than a guy on a horse, a whole lot of people went a lot less hungry in a hurry.

If you want to measure the change in cooperative capabilities illustrated here, you can start with the 18,800-fold increase in horsepower between a horse and a Hercules. Total carrying capacity over time is perhaps a better metric, and there the gains are larger. A horse can lug two hundred pounds more than thirty miles in a day, but a C-130 carries forty-two thousand pounds over eight thousand miles during those same twenty-four hours. This makes for a 56,000-fold improvement in our ability to cooperate with one another.

The second cooperative tool is the information and communication technology (ICT) revolution we've already documented. This has produced even larger gains during this same two-hundred-year period. In his book *Common Wealth: Economics for a Crowded Planet,* Columbia University economist Jeffery Sachs counts eight distinct contributions ICT has made to sustainable development—all of them cooperative in nature.

The first of those gains is connectivity. These days, there's no way to avoid the world. We are all part of the process, as we all know one another's business. "In the world's most remote villages," writes Sachs, "the conversation now often turns to the most up-to-date political and cultural events, or to changes in commodity prices, all empowered by cell phones even more than radio and television." The second contribution is an increased division of labor, as greater connectivity produces greater specialization, which allows all of us to participate in the global supply chain. Next comes scale, wherein messages go out over vast networks, reaching millions of people in almost no time at all. The fourth is replication: "ICT permits standardized processes, for example, online training or production specifications, to reach distant outlets instantaneously." Fifth is accountability. Today's new platforms permit increased audits, monitoring, and evaluation, a development that has led to everything from better democracy, to online banking, to telemedicine. The sixth is the Internet's ability to bring together buyers and sellers—what Sachs calls "matching"—which, among many other things, is the enabling factor behind author and *Wired* magazine editor in chief Chris Anderson's "long-tail" economics. Seventh is the use of social networking to build "communities of interest," a gain that has led to everything from Facebook to SETI@home. In the eighth spot is education and training, as ICT has taken the classroom global while simultaneously updating the curriculum to just about every single bit of information one could ever desire.

Obviously, the world is a significantly better place because of these new tools of cooperation, but ICT's impact doesn't end with novel ways to spread information or share material resources. As Rob McEwen discovered when he went looking for gold in the hills of northwestern Ontario, the tools of cooperation can also create new possibilities for sharing mental resources—and this may be a far more significant boost for abundance.

Gold in Dem Hills

A dapper Canadian in his midfifties, Rob McEwen bought the disparate collection of gold mining companies known as Goldcorp in 1989. A decade later, he'd unified those companies and was ready for expansion—a process he wanted to start by building a new refinery. To determine exactly what size refinery to build, McEwen took the logical step of asking his geologists and engineers how much gold was hidden in his mine. No one knew. He was employing the very best people he could hire, yet none of them could answer his question.

About the same time, while attending an executive program at MIT's Sloan School of Management, McEwen heard about Linux. This open-source computer operating system got its start in 1991, when Linus

Torvalds, then a twenty-one-year-old student at the University of Helsinki, Finland, posted a short message on Usenet:

> I'm doing a (free) operating system (just a hobby, won't be big and professional like gnu) for 386(486) AT clones. This has been brewing since April, and is starting to get ready. I'd like any feedback on things people like/dislike in minix . . .

So many people responded to his post that the first version of that operating system was completed in just three years. Linux 1.0 was made publicly available in March 1994, but this wasn't the end of the project. Afterward, support kept pouring in. And pouring in. In 2006 a study funded by the European Union put the redevelopment cost of Linux version 2.6.8 at $1.14 billion. By 2008, the revenue of all servers, desktops, and software packages running on Linux was $35.7 billion.

McEwen was astounded by all this. Linux has over ten thousand lines of code. He couldn't believe that hundreds of programmers could collaborate on a system so complex. He couldn't believe that most would do it for free. He returned to Goldcorp's offices with a wild idea: rather than ask his own engineers to estimate the amount of gold he had underground, he would take his company's most prized asset—the geological data normally locked in the safe—and make it freely available to the public. He also decided to incentivize the effort, trying to see if he could get Torvald's results in a compressed time period. In March 2000 McEwen announced the Goldcorp Challenge: "Show me where I can find the next six million ounces of gold, and I will pay you five hundred thousand dollars."

Over the next few months, Goldcorp received over 1,400 requests for its 400 megabytes of geological data. Ultimately, 125 teams entered the competition. A year later, it was over. Three teams were declared winners. Two were from New Zealand, one was from Russia. None had ever visited McEwen's mine. Yet so good had the tools of cooperation become and so ripe was our willingness to use them that by 2001, the gold pinpointed by these teams (at a cost of $500,000) was worth billions of dollars on the open market.

When McEwen couldn't determine the amount of ore he had underground, he was suffering from "knowledge scarcity." This is not an uncommon problem in our modern world. Yet the tools of cooperation have become so powerful that once properly incentivized, it's possible to bring the brightest minds to bear on the hardest problems. This is critical, as Sun Microsystems cofounder Bill Joy famously pointed out: "No matter who you are, most of the smartest people work for someone else."

Our new cooperative capabilities have given individuals the ability to understand and affect global issues as never before, changing both their sphere of caring and their sphere of influence by orders of magnitude. We can now work all day with our hands in California, yet spend our evenings lending our brains to Mongolia. NYU professor of communication Clay Shirky uses the term "cognitive surplus" to describe this process. He defines it as "the ability of the world's population to volunteer and to contribute and collaborate on large, sometimes global, projects."

"Wikipedia took one hundred million hours of volunteer time to create," says Shirky. "How do we measure this relative to other uses of time? Well, TV watching, which is the largest use of time, takes two hundred billion hours every year—in the US alone. To put this in perspective, we spend a Wikipedia worth of time every weekend in the US watching advertisements alone. If we were to forgo our television addiction for just one year, the world would have over a trillion hours of cognitive surplus to commit to share projects." Imagine what we could do for the world's grand challenges with a trillion hours of focused attention.

An Affordable Android

Until now, we've kept our examination of the tools of cooperation rooted in the past, but what's already been is no match for what's soon to arrive. It can be argued that because of the nonzero nature of information, the healthiest global economy is built upon the exchange of information. But this becomes possible only when our best information-sharing devices—specifically devices that are portable, affordable, and hooked up to the Internet—become globally available.

That problem has now been solved. In early 2011 the Chinese firm Huawei unveiled an affordable $80 Android smart phone through Kenya's telecom titan Safaricom. In less than six months, sales skyrocketed past 350,000 handsets, an impressive figure for a country where 60 percent of the population lives on less than $2 a day. Even better than the price are the 300,000-plus apps these users can now access. And if that's not dramatic enough, in the fall of 2011 the Indian government partnered with the Canada-based company Datawind and announced a seven-inch Android tablet with a base cost of $35.

But here's the bigger kicker. Because information-spreading technology has traditionally been expensive, the ideas that have been quickest to spread have usually emerged from the wealthier, dominant powers—those nations with access to the latest and greatest technology. Yet because of the cost reductions associated with exponential price-performance curves, those rules are changing rapidly. Think about how this shift has impacted Hollywood. For most of the twentieth century, Tinseltown was the nexus of the entertainment world: the best films, the brightest stars, an entertainment hegemony unrivalled in history. But in less than twenty-five years, digital technology has rearranged these facts. On average, Hollywood produces five hundred films per year and reaches a worldwide audience of 2.6 billion. If the average length of those films is two hours, then Hollywood produces one thousand hours of content per year. YouTube users, on the other hand, upload forty-eight hours' worth of videos every minute. This means, every twenty-one minutes, YouTube provides more novel entertainment than Hollywood does in twelve months. And the YouTube audience? In 2009 it received 129 million views a day, so in twenty-one days, the site reached more people than Hollywood does in a year. Since content creators in the developing world now outnumber content creators in the developed world, it's safe to say that the tools of cooperation have enabled the world's real silent majority to finally find its voice.

And that voice is being heard like never before. "The global deployment of ICT has utterly democratized the tools of cooperation," says Salim Ismail, SU's founding executive director and now its global ambassador. "We saw this in sharp relief during the Arab Spring. The aggregated self-publishing capabilities of the everyman enabled radical transparency and transformed the political landscape. As more and more people learn how to use these tools, they'll quickly start applying them to all sorts of grand challenges." Including, as we shall see in the next chapter, the first stop on our abundance pyramid: the challenge of water.

CHAPTER EIGHT
WATER

Water for Water

Peter Thum didn't intend to become a social entrepreneur. In 2001 he was consulting for McKinsey & Company on a bottled water project in South Africa—a country with an ongoing water crisis. Every day, he would watch women and children load up with empty jugs and set out on what was often a four-hour journey to bring back enough water for their families to survive. One afternoon, while driving on an empty dirt road miles and miles from the closest town, Thum came across a solitary woman struggling to carry a forty-pound jug on her head. "This was the middle of nowhere," he recounts. "It was pretty clear this woman had been walking for a long time and was going to have to keep walking for a long time. Even though I had seen evidence of the water crisis around South Africa, at that moment it became crystal clear: something had to be done about the problem."

Thum decided the easiest way to facilitate change was to connect bottled water, then becoming one of the world's hottest commodities, with the water shortage that was becoming one of the world's biggest crises. He returned to the States, partnered with old friend Jonathan Greenblatt, and created Ethos Water: a superpremium brand of bottled water that would donate a portion of its proceeds to help children of the world get clean water and raise awareness around the issue. In 2005 Howard Schultz, CEO of Starbucks, decided to acquire Ethos, putting its water in about 7,000 stores in America. With Starbucks' help, by donating five cents per bottle sold to water-related projects, Ethos has since made grants in excess of $10 million and brought water and sanitation to half a million people.

That said, the global water crisis affects a billion people, so let's be clear: $10 million isn't going to get it done. But Ethos's arrival marked something of a turning point. Historically, because of the huge amount of infrastructure required by most water projects, this space had been the domain of World Bank–style institutions. Ethos was one of the first companies to prove that social entrepreneurship could have a role in addressing water challenges. The company also helped raise awareness around the issue, and this created a snowball effect. Within a decade, water had become a top growth category for social entrepreneurs, and, as inventor Dean Kamen points out, there's still plenty of room for growth:

> When you talk to the experts about developing new technology to provide clean drinking water for the developing world, they'll tell you that—with four billion people making less than two dollars a day—there's no viable business model, no economic model, and no way to finance development costs. But the twenty-five poorest countries already spend twenty percent of their GDP on water. This twenty percent, about thirty cents, ain't much, but do the math again: four billion people spending thirty cents a day is a $1.2 billion market *every* day. It's $400 billion a year. I can't think of too many companies in the world that have $400 billion in sales a year. And you don't have to do a market study to find out whether there's a need. It's water. There's a need!

Filling that need, however profitable, won't be easy. The issue isn't just the amount of water required for hydration and sanitation, it's that water is thoroughly embedded in our lives, woven through most everything we manufacture or consume. The reason that 70 percent of the world's water is used for agriculture is because one egg requires 120 gallons to produce. There are 100 gallons in a watermelon. Meat is among our thirstiest commodities, requiring 2,500 gallons per pound or, as *Newsweek* once explained, "the water that goes into a 1,000-pound steer would float a destroyer."

And sustenance is just the beginning. In fact, everything in our abundance pyramid is affected by issues of hydrology. Beyond food, education takes a hit, as 443 million school days a year are lost to water-related disease. Thirty-five gallons of water are used to make one microchip—and a single Intel plant produces millions of chips each month—so information abundance suffers too. Then there's energy, where every step in the power production chain makes the world a dryer place. In the United States, for example, energy requires 20 percent of our nonagricultural water. At the pyramid's peak, threats to freedom have also been correlated to scarcity. In 2007 UC Berkeley professor of economics Edward Miguel found "strong evidence that better rainfall makes conflict less likely in Africa." So far, these conflicts have remained civil wars played out within countries, but some two hundred rivers and three hundred lakes share international boundaries, and not all these neighbors are friendly. (Israel and Jordan, for example, share the Jordan River.) Finally, with 3.5 million people dying annually from water-related illnesses, nothing is clearer than the direct ties between health and hydration.

Beyond the humancentric requirements of our abundance pyramid, there are even more problematic environmental concerns. Let's return to bottled water for a moment. Every year, we humans consume almost 50 billion liters of bottled water. Much of this water is what's known as "fossil water," meaning that it took tens of thousands of years to accumulate in aquifers and is not easily replenished. But fossil water also anchors the world's most delicate ecosystems. The thirst of modern agricultural practices, industrial practices, and the bottled water industry have pushed those systems toward collapse. We cannot risk further degradation. Simply put: no ecosystems means no ecosystem services, and that's a loss our species cannot survive.

Thus, addressing all of these concerns will require every tool in the toolbox. Our agricultural practices must be totally revamped, our industrial practices as well. We'll need waterwise appliances, novel infrastructure solutions, and a lot of honesty about a planetary population pushing toward nine billion. What that figure really tells us is what's really needed: a change measurable in orders of magnitude. With 97.3 percent of the water on the planet too salty for consumption, and another 2 percent locked up as polar ice, an orders-of-magnitude change does not come from bickering over the remaining .5 percent. This is not to say that we should ignore conservation and efficiency, but if our ultimate goal is abundance, then that requires an entirely new approach. Fresh water must go the route of aluminum, from one of the scarcest resources on Earth to one of the most ubiquitous. Pulling this off requires a significant amount of innovation: the type of significant innovation being unleashed by Moore's law—which, as we shall soon see, is exactly what DIY innovators like Dean Kamen are bringing to the table.

Dean vs. Goliath

Dean Kamen is a self-taught physicist, multimillionaire entrepreneur, and—with his 440 patents and National Medal of Technology—one of the greatest DIY innovators of our time. Like most DIY-ers, Kamen loves solving problems. Back in the 1970s, while he was still in college, Kamen's brother (then a medical student and now a renowned pediatric oncologist) mentioned there was no reliable way to give babies small and steady doses of drugs. Without such technology, infants were stuck with extended hospital stays, and nurses were stuck with inflexible time schedules.

So Kamen got curious. He started tinkering. One thing led to another, and pretty soon he'd invented the first portable infusion pump capable of automatically delivering the exact same drug dosages that had once required round-the-clock hospital supervision. Afterward, the miniaturization of medical technology became something of a specialty. In 1982 Kamen founded DEKA Research and Development, which soon created a portable kidney dialysis machine the size of a VCR, rather than the previous dishwasher-esque model. Then came the iBot: a motorized wheelchair that climbs stairs; the Segway, Kamen's attempt to reinvent local transportation; and the "Luke" Arm—a radical step forward in the development of prosthetic limbs.

Throughout all of this, Kamen never lost his interest in the challenges surrounding dialysis. "Every day," he says, "dialysis patients flush five gallons of sterilized water through their system. Getting this much clean water is a hassle. Often it means backing up delivery trucks to patients' homes once a week, and filling their garages with hundreds of bags of sterile water. I kept thinking there's gotta be a better way."

Kamen's first idea was to recycle the sterile water, but after consulting with biologists, he realized that there was no way to filter out mechanically what the kidney takes out naturally. "There's ammonia, urea, all these middle molecules. What the kidney takes out, you just can't filter." So if he couldn't recycle the water, perhaps there was a way to make tap water clean enough for injection.

That adventure took a few more years. "Turns out going from potable water to sterile water using filters was impossible," he explains. "Osmosis membranes don't work. The gold standard was pure, distilled deionized water, but there were no miniature distillers that could meet that standard." So Kamen decided to build one. Unfortunately, after doing the calculations, he realized that the amount of electrical power needed to run even a small unit would require rewiring most homes.

Next came a crazier idea: build a distiller capable of recycling its own energy. "A couple of years later, we finally got this little box that had 98 percent energy recovery and produced a reasonable amount of sterile water. We tested it with all these different tap waters, and it worked perfectly. It was so good that we didn't need to use tap water: we could use gray water instead. Then it hit me: if I can make gray water sterile enough for injection with 98 percent energy recovery, why am I trying to optimize a device to produce five to ten gallons a day? That machine could help a few tens of thousands of dialysis patients. But if I made a different machine [with a greater output] it might help a few billion people. Instead of creating an alternative to a minimally difficult problem [water delivery], I can stop people from dying [from water-related illness]."

That different machine was finished in 2003. As this is the technology that Kamen wants to use to bring down the giant problem of waterborne illness, he named it the Slingshot, for the technology that David used to bring down Goliath. It's the size of a dorm-room refrigerator, with a power cord, an intake hose, and an outflow hose. According to the inventor, "Stick the intake hose into anything wet—arsenic-laden water, salt water, the latrine, the holding tanks of a chemical waste treatment plant; really, anything wet—and the outflow is one hundred percent pure pharmaceutical-grade injectable water."

The current version can purify 1,000 liters (250 gallons) of water a day using the same amount of energy it takes to run a hair dryer. The power source is an updated version of a Stirling engine, designed to burn almost anything. Over a six-month field trial in Bangladesh, the engine ran only on cow dung and provided villagers with enough electricity to charge their cell phones and power their lights. And because Kamen wants to deploy the system in some of the remotest villages in the world, it's also designed to run maintenance free for at least five years.

"It better work that well," says Greenblatt, "because the world is littered with water pumps and purifiers that were not sustainable. I was in a village in Ethiopia that had made a water pump out of bicycle parts, and it worked because, when it broke down, people could fix it; they could get bicycle parts. That's the kind of supply chain you want."

Greenblatt is not alone in this assertion. Many believe water is an issue of money and will be best solved locally, and without the aid of techie gizmos. It's an opinion based on hindsight. The last century saw governments dithering while they searched for a high-tech, silver-bullet solution. Millions died in the interim, and the world is full of gadgets either unsuitable for the ruggedness of their deployment area or impossible to maintain because supply chains did not extend far enough. A great many of these bright ideas, because no one bothered to have an open discussion ahead of time, simply violated cultural barriers. Rob Kramer, chairman of the Global Water Trust, likes to tell an apocryphal story of a trunk line extension project in remote Africa, where pipe was run to within a quarter mile of a village in need—but the pipe kept getting vandalized. "Turns out," he says, "the four hours every other day that the women spent hiking out to gather water was the only time they got away from their husbands. They cherished this privacy, so they kept sabotaging the pipe."

All of these facts are correct, but they overlook others. As admirable as the bicycle-parts pump's ingenuity, it's not a long-term solution. The bicycle-parts pump is a transition technology, not unlike the early copper-wire phone systems that led to wireless 3G networks. For long-term sustainability, we still need massively disruptive Slingshot-like solutions.

Secondly, we can learn from our mistakes. Certainly we screwed up water (and not just in the developing world: America's infrastructure is so old that wooden pipes still run beneath the city of Philadelphia), but issue awareness is at an all-time high. And thanks to the wireless revolution, we're communicating best practices better than ever. Moreover, we now understand that community support is the most critical component for any water solution; without it, all of these efforts are sunk. We also know that parts must be readily available, that maintenance workers need to be incentivized, and, ideally, that these technologies are assembled and maintained locally. But we've learned this is true for all solutions, both high tech and low tech. Moreover, the idea that high-tech solutions won't work in rural environments went away with the cell phone. What's more high tech than a Nokia mobile phone? Yet there are nearly a billion of them working all over Africa.

Energy and infrastructure capitalization are the two main issues with most technological solutions to our water problems. With abundant energy, half of this problem is solved. How we'll generate that energy is a topic saved for a later chapter, so let's now turn to capitalization. April Rinne, the director of WaterCredit, says, "The average microfinance loan in the water space is between $200 and $800." Currently the cost of producing a single Slingshot is $100,000. According to Kamen, building them commercially, at volume, brings it to $2,500 per unit, plus another $2,500 for the Stirling engine to power the device. If the system really works for five years, then the cost of producing one thousand liters of drinking water per day is $0.002 per liter. Even if you tripled that to cover interest and labor, the price of five liters is only four cents—compared to today's thirty cents for the same supply.

Kamen, though, has decided there's another way to settle the matter. He's entered into negotiations with Coca-Cola to build, distribute, and, most importantly, use its enormous supply chain (the largest in Africa) to help maintain the Slingshot. "That's not the end of the road," he says. "I do think there needs to be a third party involved; someone making the whole process transparent, making it safe, educating people about it. But I also think Coca-Cola could do the major lifting, the major capitalization, the major distribution channel, development, support, education, and maintenance. It's one-stop shopping. Most of what needs to be done, I think they could do it."

And Coca-Cola has agreed to try. In May 2011 the world's biggest soda manufacturer launched a series of Slingshot field trials. Success could provide salvation for rural communities everywhere, but there are limits. According to Kamen, the Slingshot is built to serve one hundred people. Multiple machines could provide water for much larger communities, but they're not designed for large-scale urban deployment, nor can they satisfy our agricultural or industrial needs. But before we look at solutions to these problems, let's examine how the Slingshot dents another fundamental issue that many have with abundance: our current population explosion.

Prophylaxis

Malthusians often use the word *cornucopians* to describe people lobbying for abundance. It's not meant as a term of endearment. Central to their stance is the issue of population growth. Cornucopians feel that the rate of technological growth will outpace the rate of population growth, and that will solve all our problems. Malthusians believe that we've already exceeded the planet's carrying capacity, and if population growth continues unchecked, nothing we invent will be powerful enough to reverse those effects. But Kamen's technology provides a much-needed middle path.

Population is linked directly to fertility. Today the majority of developed countries have fertility rates at or below replacement levels—meaning that population is either stable or declining. The issue lies in the developing world, where the number of babies born is much higher. And the problem isn't in cities. Urbanization actually lowers fertility rates. The issue is in the country, as the most fecund population on the planet is the rural poor. It takes lots of hands to do farm work, so farmers have large families. But they want boys—usually three at the minimum. Their logic is heartbreaking. Three boys are desirable because one will probably die, while the second will stay home to tend the farm, providing for parents as they age as well as making enough money to send the third child to school so that he can get a better job and end this cycle. Thus child mortality among the rural poor is one of the largest factors driving population growth, and dirty water is often the root of this problem.

Of the 1.1 billion people in the world without access to safe water, 85 percent of them live in the countryside. Of the 2.2 million children that die each year from drinking contaminated water, the vast majority are rural as well. So a machine capable of providing clean drinking water for these communities, by boosting health and child survival rates, actually reduces fertility in the one place where it matters most. Beyond being a water purifier, the Slingshot is an extremely well-targeted family planning device: a prophylactic disguised as a drinking fountain.

Getting Roomier at the Bottom

As great as the Slingshot sounds, the solution to water is not any one technology; rather it will be a combination of technologies built for a combination of needs. One of those needs is disaster readiness. Even in the developed world, our relief systems are no match for the devastation of earthquakes, tidal waves, and tropical storms. When Hurricane Katrina hit New Orleans in 2005, it took five days to get water to refugees in the Superdome.

An English engineer named Michael Pritchard was stunned by Katrina, less than a year after he'd been stunned by the Asian tsunami. Pritchard was an expert in water treatment, an issue at the heart of both tragedies. Not only were survivors unable to get clean water immediately after the disaster, the solution to that problem only exacerbated others. "Traditionally," Pritchard told a TED audience, "in a crisis, what do we do? We ship water. After a few weeks, we set up camps, and people are forced to come into these camps to get their safe drinking water. What happens when twenty thousand people congregate in a camp? Diseases spread, more resources are required, the problem just becomes self-perpetuating."

So Pritchard decided to do something. A few years later, in 2009, he'd completed the Lifesaver bottle. With a hand pump on one end and a filter on the other, the bottle doesn't look especially high tech, but that filter is unlike any other. Researchers in nanotechnology work at miniscule scales, where distances are measured in atoms. One billionth of a meter—a nanometer, in technical parlance—is their baseline. Before Pritchard came along, the best hand-pumped water filters on the market worked down to the level of 200 nanometers. That's small enough to capture most bacteria, but viruses, which are considerably more microscopic, still slipped through. So Pritchard designed a membrane with pores 15 nanometers wide. In seconds, it removes everything there is to remove: bacteria, viruses, cysts, parasites, fungi, and other waterborne pathogens. One filter lasts long enough to produce six thousand liters of water, and the system automatically shuts off when the cartridge is expired, preventing the user from drinking contaminated water.

Lifesaver was designed for disaster relief, but why wait? A jerry can version of the system produces twenty-five thousand liters of water—enough for a family of four for three years. Even better, it costs half a cent a day to run. "For eight billion dollars," says Prichard, "we can hit the Millennium Goals' target of halving the number of people without access to safe drinking water . . . For twenty billion, everyone can have access to safe drinking water."

And Lifesaver is just the beginning. The nanotechnology industry is exploding. Between 1997 and 2005, investment rose from $432 million to $4.1 billion, and the National Science Foundation predicts that it will hit $1 trillion by 2015. We are entering the era of molecular manufacturing, and when you work at this scale, rearranging atoms leads to entirely new physical properties.

To return to water, there are now nanomaterials with increased affinity, capacity, and selectivity for heavy metals, among other contaminants. This means that heavy metals are drawn to these particles, and these particles can better transform those metals into harmless compounds, thus helping to clean up polluted waterways, contaminated aquifers, and Superfund sites.

Meanwhile, researchers at IBM and the Tokyo-based company Central Glass have developed a nanofilter capable of removing both salt and arsenic—which was, until fairly recently, an all but impossible trick. On the sanitation front, plumbing fixtures are now being built with self-cleaning nanomaterials that remove clogs and eliminate corrosion; while further back in development are nano-based self-sealing pipes that repair leaks on their own accord. Out on the wild frontier, German scientist Helmut Schulze and researchers at DIME Hydrophobic Materials, a company based in the United Arab Emirates, have an idea straight out of *Dune.* They've developed a nano-based hydrophobic sand, a ten-centimeter layer of which, when placed beneath desert topsoil, decreases water loss by 75 percent. In the Middle East, where 85 percent of all water is used for irrigation, this could be used both to grow crops and combat desertification.

With 40 percent of the Earth's population living within 100 kilometers (62 miles) of a coast, it's the combination of nanotech and desalination that holds even greater promise. Currently the majority of the world's seven thousand desalination plants rely on thermal desalination (often called "multistage flash") or reverse osmosis. The former means to boil water and condense the vapor; the latter feeds water through semipermeable membranes. Neither is the solution we need.

Thermal desalination consumes too much energy for large-scale deployment (about 80 megawatt hours per megaliter) and the brine by-product fouls aquifers and is devastating to aquatic populations. Reverse osmosis, on the other hand, uses comparatively less energy, but toxins such as boron and arsenic can still sneak through, and membranes clog frequently, reducing the lifetime of the filter. But the Los Angeles–based company NanoH_2O won a spot on the 2010 Cleantech 100 list for a novel filter that uses 20 percent less energy while producing 70 percent more water.

Of course, we could continue on like this for the rest of the book. There are dozens and dozens of nanotechnologies currently in development that will impact water. And for every amazing nanotech solution, there are mirroring developments in biotech. For every biotech solution, there's a wastewater recycling solution equally as exciting. But many believe the most promising line of development isn't even in the water space; it's in the metatechnologies surrounding this space.

The Smart Grid for Water

When IBM "Distinguished Scientist" and chief technology officer for Big Green Innovations Peter Williams says, "The biggest opportunity in water, isn't in water: it's in information," what he's talking about is waste. Right now, in America, 70 percent of our water is used for agriculture, yet 50 percent of the food produced gets thrown away. Five percent of our energy goes to pump water, but 20 percent of that water streams out holes in leaky pipes. "The examples are endless," says Williams, "the bottom line is the same: Show me a water problem and I'll show you an information problem."

The solution to this information problem is to create intelligent networks for all of our waterworks, what's being called the "Smart Grid for Water." The plan is to embed all sorts of sensors, smart meters, and AI-driven automation into our pipes, sewers, rivers, lakes, reservoirs, harbors, and, eventually, our oceans. Mark Modzelewski, executive director of the Water Innovations Alliance, believes a smart grid could save the United States 30 percent to 50 percent of its total water use.

IBM believes that the smart grid for water will be worth over $20 billion in the next five years, and the company is determined to get in on the ground floor. In the Amazon basin, it has partnered with the Nature Conservancy to build a new computer-modeling framework that allows users to simulate the behaviors of river basins and make significantly better decisions about currently unsolvable problems, such as determining in advance whether or not clear-cutting an upstream forest would destroy fish stocks in downstream watersheds. In Ireland, Big Blue has teamed up with the Marine Institute for the Smart Bay project, monitoring wave conditions, pollution levels, and marine life in Galway Bay. There's also a "smart levee" project in the Netherlands, a sewer system analytic upgrade in Washington, DC, and several dozen more efforts spattered around the globe.

Other companies are following suit. Working in Detroit, Hewlett-Packard has implemented a smart metering system that has already increased productivity by 15 percent. In the academic sector, researchers at Chicago's Northwestern University have created a "Smart Pipe"—a multi-nanosensor array that measures everything from water quality to water flow. Internationally, efforts are also increasing. Spain just installed a nationwide computer-assisted irrigation system designed to save farmers 20 percent of the nine hundred *billion* gallons of water they annually use.

Computer-assisted irrigation is a subcategory of "precision agriculture," which is a big part of the smart grid's potential. The full complement blends computer-assisted irrigation with GPS tracking and remote sensing technologies to get, as the saying goes, more crop per drop. This combination allows farmers to know everything going on in their fields: temperature, transpiration, moisture content in the air and soil, the weather forecast, how much fertilizer has been applied to every plant, how much water each plant has received, and so forth. An unsustainable 70 percent of the water on Earth is now used for growing food. "With precision agriculture," says Doug Miell, a water management consultant who advises the state of Georgia, "farmers can lower their water use by thirty-five percent to forty percent, and increase their yields by twenty-five percent."

And the massive savings talked about in this section are the starting point of this discussion, not its closing arguments. Once our waterworks are turned into an intelligent network, water truly becomes an information science—thus strapping itself to the rising tide of exponential growth. What is now being discussed as the smart grid for water is really a beta-level deployment. This grid will beget the next and the next, and—as we humans are lousy at anticipating the results of exponential growth—there's really no telling exactly where we'll end up. One thing for certain, though, it'll be a place with a whole lot more water.

Solving Sanitation

It's an open debate: Who invented the modern toilet? Apocrypha holds that it was Thomas Crapper, a nineteenth-century English plumber, but the real story actually begins much earlier. In the West, while his technology was never commercialized, credit is now given to Sir John Harington, who invented a water closet in 1596 for his godmother, Queen Elizabeth I. In the East, innovation stretches back much further. Archaeologists recently unearthed a Han dynasty latrine dating to 206 BC. Complete with a running water supply, stone bowl, and an armrest, this 2,400-year-old Chinese technology looks downright modern. And that's the problem: when it comes to our indoor plumbing, not much has changed in a very long time.

But imagine the potential upgrades. Imagine toilets that require no infrastructure. No pipes under the floor, no leach field under the lawn, no sewer systems running down the block. These high-tech outhouses powder and burn the feces and flash evaporate the urine, rendering everything sterile along the way. Rather than wasting anything, these toilets give back: packets of urea (for fertilizer), table salt, volumes of freshwater, and enough power that you can charge your cell phone while taking a crap, should the need arise. Tie these toilets into the smart grid, and the electricity can be sold back to the utility company, marking the first time in history that anyone has been paid to poop. As a final component, do all this at a cost to the consumer of five cents a day. Now, that's not just an upgrade, it's a revolution.

It's also the goal of a recently announced Bill & Melinda Gates Foundation program. Eight universities have received funding to help bring toilet technology into the twenty-first century, which is how Lowell Wood got involved in the effort. Wood is not your typical sanitation expert. He's an astrophysicist at Lawrence Livermore National Laboratory, with a background in thermonuclear fusion, computer engineering, X-ray lasers, and, most famously, President Ronald Reagan's "Star Wars" missile defense program.

"The thrust of the Gates project," says Wood, "is to upgrade a system that hasn't really evolved in 130 years, since Victorian England. In the developing world, where sanitation issues cause tremendous death and disease, this will obviously save millions and millions of lives, but in the developed world, three-quarters of our water bill is the cost of hauling away waste and running sewage treatment plants. So the goal is to solve both problems: to find a way for people to go to the bathroom that doesn't involve running water or sewage, while still rendering human waste completely harmless."

This may sound like fantasy, but no magic is required. "You can burn the fecal portion of the waste and use that energy to completely clean up the urine, turning it back into water and solids," explains Wood. "There's over a megajoule per day of energy in human feces, which is enough to do everything the toilet needs to do, with plenty left over for cell phones and lights. And we have the technology already; we can literally do this with off-the-shelf parts. The biggest challenge is it has to be done at a cost of five cents a day because that's the cost that's affordable in the developing world."

The upside of this toilet is almost incalculable. For starters, removing human feces from the equation solves an enormous portion of the global disease burden (which also slows population growth). Doing so in a way that is distributed (so that it doesn't require massive upfront infrastructure investment) and net positive for water and power makes this technology radically disruptive. Moreover, the efficiencies provide a much-needed savings. Toilets account for 31 percent of all water use in America. The US Environmental Protective Agency (EPA) estimates 1.25 trillion gallons of water—the combined annual usage of LA, Miami, and Chicago—leaks from US homes each year, with toilets being the biggest waster. Lastly, in addition to feces and urine, this technotoilet processes all organic wastes, including table scraps, garden cuttings, and farm refuse, thus closing all the loops while providing a family with all the water they might require.

The Pale Blue Dot

In 1990, in one of the most celebrated acts of an extremely illustrious career, astronomer Carl Sagan decided it might be interesting to have the Voyager 1 spacecraft, after completing its mission at Saturn, spin around and take a snapshot of the Earth. Viewed across this vast distance, the Earth is inconsequential, a nondescript speck among specks—or, as Sagan says, "a mote of dust suspended on a sunbeam." But it's a blue mote; thus the photograph's famous name: "the pale blue dot."

Our planet is a pale blue dot because it's an aqueous world, two-thirds of its surface covered by oceans. Those oceans are our backbone and our lifeblood. There is no question that a billion people now lack access to safe drinking water, but our oceans hold the secret to a better future. To return to an earlier theme: abundance is not a cornucopian vision. While the innovations just explored share the potential to tap these oceans—recycle their contents and change their chemistry, providing us with all the water we need and then some—it will not happen automatically. We have much work ahead. Yet because these waterwise technologies are all on exponential growth curves, they represent the greatest leverage available. They are the easiest path from A to B, but—and it's a critical "but"—we still must commit ourselves to the path.

Of his famous photograph, Sagan once said: "This distant image of our tiny world . . . underscores our responsibility to deal more kindly with one another, and to preserve and cherish the pale blue dot, the only home we've ever known." And we couldn't agree more. So today, right now, bring on the efficiencies, take shorter showers, eat less beef, do all that we can to preserve a currently limited resource. But for tomorrow, know that a world of watery plenty is a very real possibility, and putting our energy behind exponentials puts us on the fast track. The technologies explored in this chapter and the fields of research they represent are the very best way to preserve the only home we've ever known: this pale blue dot.

CHAPTER NINE

FEEDING NINE BILLION

The Failure of Brute Force

It's been said that feeding the hungry is the world's oldest philanthropic aim, but that doesn't mean we've gotten good at it. According to the UN, 925 million people currently don't have enough to eat. That's almost 1 out of every 7 of us, with the young being the most visible victims. Each year, 10.9 million children die—half because of issues related to undernourishment. In developing nations, 1 out of 3 children show stunted growth resulting from malnutrition. Iodine deficiency is the single leading cause of mental retardation and brain damage; a lack of vitamin A kills a million infants annually. And this is where we are today, right now, before the world's population balloons by billions, before global warming reduces arable land, before—that is to say—an already unfathomable problem becomes downright ineffable.

That said, the situation brings to mind the story of two shoe salesmen from Britain circa 1900. Both go to Africa to explore new markets. After a week, each writes a letter home. The first salesman reports: "Prospects are terrible, no one here wears shoes, I'm on the next boat out." But the second sees things differently: "This place is amazing. Market potential is almost unlimited. I may never leave." In other words, when it comes to food, there's ample opportunity for improvement.

Over the past one hundred years, agriculture has mainly been a brute force equation. First we industrialized our farms, next we industrialized our food. We backboned our food production and distribution systems with petroleum products. These days, it takes 10 calories of oil to produce 1 calorie of food. In a world facing energy shortages, this alone makes the process untenable. Irrigation systems have pumped our reservoirs dry. Major aquifers in both China and India are almost gone, resulting in dust bowls far worse than the American Midwest suffered in the 1930s. Toxic herbicides and pesticides have destroyed our waterways. Runoff from nitrogen-laden fertilizer has turned our coastal waters into dead zones so severe that the United States, a nation surrounded by oceans, must now import 80 percent of its seafood from abroad.

But even that bizarre practice can't last. Modern fishing practices are another part of this brute force equation. Bottom trawling destroys about six million square miles of that sea floor every year—that's an area the size of Russia. So forget about importation. A 2006 report in the journal *Science,* written by an international group of ecologists and environmentalists, showed that at our current pace of exploitation, the world will run out of seafood by 2048.

Moreover, we seem to be exhausting the potential of many of the technologies that have produced the greatest gains in food production over the past half century. According to Lester Brown, founder of the Worldwatch Institute and the Earth Policy Institute, "The last decade has witnessed the emergence of yet another constraint on growth in global agricultural productivity: the shrinking backlog of untapped technologies." Japan, for example, has used just about every technology available, and rice yields have flatlined for fourteen years. South Korea and China are facing similar situations. Production of wheat in France, Germany, and Britain, the three countries that account for one-eighth of the world's wheat, has similarly plateaued. And industrial farming has left poorer nations in even more precarious shape. Writing about the Punjab region in India—which many claim was transformed by the Green Revolution from "begging bowl" into "bread basket"—the celebrated environmentalist Vandana Shiva points out: "[F]ar from bringing prosperity, two decades of the Green Revolution have left the Punjab riddled with discontent and violence. Instead of abundance, the Punjab is beset with diseased soils, pest-infested crops, waterlogged deserts, and indebted and discontented farmers."

Yet, despite all of this devastation, the past century has also seen a miraculous change in our ability to produce food. We've managed to feed more people using less space than ever before. Currently we farm 38 percent of all the land in the world. If production rates had remained as they were in 1961, we would have needed 82 percent to produce the same amount of food. This is what petrochemical-backed agricultural intensification has made possible. The challenge going forward is to replace this unsustainable brute force with a considerably more nuanced approach. If we can learn to work with our ecosystems rather than run roughshod over them, while simultaneously optimizing our food crops and food systems, we could easily find ourselves in the place of that second shoe salesman: with a wide-open market and an infinite potential.

Cooking for Nine Billion

Many feel the question of how to best improve our food crops has been reduced to a binary—to GMO (genetically modified organism) or not to GMO. Truthfully, though, that's no longer the question. In 1996 there were 1.7 million hectares of biotech crops in the world; by 2010, the number had jumped to 148 million hectares. This 87-fold increase in hectares makes genetically engineered seeds (GEs) the fastest-adopted crop technology in the history of modern agriculture. Seriously, that horse has already left the barn.

Furthermore, the idea that GE crops are a Frankenfood sin against nature is, to be blunt, pretty ridiculous. It rests on the proposition that there's something natural about agriculture. As idyllic as it seems, farming is just a 12,000-year-old way of optimizing lunch. In fact, as Matt Ridley explains:

> [A]lmost by definition, all crop plants are "genetically modified." They are monstrous mutants capable of yielding unnaturally large, free-threshing seeds or heavy, sweet fruits and dependent on human intervention to survive. Carrots are orange thanks only to the selection of a mutant first discovered perhaps as late as the sixteenth century in Holland. Bananas are sterile and incapable of setting seed. Wheat has three whole diploid (double) genomes in each of its cells, descended from three different wild grasses, and simply cannot survive as a wild plant—you never encounter wild wheat.

The lineage of agriculture is a lineage of humans rearranging plant DNA. For a very long time, crossbreeding was the preferred method, but then came Mendel and his peas. As we began to understand how genetics worked, scientists tried all kinds of wild techniques to induce mutations. We dipped seeds in carcinogens and bombarded them with radiation, occasionally inside of nuclear reactors. There are over 2,250 of these mutants around; most of them are certified "organic."

GE, on the other hand, allows us to be more precise in our search for new traits. For the first time in the history of plant breeding, the tools of genetic engineering allow us to understand what it is that we're doing. That's the real difference. That's what all this fuss has been about: a radical change in the quality and quantity of information available to us, a move from evolution by natural selection to evolution by intelligent direction.

This is not to say there aren't interesting non-GE techniques of seed optimization in development. The Kansas-based Land Institute is attempting to turn annual food crops like wheat and corn into perennials. The results could be fantastic. Natural ecosystems are far better than human-managed agricultural systems at converting sunlight into living tissue. Perennials—and mainly polyculture perennials (meaning a mixture of perennials growing side by side)—anchor those ecosystems. These plants have long roots and diverse architectures, making them weather tolerant, pest resistant, disease resistant, and able to produce more biomass per acre than human agriculture without requiring any fossil fuel inputs or degrading the soil and water. The issue is one of time. The Land Institute expects it to take another twenty-five years until these perennials are profitable and productive. Biocrops, meanwhile, are here today.

Moreover, after thirty years of research, a great many of our GE fears have been quieted. Health concerns appear to be a nonstarter. More than a trillion GE meals have been served, and not a single case of GE-induced illness has turned up. Ecological devastation was another worry, but, overall, GE appears to be good for the environment. The seeds don't require plowing, so soil structure remains intact. This halts erosion, improves carbon sequestration and water filtration, and massively reduces the amount of petrochemical inputs needed to grow our food. Herbicide use is also down, while yield increases are up.

"[W]hen farmers in India adopted Bt cotton in 2002," writes Stewart Brand in the *Whole Earth Discipline: An Ecopragmatist Manifesto,* "the nation went from a cotton importer to an exporter, from 17 million bales to 27 million bales. What was the social cost of that? The main event was that Bt cotton increased yields by 50 percent and decreased pesticide use by 50 percent, and the Indian grower's total income went from $540 million to $1.7 billion."

This is a present-tense progress report. The agricultural portion of the biotech industry is growing at 10 percent a year; the technology itself, on a faster curve. In 2000, when the first plant genome was sequenced, it took seven years, $70 million, and five hundred people. The same project today takes about three minutes and costs about $100. This is good news. More information means better targeted approaches. Right now we're enjoying first generation GE crops; soon we'll have versions that can grow in drought conditions, in saline conditions, crops that are nutritionally fortified, that act as medicines, that increase yields and lower the use for pesticides, herbicides, and fossil fuels. The best designs will do many of these things at once. The Gates Foundation–led effort BioCassava Plus aims to take cassava, one of the world's largest staple crops, fortify it with protein, vitamins A and E, iron, and zinc; lower its natural cyanide content, make it virus resistant, and storable for two weeks (instead of one day). By 2020, this one genetically modified crop could radically improve the health of the 250 million people for whom it is a daily meal.

Sure, there are issues with GE. No one wants to see a few companies in charge of the world's food supply, so who owns the seed is a real concern. But this too won't last. As the wife-and-husband team of University of California at Davis plant pathologist Pamela Ronald and UC Davis organic farming expert Raoul Adamchak described in their book *Tomorrow's Table: Organic Farming, Genetics, and the Future of Food:* "It [GE] is a relatively simple technology that scientists in most countries, including many developing countries, have perfected. The product of GE technology, a seed, requires no extra maintenance or additional farming skills." This means that GE is already democratic, provided that we can learn to share the intellectual property. This hasn't happened yet (or not in any great measure), but in a recent speech given at the Long Now Foundation, author and organic activist Michael Pollan called for an open source movement for GE crops. Stewart Brand agrees, arguing that "if Monsanto throws a fit, tell them that if they're polite, you might license back to them the locally attuned tweaks you've made to their patented gene array."

But even with open-sourced GE crops, feeding the world isn't just about the production side of the equation—there's also distribution to consider.

So consider this: we live on a planet where nearly one billion people are hungry, yet we already produce more than enough food to feed the world. According to the Institute for Food and Development Policy/Food First, there are 4.3 pounds for every person every day: 2.5 pounds of grain, beans, and nuts; about a pound of meat, milk, and eggs; and another pound of fruits and vegetables. Many believe the incredible waste in our distribution system is the issue. While that's true, if we're really serious about feeding the world, the solution isn't to find new ways to move food around more efficiently. It's time to move the farm.

Vertical Farming

This isn't the first time we've been forced to move the farm. During the tail end of the Second World War, the US military was having trouble feeding itself. This too was a distribution problem. With troops strung out all over the world, not only was it prohibitively expensive to transport perishables hither and yon but also supply ships tended to be easy prey for submarine attacks. The obvious answer was also to grow food locally, but with soldiers stationed on barren islands in the Pacific and in arid deserts in the Middle East, fertile soil was not readily available. Then again, who needs soil when there's water?

The idea of growing food in water dates back, at least, to the Hanging Gardens of Babylon. But hydroponics, the growing of food in a nutrient-rich solution, is a more modern development. The first published work on the subject was Francis Bacon's 1627 *Sylva Sylvarum: or, a Natural History, in Ten Centuries,* but the tech didn't come of age until the 1930s, when scientists perfected the chemical composition of the growth medium. Yet beyond the occasional odd application—Pan American Airways grew veggies on Wake Island in the 1930s so that passengers could enjoy leafy greens with their midflight meal—no one had tried to farm this way at scale.

World War II changed all of this. In 1945 the US military began building a series of large-scale hydroponic experiments, first on Ascension Island in the South Atlantic, and later on Iwo Jima and in Japan—including what was then the world's largest hydroponic facility: a twenty-two-acre farm in Chofu. Simultaneously, because we had troops guarding our oil supply, more hydroponic farms were built in Iraq and Bahrain. All were incredibly successful. In 1952 alone, the army's hydroponic division grew over eight million pounds of fresh produce.

After the war, most people forgot about these successes. Food production went back to the soil. The Green Revolution occurred, and hydroponics was further sidelined for petrochemical solutions. A trickle of research continued. NASA, which wanted to know how to feed astronauts on Mars, stuck with it. A few others did as well. In 1983 Richard Stoner made a major breakthrough, discovering that it was possible to suspend plants in midair, delivering food through a nutrient-rich mist. This was the birth of *aero*ponics, which was when things started to get really interesting.

Traditional agriculture uses 70 percent of the water on the planet. Hydroponics is 70 percent more efficient than traditional agriculture. Aeroponics, meanwhile, is 70 percent more efficient than hydroponics. Thus, if we used aeroponics for agriculture, we could drop water use from 70 percent to 6 percent—quite the savings. With the threat of water scarcity getting more serious every day, it's hard to believe these technologies haven't been widely adopted.

"It's a PR problem," says Dickson Despommier. "When people hear *hydroponics,* they don't think NASA, they think pot grower. Hell, until about ten years ago, *I* thought pot grower."

But this is starting to change, and Dr. Despommier is somewhat responsible. A tall man with a gray beard, Despommier is a microbiologist and ecologist by training, one of the world's leading experts on intracellular parasitism, and, until his retirement in 2009, a professor of public health at Columbia University. In 1999 Despommier was teaching a class in medical ecology that included a section on climate change and its potential impact on food production.

"It was a really depressing thing to have to teach," he recalls. "The FAO [Food and Agriculture Organization of the United Nations] estimates that agricultural production needs to double by 2050 to keep up with population growth. Yet eighty percent of the arable land is already in use, and our current reports on climate change show crop production declining by ten percent to twenty percent in the next ten years. By the time I was done laying this out for my students, they wanted to throw rotten tomatoes at me."

Sick of the doom and gloom, Despommier set aside his regular curriculum and instead challenged his students to come up with a positive solution. After thinking it over, they came back to him with rooftop gardening.

"It was local," says Despommier. "It seemed doable. They wanted to know how many people they could feed by growing food on all the rooftops—no commercial buildings, just apartment complexes—in Manhattan. So I gave them the rest of the semester to figure it out."

As this was the era before Google Maps, just deducing the available rooftop space took three weeks in the New York Public Library. "What to grow?" was the next question. Their crop needed to be capable of dense production but pack a large nutritional punch. They settled on rice. But then they did the math. Growing rice on all the rooftops in New York would feed only 2 percent of the city's population.

"They were pretty upset," recalls Despommier. "All that work, and all they could feed was two percent of New York. I tried to mollify them, saying, 'Well, if you can't grow food on the rooftops, what about all those apartment buildings that are abandoned? What about Wright-Patterson Air Force Base? What about skyscrapers? Imagine how much food we could grow if we just stuck it inside tall buildings.' "

At the time, for Despommier, it was mostly a throwaway notion, something said quickly to appease his students. But the idea stayed with him. His wife wanted to know how it would work, so he found himself looking up hydroponics on the Internet. "I read about what the military accomplished during WWII and realized two things: Hydroponics wasn't just for pot growers. And my crazy vertical farming idea—it wasn't so crazy."

His students were equally enthralled. They went right back to work. Within a year, a rough design was hashed out, and their vertical farm could feed a heck of a lot more than just 2 percent of New York's population. "One thirty-story building," says Despommier, "one square New York block in footprint, could feed fifty thousand people a year. One hundred fifty vertical farms could feed everyone in New York City."

And they have astounding advantages. Vertical farms are immune to weather, so crops can be grown year-round under optimal conditions. One acre of skyscraper floor produces the equivalent of ten to twenty traditional soil-based acres. Employing clean-room technologies means no pesticides or herbicides, so there's no agricultural runoff. The fossil fuels now used for plowing, fertilizing, seeding, weeding, harvesting, and delivery are gone as well. On top of all that, we could reforest the old farmland as parkland and slow the devastating loss of biodiversity.

So how does this all work? Nutrition, obviously, is hydroponically or aeroponically delivered. Plants also need sunlight, so vertical farms are designed for maximum shine. Parabolic mirrors bounce light around the building's interior, while the exterior is skinned in ethylene tetrafluoroethylene, a revolutionary polymer that is extremely light, nearly bulletproof, self-cleaning, and as transparent as water. Grow lights are also used, both at night and during cloudy conditions, and the electricity needed to run them will be generated by capturing the energy we now flush down our toilets. That's right: we will recycle our own dung. "New York City alone," says Despommier, "is shitting away nine hundred million kilowatts of electricity each year."

Perhaps most importantly, the average American foodstuff now travels 1,500 miles before being consumed. That's only the average. The typical US meal contains five ingredients grown in other countries. Dinner in LA could easily include beef from Chile (5,585 miles), rice from Thailand, (8,263), olives from Italy (6,353), mushrooms from New Zealand (6,508), and a nice shiraz from Australia (7,487). As 70 percent of a foodstuff's final retail price comes from transportation, storage, and handling, these miles add up quickly.

Vertical farms change all this. They reduce the number of days it takes sustenance to reach our plates to the number of minutes it takes to walk a head of lettuce down ten flights of stairs. And despite their futuristic feel, there are no new technologies involved, so vertical farms are already cropping up. There are a number of pilot projects in the United States, and more substantial efforts overseas. Japan, while it hasn't switched yet from horizontal to vertical production, is attempting to build several hundred "plant factories" to increase domestic food security. Using clean-room techniques and employing senior citizens to tend the plants, they can now harvest twenty lettuce crops a year instead of one or two, using traditional practices. Meanwhile, Sweden's Plantagon is already working on five vertical farming projects: two in Sweden, two in China, and one in Singapore. Its standard model, a huge glass sphere with planting boxes arranged in a giant spiral, allows a greenhouse of 10,000 square meters to grow 100,000 square meters' worth of produce.

Yet the real promise of vertical farms comes from adding tomorrow's technologies to today's ideas. Imagine ubiquitous embedded sensors perfecting temperature, pH balance, and nutrient flows. Add in AI and robotics that maximize planting, growing, and harvesting of every square meter.

Since food production is limited by a plant's ability to convert sunlight into fuel, how about using GE to improve this as well? Researchers at the University of Illinois have been working on this idea for a while now. They believe that over the next ten to fifteen years, photosynthetic optimization could increase crop yields by as much as 50 percent. By growing these optimized crops inside of vertical farms—and optimizing our LED lights to the plants' preferred spectrum—we could save even more energy (by removing the bandwidths that plants don't use) and push those yields significantly higher.

What all of this means is that for the 70 percent of us who will soon live in cities, vertical farms offer the clearest path toward ending hunger and malnutrition. These farms already have the ability to increase the amount of food grown per harvest by orders of magnitude and increase the number of possible harvests by factors of ten. They have the potential to produce all of this food while simultaneously requiring 80 percent less land, 90 percent less water, 100 percent fewer pesticides, and nearly zero transportation costs. Integrate a few emerging technologies—aquaponics for closed-loop protein production; robotic crop harvesting to lower labor costs; AI systems attached to biosensors for better environmental regulation; the continued development of biomass energy systems (so that the parts of the plant that are not eaten can be recycled as a fuel); the betterment and continued integration of waste recycling systems (to further close the loop and drop energy costs)—and we end up with the gold standard of sustainable agriculture: an entirely local food production and distribution system with no waste, zero environmental impact, and the scalable potential to feed the world.

Protein

We still have a problem. The strategies discussed so far in this chapter all improve crop production, but optimal health means 10 to 20 percent of one's total calories must come from protein. We can eat more tofu, but for much of the world, meat is the preferred choice. Unfortunately, while meat might not be murder, it's certainly murdering the planet.

Cattle, for starters, are energy hogs, with the standard ratio of energy input to beef output being 54:1. They're also a land hog, with livestock production accounting for 70 percent of all agricultural lands and covering 30 percent of all land surface on the planet. Ranching produces more greenhouse gases than all the cars in the world, and is the leading cause of soil erosion and deforestation. Disease is another issue. Tightly packed herds of animals are breeding grounds for pandemics. The global demand for meat is expected to double by 2050, so unless something changes, the threat of pandemics can only increase.

And the danger is increasing. As people rise out of poverty, their taste for meat rises too. Between 1990 and 2002, China's level of carnivorous consumption doubled. Back in 1961 the Chinese consumed 3.6 kilograms per person per year. By 2002, that had jumped to 52.4 kilograms. This same pattern can be seen emerging globally.

But something is changing—actually, two things. In the near term, there's aquaculture; in the long term, there's in-vitro meat. Aquaculture is nothing new. How old is another question. Manuscripts from the fifth century BC show fish farming was practiced in ancient China. Both the Egyptians and the Romans cultivated oysters as well. The more modern incarnation was a post–World War II innovation that's been pretty unstoppable ever since. From 1950 through 2007, global aquaculture yields increased from two million metric tons to fifty million metric tons. So while natural fisheries have been in decline during this same period (the global fish catch peaked in the 1980s), fish farming has allowed human consumption to keep on rising. Aquaculture is now the fastest-growing animal food production system, supplying nearly 30 percent of our seafood.

And that number needs to climb significantly higher. Back in 2003, the journal *Nature* reported that 90 percent of all large fish in the sea are gone, taken either for direct human consumption or for animal food, fertilizers, and oil. This list includes tuna, swordfish, marlin, and the large groundfish such as cod, halibut, skates, and flounder, all threatened by the downstream effect of overfishing and industrial fishing practices. As fabled oceanographer Sylvia Earle (often called "Her Deepness") explained in the pages of *National Geographic*:

> Trawling takes huge amounts of bycatch, birds, mammals, and a whole host of life. Many creatures we don't even have names for yet get lost, killed in the process of dragging nets across the sea floor to catch shrimp and flounder and other bottom dwellers. And longlines—with baited hooks every

> few feet—may run 50 or 60 miles through the ocean and just catch whatever's there. There's no sign on the hook that says it shouldn't be swordfish or tuna, and those are two that shouldn't be caught right now. If we want to have recovery take place, we should be giving them a break.

Aquaculture is a large part of that break. The practice is renewable and scalable. And besides helping to protect our oceans, the National Oceanic and Atmospheric Administration (NOAA) believes that fish farming can reduce America's need for seafood imports ($10 billion worth a year), create jobs, reduce the trade deficit, and improve food security. Others are more cautious. For carnivorous fish such as salmon, aquaculture requires two pounds of wild-caught fish to feed one pound of farmed fish. Breeding farms suffer all the issues of factory farming: concentrate thousands of fish, and waste and disease become a problem. Another is the destruction of natural habitats. Shrimp farming, for example, has devastated coastal mangrove forests around the world.

But here too we are learning from our mistakes. Thanks to a considerable amount of international pressure, the shrimp industry is starting to clean up its act. Improved vegetable proteins and rendered animal by-products, fortified with amino acids, are replacing wild-caught fish in most salmon farming operations. There are even bigger gains found in combining integrative agriculture with aquaculture.

On a smaller scale, Asian rice farmers use fish to fight rice pests such as the golden snail, both boosting rice yields and protein consumption (as they also get to harvest the fish). In Africa, farmers are installing fish ponds in home gardens, as the mud from the bottom of the pond makes a great mineral-rich fertilizer. On a larger scale, the most exciting innovation may belong to Will Allen, the MacArthur Genius Award–winning force behind Growing Power, a Milwaukee-based organization building one of the United States's first vertical farms. Allen, a pioneer in urban aquaculture, aims to devote the first floor of his vertical farm to the process. Some 110,000 gallons of water will produce 100,000 tilapia, lake perch, and, possibly, bluegill a year. The fish feces will be recycled to fertilize plants on higher levels of the greenhouse.

But this is just a starting point. If we're really serious about protecting our oceans and preserving seafood as a source of protein, integrated aquaculture needs to be a significant part of our entire food chain. "If we value the ocean and the ocean's health at all," continues Earle, "we have to understand that fish are critical to maintaining the integrity of ocean systems, which in turn make the planet work. We have been so single-minded about fish, thinking that the only good fish is a cooked fish, rather than recognizing their importance to the ecosystem that also has a great value to us."

Cultured Meat

In 1932 Winston Churchill said, "Fifty years hence, we shall escape the absurdity of growing a whole chicken in order to eat the breast or wing by growing these parts separately under a suitable medium." As it turns out, it took a few extra decades for biotechnologists to deliver on Churchill's promise, but more and more, it looks like it was worth the wait.

Cultured meat (or in-vitro meat, as some prefer) is meat grown from stem cells. The process was pioneered by NASA in the late 1990s, as the agency suspected this might be a good way to feed astronauts on long space flights. By 2000, goldfish cells were being used to create edible muscle protein, and research began in earnest. By 2007, there had been enough progress that a collection of international scientists formed the In Vitro Meat Consortium to promote large-scale cultured meat production. The following year, an economic analysis presented at the In Vitro Meat Symposium in Norway showed that meat grown in giant tanks known as bioreactors could be cost competitive with European beef prices, and the People for the Ethical Treatment of Animals (PETA) created a $1 million incentive prize to move things along. By 2009, scientists in the Netherlands had succeeded in turning pig cells into pork inside a petri dish. More work has been done since then, and while we'll still a decade away from bringing this technology to market, we are definitely heading in that direction.

Providing people with protein is not all that will drive this change. "Cattle ranching is always going to be an environmental disaster, and ground beef is always going to be bad for you," says Jason Matheny, director of New Harvest, a nonprofit that funds research into cultured meat. "On reducing greenhouse gas emissions alone, switching to cultured meat is the equivalent of everyone in America suddenly driving hybrids. And, healthwise, real beef is always going to have fatty acids that contribute to heart disease.

You just can't turn a cow into a salmon, but cultured meat allows us to do just that. With in vitro meat, we can create a hamburger that prevents heart attacks, rather than one that causes them."

By growing beef in bioreactors, we also become less vulnerable to emerging diseases (70 percent of emerging diseases come from livestock) and contamination—something that occurs when workers in slaughterhouses accidentally slice open an animal's intestinal tract. Cultured meat has no gastrointestinal tract, so there's no danger of harmful bacteria spilling into our food supply. There are, of course, concerns that the same hostility facing GE crops will be encountered with cultured meat, but the medical establishment is in hot pursuit of organ regeneration. If we're willing to live with a lab-grown kidney permanently inside our bodies, then what concerns could we possibly have with cultured beef spending a few hours in our stomachs?

Beyond the increased health benefits, both from nutritionally fortified meat and from the reduced chance of pandemic, the 30 percent of the world's surface that is currently used for livestock can be reforested. The Belgium-sized chunk of Amazonian rain forest razed annually for cattle production can now be kept intact, the 40 percent of the world's cereal grains now devoured by livestock can be repurposed for human consumption, and the forty *billion* animals killed each year (in the United States alone) no longer have to suffer for our benefit. As PETA president Ingrid Newkirk told the *New Yorker*: "If people are unwilling to stop eating animals by the billions, then what a joy to be able to give them animal flesh that comes without the horror of the slaughterhouse, the transport truck, and the mutilations, pain, and suffering of factory farming."

Between Now and Then

The three technologies presented in this chapter so far have world-feeding potential, but there are still issues to be discussed. While aquaculture is here today, the GE industry is dominated by three seeds (cotton, corn, soybean) and has yet to penetrate deep into the food crop market. That said, golden rice (rice fortified with vitamin A) is about to clear regulatory hurdles and enter the food chain. As many believe that this technology will save millions of lives, its arrival could bring a much-needed shift in public opinion and speed the acceptance of other biocrops. But, between GE's developmental timetables and regulatory hurdles, we're still five to ten years away from significant change.

Cultured meat, meanwhile, is probably ten to fifteen years out, and the same appears true for widespread deployment of vertical farms. Moreover, vertical farms are designed to be built within cities or just outside of them, but the majority of the world's hungry and malnourished now live in rural poverty. In light of these facts, this does raise the issue of stopgap measures.

While no blanket technology fits this bill, there's now an emerging set of agricultural practices that blends the best of agronomy, forestry, ecology, hydrology, and a number of other sciences. Known as agroecology, the basic idea is to design food systems that mimic the natural world. Instead of striving for zero-environmental impacts, agroecologists want systems that produce more food on less land while simultaneously enhancing ecosystems and promoting biodiversity.

And they're getting them. A recent UN survey found that agroecology projects in fifty-seven countries have increased crop yields an average of 80 percent, with some being pushed up to 116 percent. One of the most successful of those is the push-pull system, developed to help Kenyan maize farmers deal with pestilence, invasive parasitic weeds, and poor soil conditions. Without getting too technical, push-pull is an intercropping system in which farmers plant specific plants between rows of corn. Some plants release odors that insects find unpleasant. (They "push" insects away.) Others, like sticky molasses grass, "pull" the insects in, acting as a kind of natural flypaper. Using this simple process, farmers have increased crop yields by 100 to 400 percent.

More importantly, while these agroecological techniques are widely available today (three hundred thousand African farmers have already adopted push-pull), we are only beginning to understand their real potential. Although the practices themselves look decidedly low tech, all the fields they're informed by are information-based sciences and thus on exponential growth curves. Moreover, there's no anti-GE bias permeating agroecology, so as better and better biotech becomes available, these new seeds can be quickly integrated into these sustainable systems. As UC Davis plant pathologist Pamela Ronald explained in an article for the *Economist,* this may be the very best way forward:

> A premise basic to almost every agricultural system (conventional, organic, and everything in between) is that seed can only take us so far. The farming practices used to cultivate the seed are equally important. GE crops alone will not provide all the changes needed in agriculture. Ecologically based farming systems and other technological changes, as well as modified government policies, undoubtedly are also required. Yet . . . there is now a clear scientific consensus that GE crops and ecological farming practices can coexist, and if we are serious about building a future sustainable agriculture, they must.

A Tough Row to Hoe

So there you have it: a long chain of sustainable intensification backed up by agroecological principles, GE crops, synthetic biology, perennial polycultures, vertical farms, robotics and AI, integrated agriculture, upgraded aquaculture, and a booming business in cultured meat. This is what it's going to take to feed a world of nine billion. It won't be easy. All these technologies will need to be scaled up simultaneously, and the sooner the better. This last point is key. We have a measure for the amount of plant mass-produced each year: it's called primary productivity. As every animal on Earth eats either plants or animals that eat plants, this number is a good metric for examining the impact that human food consumption is having on the planet. Right now we're consuming 40 percent of the planet's primary productivity. That's a dangerously high number. What's the tipping point? Perhaps 45 percent could be enough to start a catastrophic loss of biodiversity from which our ecosystems cannot recover. Perhaps it's 60 percent. No one knows for sure. What is known is that unless we figure out how to better the system and lower our impacts, then, with our ever-burgeoning population, we have little hope of a sustainable future. But if we follow the blueprint outlined in this chapter, we can radically increase the planet's primary productivity, protect its biodiversity, and concurrently make good on mankind's oldest humanitarian pledge: to feed the hungry. And we can do so in a truly abundant fashion.

PART FOUR
THE FORCES OF ABUNDANCE

CHAPTER TEN
THE DIY INNOVATOR

Stewart Brand

In the opening pages of *The Electric Kool-Aid Acid Test,* Tom Wolfe describes "a thin blond guy with a blazing disk on his forehead too, and a whole necktie made of Indian beads. No shirt, however, just an Indian bead necklace on bare skin and a white butcher's coat with medals from the King of Sweden on it." This guy is Stewart Brand: a Stanford-trained biologist, ex-army paratrooper, turned Ken Kesey cohort and fellow Merry Prankster who was about to become the voice of one of the most potent forces for abundance the world had yet seen: the Do-It-Yourself (DIY) innovator.

The story goes like this: a few months after Wolfe's book was published, in March 1968, Brand was reading a copy of Barbara Ward's *Spaceship Earth* and trying to answer a pair of questions: How can I help all my friends who are currently moving back to the land? And, more importantly, how can I save the planet?

His solution was pretty straightforward. Brand would publish a catalog in the vein of L. L. Bean, blending liberal social values, ideas about appropriate technology, ecological notions of whole systems thinking, and—perhaps most importantly—a DIY work ethic. This ethic has a long history, dating back at least as far as Ralph Waldo Emerson's 1841 essay "Self-Reliance," resurfacing again in the Arts and Crafts renaissance of the early twentieth century, then gaining even more steam with the hot-rodding and home improvement movements of the 1950s. But the late 1960s marked the largest communal uprising in American history, with conservative estimates putting the number at ten million Americans moving back to the land. All of these transplants soon learned the same lesson: agrarian success depended on one's DIY capabilities, and those capabilities, as Brand so clearly realized, depended on one's access to tools—and here tools mean anything from information about windmills to ideas about how to start a small business. "I was in the thrall of Buckminster Fuller," Brand recalls. "Fuller had put out this idea that there's no use trying to change human nature. It's been the same for a very long time. Instead, go after the tools. New tools make new practices. Better tools make better practices."

Out of all of this was born the *Whole Earth Catalog* (*WEC*). The first version, published in July 1968, was a six-page mimeograph that began with Brand's now-legendary DIY statement of purpose: "We are as gods and we might as well get good at it," and then a selection of tools and ideas to facilitate exactly this kind of personal transformation. Because so many people were then interested in such ideas, the catalog had the downstream effect of uniting once-disparate DIY-ers into a potent force. As TED founder Richard Saul Wurman explains: "This was a catalog for hippies that won the National Book Award. It was a paradigm shift in information distribution. I think you can draw a pretty straight line from the *WEC* to a lot of today's culture. It created an aroma that was sniffed by an awful lot of people. It's so pervasive that most don't even know the source of the smell."

At the center of that scent was the *WEC*'s embrace of personal technology: most importantly the PC. Brand is credited with inventing the term "personal computer," and while some of this had to do with his scientific background, more had to do with the Stanford Research Institute. In 1968 SRI was both at the cutting edge of computer research and located just around the corner from the Menlo Park offices of the *WEC*. Brand was a frequent visitor. On these trips, he was exposed to the computer mouse, interactive text, videoconferencing, teleconferencing, email, hypertext, a collaborative real-time editor, video games, and more. Brand saw the amazing potential of these tools and, in the pages of the *WEC*, told the world about what he'd seen.

"Stewart is singlehandedly responsible for American culture's acceptance of the personal computer," says Kevin Kelly (who was a *WEC* editor before founding *Wired* magazine). "In the sixties, computers were Big Brother. The Man. They were used by the enemy: massive, gray-flannel-suit corporations and the government. But Brand saw what was possible with computers. He understood that if these tools became personal, it flipped the world around into a place where people were gods."

Brand's marriage of self-reliance and technology helped shape the DIY innovator into a force for abundance, but just as important was the movement's adoption of two more *WEC* principles. The first was what would later become known as the "hacker ethic," the idea—as Brand famously put it—that "information wants to be free." The second was the then-strange notion that business could be a force for good. "Brand united the idea that you can do it yourself with new Utopian society," explains technology writer Howard Rheingold. "He really believed that given the right tools, any change was possible." And, as a man named Fred Moore discovered, the personal computer was exactly the right tool.

Homebrew History

The DIY innovator did not become a force for abundance overnight. The notion took some coaxing. It took a serious equipment upgrade. And, mostly, it took the help of a longtime political activist turned DIY innovator named Fred Moore.

In the early 1970s, Moore realized there was power in networking. If he could find a way to connect all the key players in all the various left-leaning movements operating in America, perhaps those movements could really become a force for reckoning. He started keeping records of the players and their contact information on three-by-five-inch note cards, but there were so many of them that he was soon overwhelmed. He suspected that his database would be significantly more effective if he could use a computer to manage it, but how to afford a computer was the real issue. Because Moore didn't have enough money to buy a machine of his own, in 1975 he decided to start a hobbyists club to help him build one.

This was the birth of the Homebrew Computer Club, a collection of tech hobbyists who gathered at the Community Computer Center in Menlo Park to swap circuits and stories. Early members included fabled hackers such as John Draper (Captain Crunch), Osborne 1 creators Adam Osborne and Lee Felsenstein, and Apple cofounders Steve Wozniak and Steve Jobs. Moore never lost sight of his activist past and was constantly reminding people to "give more than you take"—which was a fancy way of saying "Share your trade secrets"—but his members took it to heart. The Homebrew Club believed in building amazing machines, selling its creations (hardware), and sharing its intellectual property (software). As John Markoff explains in *What the Dormouse Said: How the 60s Counterculture Shaped the Personal Computer Industry,* nothing has been the same since:

> The Homebrew Computer Club was fated to change the world . . . At least twenty-three companies, including Apple Computer, were to trace their lineage directly to Homebrew, ultimately creating a vibrant industry that, because personal computers became such all purpose tools for both work and play, transformed the entire American economy. With Ted Nelson's computing power-to-the-people rallying cry echoing across the landscape, the hobbyists would tear down the glass-house computing world and transform themselves into a movement that emphasized an entirely new set of values from traditional American business.

With his championing of the DIY innovator, Stewart Brand had sparked a match, and the Homebrew Computer Club was part of the resulting conflagration. But it was not the only part. As we shall see in the next section, because I came of age at a time when DIY innovators had already transformed big business and big science, the idea of taking the space race out of the hands of government didn't seem entirely impossible. "The *WEC* not only gave you permission to invent your life," Kevin Kelly once said, "it gave you the excuses and the tools to do just that. And you believed you could do it, because on every page of the catalog were other people doing it." So while making off-world travel a DIY enterprise might not be easy, the reverberations of the *WEC* gave me exactly what they gave so many other people: the courage to try.

The Power of Small Groups (Part I)

The argument that sits at the core of this chapter is that because of people like Stewart Brand and Fred Moore—and because the quality of our tools has finally caught up to the scope of their vision—small groups of dedicated DIY innovators can now tackle problems that were once solely the purview of big governments and large corporations. While I've seen this happen repeatedly, no example is more illustrative than the story of Burt Rutan.

Rutan is a tall man, with a wide brow, gray hair, and a pair of mutton-chops to rival Neil Young. Before he retired in 2010, he ran a design and test flight facility called Scaled Composites. In 2004 Scaled responded to the Ansari X PRIZE (more on this later) and did something that every major aerospace company and government agency thought impossible: changed the paradigm of human spaceflight.

In America, our relationship with the final frontier began in the spring of 1952, when the National Advisory Committee for Aeronautics (NACA)—which would later become NASA—decided it was time to go up, up, and away. The aim was to fly an airplane faster and higher than anyone had ever gone before, with an official goal of Mach 10 (ten thousand feet per second) and one hundred kilometers straight up (into the middle of the mesosphere). The result was the X-series of experimental aircraft, including the X-1, which carried pilot Chuck Yeager through the sound barrier, and the X-15, which carried Joe Walker so much farther.

The X-15 was an extreme machine. Built from a nickel-chrome alloy called Inconel X, the plane could withstand temperatures hot enough to melt aluminum and render steel useless. It "took off" from California's Edwards Air Force Base, strapped beneath the wing of a B-52. The bomber carried the X-15 some forty-five thousand feet into the air, then dropped it like a rock. After falling a safe distance away, the rocket plane fired up its engines and went bat out of hell through the sky—which is what it took to get pilot Joe Walker off this planet.

Walker's departure took place on July 19, 1963, the date he flew the X-15 past the one-hundred-kilometer mark, becoming the first man to fly a plane into space. It was an incredible feat, and one that required an incredible effort. It took two major aerospace contractors employing thousands of engineers to build the X-15. By 1969, the program had cost about $300 million—more than $1.5 billion today. But this was the cost of flying to the edge of space until Burt Rutan came along.

Rutan didn't start out wanting to build spaceships, he started out building airplanes. He built a lot of them. Extremely lucky airplane designers work on three or four machines over the course of a career. Rutan, on the other hand, is prolific. Since 1982, he's designed, built, and flown an unprecedented forty-five experimental aircraft, including the Voyager, which made the first nonstop, non-refueled flight around the world, and the Proteus, which holds the world record for altitude, distance, and payload lift. Along the way, Rutan also developed a serious frustration with NASA's inability to truly open the space frontier.

In his mind, the problem was one of volume. "The Wright Brothers lifted off in 1903," he says, "but by 1908, only ten pilots had ever flown. Then they traveled to Europe to demonstrate their aircraft and inspired everyone. The aviation world changed overnight. Inventors began to realize, 'Hey, I can do that!' Between 1909 and 1912, thousands of pilots and hundreds of aircraft types were created in thirty-one countries. Entrepreneurs, not governments, drove this development, and a $50 million aviation industry was created."

Now contrast this with human spaceflight. Since Soviet cosmonaut Yuri Gagarin in 1961, only one spaceplane and a handful of rockets have carried humans into space: X-15, Redstone, Atlas, Titan, Saturn, Shuttle, Vostok, Voskhod, and Soyuz. All government owned and operated. As of April 2010, forty-nine years since spaceflight became possible, about three hundred manned flights have taken a total just over five hundred people into space—an unacceptable total, in Rutan's mind.

"When Buzz [Aldrin] first walked on the Moon," he says, "I'll bet he was thinking that in forty years we'll be walking on Mars. But we're not, and we're not close. Space travel is still primitive. Our rate of spaceflight is pathetically low: less than one flight every two months. Rather than go on to Mars, we have retreated to low Earth orbit. We serially abandoned former launch capabilities, and now the only spaceship we have, the Space Shuttle [the Shuttle program ended in 2011], is the most complex, most costly, and most dangerous. Why is the space program making acronyms for engineering welfare programs instead of having the courage to fly hardware? We have the courage here at Scaled."

This is not just egotistical chatter. Rutan backed up his words with action, beating the behemoths at their own game. His human-carrying spaceplane, imaginatively called SpaceShipOne, outperformed the government's X-15 in every measure. Rather than costing billions and requiring a workforce of thousands, in 2004 SS1 took flight with only $26 million and a team of thirty engineers. Instead of just one astronaut, SS1 boasted three seats. Forget a turnaround time measured in weeks, Rutan's vehicle set a record flying to space twice in just five days. "The success of SpaceShip-One altered the perceptions of what a small group of developers can do," says Gregg Maryniak, director of the James S. McDonnell Planetarium in Saint Louis. "Everyone had grown to believe that only NASA and professional astronauts could travel into space. What Burt and his team did was demonstrate that all of us will have the chance to make that trip in the near future. He changed the paradigm."

The Maker Movement

A few years after Burt Rutan changed the paradigm for spaceflight, Chris Anderson did the same thing for unmanned air vehicles (UAV). Anderson is the editor in chief of *Wired* and, not surprisingly, something of a geek dad. About four years ago, he decided to spend the weekend with his kids building a LEGO Mindstorms robot and a remote control airplane. But nothing went as planned. The robots bored the kids—"Dad, where are the lasers?"—and the airplane crashed into a tree right out of the gate. While Anderson was cleaning up the wreckage, he began wondering what would happen if he used the LEGO autopilot to fly the plane. His kids thought the idea was cool—for about four hours—but Anderson was hooked. "I didn't know anything about the subject," he says, "but I recognized that I could buy a gyro from LEGO for $20 and turn it into an autopilot that my nine-year-old could program. That was mind blowing. Equally amazing was the fact that an autonomous flying aircraft is on the Department of Commerce's export control restrictions list—so my nine-year-old had just weaponized LEGO."

Curious to learn more, Anderson started a nonprofit online community called DIY Drones. In the beginning, the projects were simple, but as his community grew (currently to seventeen thousand members), so did their ambition. The cheapest military-grade UAV on the market is the Raven. Built by AeroVironment, this drone retails for $35,000, with the full system for $250,000. One of DIY Drones' first major projects was an attempt to build an autonomous flying platform with 90 percent of Raven's functionality at a radically reduced price. The members wrote and tested software, designed and tested hardware, and ended up with the QuadCopter. It was an impressive feat. In less than a year, and with almost no development costs, they created a homebrew drone with 90 percent of the Raven's functionality for just $300—literally 1 percent of the military's price. Nor is this a one-off demonstration. The DIY Drones community has developed one hundred different products in the same way, each in under a year, for essentially zero development cost.

But homebrew UAVs are only the beginning. Anderson's decision to hack his kids' toys puts him squarely amidst the burgeoning Maker Movement. Built around a desire to tinker with the objects in our daily environment, most date the origin of this movement to 1902, when the first issue of *Popular Mechanics* hit the stands. By the 1950s, tinkering had become a middle-class virtue. "Fix your house, fix up an old boat, fix up an old car," says Dale Daugherty, founder and publisher of *Make* magazine. "Tinkering was a way for a guy with a modest income to improve his life."

With the advent of the computer, hacking code became more fun than hacking objects, and the movement dropped underground, resurfacing as the bedrock ethos of punk-rock culture, later a mainstay at events like Burning Man. Over the past ten years, though, a leap from software back into hardware has occurred. "These days," says Daugherty, "there's a hands-on imperative. People are really passionate about getting access to and control of the technology in their lives. We're back to hacking the physical."

And the physical has never been more hackable. Think of it this way: less than five years after Burt Rutan spent $26 million beating the aerospace giants at their own game, DIY Drones took them down with volunteer labor, a few toys, and a couple hundred dollars' worth of spare parts. "It's radical demonetization," says Anderson, "a true DIY story about using open-source design to reduce costs a hundredfold while keeping ninety percent functionality." The aerospace industry, Anderson feels, is ripe for such demonetization, and his vision should make some of the stodgier companies very nervous. "Two orders of magnitude in cost reduction was easy," he says. "We're now going for three."

For exactly these reasons, the Maker Movement has serious abundance potential. Cheap drones can ferry supplies to places such as Bangladesh, where monsoons wash out roads, or to Botswana, where roads don't exist. Matternet, a Singularity University (SU) 10^9+ company, is planning an AI-enabled network of UAVs and recharging stations housed in shipping containers scattered throughout Africa. Orders are placed via smart phone. For villages disconnected from the global transportation network, this means that everything from replacement parts for farm machinery to medical supplies can now be shipped in via an autonomous QuadCopter—for less than six cents per kilogram-kilometer.

Conservation is another possible use for low-cost autonomous platforms. Knowing how many tigers are left in Siberia is critical to developing a protection plan, but with an area 7.5 million square miles, how do you count? A fleet of DIY drones could do the counting for us, or patrol rain forests for illegal logging, or hundreds of other suddenly affordable applications.

And UAVs are only one technology. Makers are now impacting just about every abundance-related field, from agriculture to robotics to renewable energy. Hopefully, you'll find this inspirational. One of this book's key messages is that anyone can take on a grand challenge. In less than five years, Chris Anderson went from knowing nothing about UAVs to revolutionizing the field. You too can start a community and make a contribution. And if software and hardware aren't your flavors of choice, how about wetware? As we shall see in the next section, groups of high school and college students have set out to hack the very stuff of life itself and launch the DIY bio moment.

DIY Bio

In the early 2000s, a biologist named Drew Endy was growing increasingly frustrated with the lack of innovation in genetic engineering. Endy grew up in a world where anyone could purchase transistor parts at RadioShack, snap them together, and they worked just fine. He wanted the exact same off-the-shelf reliability from DNA. In his mind—and in the minds of many genetic engineers at the time—there was no difference between cells and computers. Computers use a software code of 1s and 0s, whereas biology uses a code of *A*s, *C*s, *T*s, and *G*s. Computers use compilers and storage registries; biology uses RNA (ribonucleic acid) and ribosomes. Computers use peripherals; biology uses proteins. As Endy told the *New York Times*: "Biology is the most interesting and powerful technology platform anyone's ever seen. It's already taken over the world with reproducing machines. You can kind of imagine that you should be able to program it with DNA."

In 2002 he came to MIT as a research fellow and met a few other folks who shared this view. The following year, alongside Gerald Sussman, Randy Rettberg, and Tom Knight, Endy founded the International Genetically Engineered Machine (iGEM) competition: a worldwide synthetic biology competition aimed at high school and undergraduate students. Their goal was to build simple biological systems from standardized, interchangeable parts—essentially DNA sequences with clearly defined structures and functions—and then operate them within living cells. These standardized parts, known technically as BioBricks, would also be collected in an open-source database accessible to anyone who was curious.

IGEM may not sound all that unusual, but ever since James Watson and Francis Crick discovered the double helix in 1953, the business as usual of biotech meant mammoth companies such as Genentech or Human Genome Project–sized government efforts, both requiring billions of dollars and thousands of researchers. All Endy and his friends did was teach a monthlong class to a handful of students.

These students were divided into five teams and asked to design a version of *E. coli* bacteria that blinked fluorescent green. A number of the teams were successful. Their homemade bacteria went from a nondescript blob to a glow stick at a rave in a month's time. More successes followed. By 2008, iGEM teams were creating genetic gizmos with real-world applications. That year, a team from Slovenia took first place with immunobricks: a designer vaccine against *Helicobacter pylori,* the bacteria responsible for most ulcers. By 2010, following the BP oil spill in the Gulf of Mexico, a winning team from Delft University of Technology created the "alkanivore," which they described as a "toolkit for enabling hydrocarbon conversion in aqueous environments"—or, in plainer language, a bug able to consume oil spills.

What's more incredible than the sophistication of this work is its rapid rate of growth. In 2004 iGEM had 5 teams that submitted 50 potential BioBricks. Two years later, it was 32 teams submitting 724 parts. By 2010, it had grown to 130 teams submitting 1,863 parts—and the BioBrick database was over 5,000 components strong. As the *New York Times* pointed out: "IGEM has been grooming an entire generation of the world's brightest scientific minds to embrace synthetic biology's vision—without anyone really noticing, before the public debates and regulations that typically place checks on such risky and ethically controversial new technologies have even started."

To understand where this revolution might go, take a look at "Splice It Yourself," a DIY bio call to arms penned by University of Washington synthetic biology pioneer Rob Carlson in the pages of *Wired*:

> The era of garage biology is upon us. Want to participate? Take a moment to buy yourself a molecular biology lab on eBay. A mere $1,000 will get you a set of precision pipettors for handling liquids and an electrophoresis rig for analyzing DNA. Side trips to sites like BestUse and LabX (two of my favorites) may be required to round out your purchases with graduated cylinders or a PCR thermocycler for amplifying DNA. If you can't afford a particular gizmo, just wait six months—the supply of used laboratory gear only gets better with time. Links to sought-after reagents and protocols can be found at DNAHack. And, of course, Google is no end of help.

Certainly the media has loved this story. Between Carlson's call to arms and the success of the iGEM competition, there have been dozens of articles claiming the next Amgen was going to come out of some teenager's garage. Even more articles appeared claiming that terrorists would soon be creating bio bugs in basements—although Carlson and others believe that the situation is not as bad as many suspect. (We explore this further in the "Dangers of the Exponentials" appendix.) Whatever the case, the era of homebrew genetics has arrived. High school kids are creating new life forms. The last frontier of big science has fallen to the DIY innovator.

The Social Entrepreneur

If the DIY innovator is taking on big government science programs, then the social entrepreneur is the DIY-er taking on big government social programs. The term itself was coined in 1980 by Ashoka founder and legendary venture capitalist Bill Drayton to describe individuals who combine the pragmatic, results-oriented methods of a business entrepreneur with the goals of a social reformer. The idea was a little ahead of its time. It took another ten years for technological evolution to catch up, but with the generation of information and communication technology that arrived in the late 1990s, Drayton's idea became a real force for abundance.

After the explosion of the Internet, websites like DonorsChoose.org,

Crowdrise, and Facebook Causes began to champion issues that had once been sole property of international agencies such as the United Nations and the World Bank. Take Kiva. Launched in October 2005—and named for the Swahili word for unity—this website allows anyone to lend money directly to a small business in the developing world via a peer-to-peer microfinance model. By early 2009, the site had grown to 180,000 member entrepreneurs receiving $1 million in loans *per week.* As of February 2011, a Kiva loan was being made every seventeen seconds, for a total amount lent of more than $977 million. And while Kiva's interest rate is nonexistent, its repayment rate is over 98 percent—meaning that it is not only changing lives, but, as *Time* magazine pointed out in 2009, "Your money is safer in the hands of the world's poor than in your 401(k)."

Kiva is only one example. The movement has seen massive growth in the past ten years. By 2007, this third sector employed around 40 million people, with 200 million volunteers. And by 2009, according to B Lab, a nonprofit that certifies purpose-driven companies, there were 30,000 social entrepreneurs in the United States alone, representing some $40 billion in revenue. Later that same year, J. P. Morgan and the Rockefeller Foundation analyzed the potential of impact investing (in other words, backing social entrepreneurs) and estimated an investment opportunity between $400 billion and $1 trillion, with profit potential between $183 billion and $667 billion.

All told, this force has produced some very real results. KickStart, started in July 1991 by Martin Fisher and Nick Moon, demonstrates how two individuals can make a significant and measurable impact. Founded to give millions of people the technological means to lift themselves out of poverty, this nonprofit has developed everything from low-cost irrigation systems, to inexpensive presses for creating cooking oils, to devices to make earthen blocks for affordable home construction. These techs are then bought by African entrepreneurs who use them to establish highly profitable small businesses. In 2010, KickStart-backed businesses accounted for 0.6 percent of Kenya's GDP and 0.25 percent of Tanzania's GDP.

An even bigger example is Enterprise Community Partners, which the magazine *Fast Company* called "one of the most influential organizations you've never heard of." This organization is a for-profit/nonprofit social entrepreneurial hybrid specializing in financing affordable housing for the poor. Over the past twenty-five years, it has helped revitalize some of America's poorest neighborhoods, including Fort Apache in the Bronx and San Francisco's Tenderloin, but its bigger accomplishment was creating a low-income housing credit that accounts for some 90 percent of affordable rental housing in the United States. One reason that social entrepreneurs are considered an end to big government social programs is because, with this single credit, Enterprise has outperformed the Department of Housing and Urban Development (HUD) on its core issue for more than two decades.

And these are only a few of the grand challenges that DIY innovators are now beginning to solve. Currently their impact is being felt at every level of our pyramid, but before telling the rest of that story, let's first turn our attention to the next force for abundance: the technophilanthropists.

CHAPTER ELEVEN
THE TECHNOPHILANTHROPISTS

The Robber Barons

It's the morning of April 16, 2011, and the X PRIZE Foundation is holding its annual Visioneering meeting. This, in our parlance, is the process of brainstorming incentive competitions to solve the world's grand challenges. To help us do the big thinking, we invite top entrepreneurs, philanthropists, and CEOs for a weekend best described as a cross between a mini-TED and Mardi Gras.

This year the meeting is being hosted by the chairman of Fox Filmed Entertainment, Jim Gianopulos, at its Los Angeles studios. The only room large enough to hold everyone is the commissary. The walls are flat white, decorated with photographs of film icons from Cary Grant to Luke Skywalker, but it's a different kind of crowd, and few pay these images much mind. Nor does anyone have much to say about box office returns or points on the back end, but there's a lot of talk about creating African entrepreneurs, reinventing the technology of health care, and increasing the energy density of batteries by an order of magnitude.

Over the years, I've been lucky enough to host many similar meetings and meet many similar people, and what seems to unify them is exactly what's on display today: a high level of optimism, a magnanimous sphere of caring, and a hearty appetite for the big and bold. Perhaps this is to be expected. These are the same captains of the digital age who, with the stroke of HTML code, have reinvented banking with PayPal, advertising with Google, and commerce with eBay. They've seen firsthand how exponential technologies and the tools of cooperation can transform industries and better lives. They now believe that the same high-leverage thinking and best business practices that led to their technological success can bring about philanthropic success. Taken together, they constitute a significant force for abundance and a new breed of philanthropist: a technophilanthropist; a young, idealistic, iPad jet-setter who cares about the world—the whole world—in a whole new way.

Where did this breed come from, what distinguishes them, and why they constitute a force for abundance is the subject of this chapter, but before we get there, some context is useful. Large-scale philanthropy, based in the private, not the public sector, is a relatively recent historical development. Going back some six hundred years, wealth was concentrated within royals whose sole goal was to keep that money in the family. This sphere of caring expanded during the Renaissance, when European merchants tried to mitigate poverty in big trading cities like London. Two centuries ago, the financial community got involved. But it was the titans of industrialization known collectively as the robber barons who really rewrote the rule book.

The robber barons were transformative. In less than seventy years, they turned America from an agricultural nation into an industrial powerhouse. What John D. Rockefeller did for oil, Andrew Carnegie did for iron and steel, Cornelius Vanderbilt did for railroads, James B. Duke for tobacco, Richard Sears for mail-order retailing, and Henry Ford for automobiles. There were dozens more. And while robber baron rapaciousness has received much attention, contemporary historians are in agreement: it was also these gilded age magnates who invented modern philanthropy.

Certainly scholars have gone back and forth about most things robber baron, including the nature of their charity. Not long ago, *BusinessWeek* wrote: "John D. Rockefeller became a major donor—but only after a public relations expert, Ivy Lee, told him that donations could help salvage a damaged Rockefeller image." Great-great grandson Justin Rockefeller, an entrepreneur and political activist, disagrees: "John David Sr., a devout Baptist, started tithing from his very first paycheck. He kept meticulous financial records. His first year in business was 1855. His income was $95, 10 percent of which he gave to the church." Either way, that $9.50 donation was only the beginning. In 1910 Rockefeller took $50 million worth of Standard Oil stock to create the foundation bearing his name. By the time of his death in 1937, half of his fortune had been given away.

Carnegie, though, was an even bigger donor, and it's to Carnegie that most of today's technophilanthropists trace their roots. When Warren Buffett wanted to inspire philanthropy in Bill Gates, he started by giving him a copy of Carnegie's essay "The Gospel of Wealth," which attempts to answer a tricky question: "What is the proper mode of administering wealth after the laws upon which civilization is founded have thrown it into the hands of the few?"

Carnegie believed that one's wealth must be used to better the world, and the best way to do so was not by leaving the money for one's children or bequeathing it to the state for public works. His interest was in teaching others how to help themselves; thus, his major contribution was to construct 2,500 public libraries. While "The Gospel of Wealth" wasn't popular in Carnegie's time, much of his philosophy is now shared by many of the technophilanthropists, though, as we'll soon see, exactly *who* to help and *how* to do so is where today's generation and yesterday's benefactors diverge.

The New Breed

In 1892, when the *New York Tribune* attempted to identify every millionaire in the United States, the newspaper came up with 4,047 names. An astonishing 31 percent of them lived in New York City. And when it came to giving back, these millionaires gave back to whence they came. There is scarcely a museum, art gallery, concert hall, orchestra, theater, university, seminary, charity, or social or educational institution in New York that does not owe its beginnings and support to these men.

Such regional myopia is to be expected. The robber barons worked in a world that was local and linear. Poverty in Africa, illiteracy in India—these were not pressing issues in their lives or businesses, and thus these industrialists kept their dollars in the neighborhood. Even Carnegie was prone to the tendency, as every library he built, he built in the English-speaking world.

This local mind-set was not restricted to the ultrawealthy in the West. Take, for example, Osman Ali Khan, known as Asaf Jah VII, the last nizam of Hyderabad and Berar, who ruled from 1911 to 1948, when these states merged with India. Khan was proclaimed the richest man in the world by *Time* magazine in 1937. He had seven wives, forty-two concubines, forty children, and a net worth of $210 billion (in 2007 dollars). During his thirty-seven-year rule, he spent a fair amount of his fortune on his people, building schools, power plants, railways, roads, hospitals, libraries, universities, museums, and even an observatory. But despite such largess, Khan focused his charity entirely in Hyderbad and Berar. Like the robber barons in America, even the richest man in the world kept his wallet close to home.

Much has changed in the past few decades. Jeff Skoll, the first president of eBay turned media mogul turned technophilanthropist, says: "Today's technophilanthropists are a different breed. While the industrial revolution focused philanthropy locally, the high-tech revolution inverted the equation. There's a different mentality now because the world is much more globally connected. In the past, things that happened in Africa or China, you didn't really know about. Today you know about them instantly. Our problems are much more interrelated as well. Everything from climate change to pandemics have roots in different parts of the world, but they affect everybody. In this way, global has become the new local."

When Skoll cashed out of eBay in 1998 for $2 billion, he too took his philanthropy global. He created a foundation to pursue a "vision of a sustainable world of peace and prosperity." The Skoll Foundation attempts to drive large-scale change by investing in social entrepreneurship. According to Skoll, social entrepreneurs are "change agents," an idea he explained further in an article for the Huffington Post:

> Whether the issue is disease and hunger in Africa; or poverty in the Middle East; or lack of education across the developing world—we all know the problems. But social entrepreneurs, I believe, have a genetic deficiency. Somehow, the gene that helps them look past the impossible is missing . . . By nature, entrepreneurs aren't satisfied until they do change the world, and let nothing get in their way. Charities may give people food. But social entrepreneurs don't just teach people to grow food—they're not happy until they've taught a farmer how to grow food, make money, pour the profits back into the business, hire ten other people, and in the process, transform the entire industry.

In its first ten years, the Skoll Foundation awarded more than $250 million to eighty-one social entrepreneurs working on five continents. These entrepreneurs, in turn, have spread their goodwill into wider spheres.

"Take Muhammad Yunus," says Skoll, "who started the Grameen Bank and helped lift a hundred million-plus people out of poverty around the world; Ann Cotton, who has educated over a quarter million African girls through her organization Camfed; and Jacqueline Novagratz, CEO of the Acumen Fund, who is affecting the lives of millions of people in Africa and Asia."

Backing social entrepreneurs is only one example of the new direction taken by today's technophilanthropists. Investing in triple-bottom-line companies, as the Rockefeller-backed Acumen Fund does, is another. Acumen is an entirely for-profit company, but it makes those profits investing in businesses that manufacture goods and services urgently needed in the developing world—reading glasses, hearing aids, mosquito nets—and selling them at very affordable prices. Then there's eBay founder Pierre Omidyar's Omidyar Network, an organization that makes for-profit investments to pursue its mission of "individual self-improvement" in key areas such as microfinance, transparency, and—of course—social entrepreneurship. "If they [the technophilanthropists] can use their donations to create a profitable solution to a social problem," writes *Economist* New York bureau chief Matthew Bishop in his book *Philanthrocapitalism: How the Rich Can Save the World* (coauthored with Michael Green), "it will attract more capital, far faster, and thus have a far bigger impact, far sooner, than would a solution based entirely on giving the money away."

In choosing to blur the border between nonprofit and for-profit, they are also attempting to redefine charity. "The new philanthropists," continues Bishop, "believe they are improving philanthropy, equipping it to tackle the new set of problems facing today's changing world; and to be blunt, it needs improvement—much philanthropy over the centuries has been ineffective. They think they can do a better job than their predecessors. Today's new philanthropists are trying to apply the secrets behind that money-making success to their giving."

One concept lately gaining momentum is "impact investing" or "triple-bottom-line investing," whereby investors back businesses that generate financial returns *and* meet measurable social or environmental goals. The practice often gives investors a further reach than traditional philanthropy—and this practice is growing. According to the research firm the Monitor Group, what was $50 billion in impact investments in 2009 is on pace to reach $500 billion within the decade.

Another of those secrets is a hands-on approach. "It's no longer 'I write the check and I'm done,' " says Paul Shoemaker, executive director of Social Ventures Partners Seattle. "Now it's 'I write the check and that's the start.' " And when they start, the technophilanthropists do much more than just bring financial capital to the table; they bring their human capital as well. "They bring networks, connections, and the ability to get high-level meetings," says Shoemaker. "When Gates decided to fight for vaccines, he built a team and led that team into meetings with world leaders and the World Health Organization. Most organizations can't get into those rooms, but Gates could, and it made a huge difference."

There's one last distinction between the new-breed philanthropists and the older generations, and it may be the one that has the biggest impact. The majority of the robber barons got generous in their august years, but many of the technophilanthropists were billionaires before the age of thirty-five, and they turned to philanthropy right afterward. "Traditional philanthropists have typically been an older lot," says Skoll. "They've made their fortune, retired, and then toward the end of their life started giving it away. And they were less ambitious in their philanthropy—it's easier to write a check to build the opera house than it is to go out and tackle malaria, or AIDS, or other global issues. Many of today's technophilanthropists have the energy and confidence that come from building global businesses at such a young age. They want to tackle audacious goals like nuclear proliferation or pandemics or water. They think they can really make a difference in their lifetimes."

All of these differences have compounded, turning the technophilanthropists into what Paul Schervish of the Boston College Center on Wealth and Philanthropy calls *hyperagents.* As Matthew Bishop explains, hyperagents "have the capacity to do some essential things far better than anyone else. They do not face elections every few years, like politicians, or suffer the tyranny of shareholder demands for ever-increasing profits, like CEOs of most public companies. Nor do they have to devote vast amounts of time and resources to raising money, like most heads of NGOs. That frees them to think long term, to go against conventional wisdom, to take up ideas too risky for government, to deploy substantial resources quickly when the situation demands it—above all, to try something new. The big question is, will they be able to achieve their potential?"

And as we shall see in the next few sections, more and more, the answer to Bishop's question appears to be a resounding "yes."

How Many and How Much?

Naveen Jain grew up in Uttar Pradesh, India, the son of a civil servant. He became a student of entrepreneurship at a very early age. "When you are poor," he says, "and basic survival is your concern, you have no alternative but to be an entrepreneur. You must take action to survive, just as an entrepreneur must take action to seize an opportunity." Jain's actions and opportunities ultimately put him on a trajectory to Microsoft, and then, through his founding of InfoSpace and Intelius, onto the Forbes 500 list.

"My parents drilled into me the importance of an education. It was a gift they themselves never had. I remember how my mother quizzed me in mathematics first thing in the morning and would often demand, 'Don't make me solve this for you.' Little did I know that she couldn't solve it because she had never been taught math in school. Today we have the technology, through AI, video games, and smart phones, to quiz every child on the planet and assure them access to the best education available."

Jain signed on as the cochair of X PRIZE's Education and Global Development Advisory Group, and is now focusing his wealth on incentive competitions to reinvent education and health care in the developing world. "Technology allowed me to create the capital I now use for philanthropy," he says, "and I can think of no better use of these resources than to focus on eradicating illiteracy and disease around the world. What is truly amazing is that today we actually have the tools to make this happen."

Jain is not the only one who feels this way. The 2010 Credit Suisse *Global Wealth Report* estimated that the world has over 1,000 billionaires: roughly 500 in North America, 245 in the Asia-Pacific region, and 230 in Europe. Finance professionals note that these numbers are probably off by a factor of two, since many choose to hide their wealth from public scrutiny. Taking a step down the economic ladder, the next group, known as "ultra-high-net-worth" individuals, cuts a broader swath, ranging from $30 million in liquid assets to centimillionaires. In total, in 2009 the number of ultra-high-net-worth individuals was just over 93,000 worldwide. Not only are these numbers higher than ever before, these individuals are giving like never before.

"The Internet's rich are giving it away, their way," proclaimed the *New York Times* in 2000. By 2004, charitable giving in America had increased to $248.5 billion, the highest yearly total ever. Two years later, the number was $295 billion. By 2007, CNBC had taken to calling our era "a new golden age of philanthropy" and Foundation Giving reported a record-setting 77 percent increase in new foundations established in the past decade, an addition of more than 30,000 organizations. Certainly those numbers dipped during the recent recession: 2 percent in 2008, 3.6 percent in 2009. The ten-year low was in 2010, but that was also the year Bill Gates put $10 billion toward vaccines, the largest pledge ever made by a charitable foundation to a single cause.

2010 was also the year that Gates and Warren Buffett, the two richest men in the world, announced the "Giving Pledge," which asks the nation's billionaires to give away half their wealth to philanthropic and charitable groups within their lifetimes or at the time of their deaths. George Soros, Ted Turner, and David Rockefeller signed up almost immediately. Skoll too was an early joiner, as was Pierre Omidyar. Oracle cofounder Larry Ellison, Microsoft cofounder Paul Allen, AOL creator Steve Case, and Facebook cofounders Mark Zuckerberg and Dustin Moskovitz have all signed on as well. As of July 2011, the total had risen to sixty-nine signatories, with more joining all the time.

That the technophilanthropists are proving to be a significant force for abundance is not a question. They've already impacted all levels of our pyramid, including those that are hard to reach. Mo Ibrahim, a Sudanese telecommunications tycoon, recently established the Ibrahim Prize for Achievement in African Leadership, which awards $5 million (and $200,000 a year for life afterward) to any African leader who serves out his or her term within the limits of a country's constitution and then leaves office voluntarily.

But the best news is that most of these technophilanthropists are still young, so they're just beginning their journey. "As some of the smartest people look at where to focus their energies next," says PayPal cofounder Elon Musk, "they are now attracted to the biggest problems facing humanity, particularly in areas such as education, health care, and sustainable energy. Without suggesting complacency, I believe it is very likely that they will solve the many challenges in those areas, and the result will be the creation of new technologies, companies, and jobs that will bring prosperity to billions on Earth."

CHAPTER TWELVE
THE RISING BILLION

The World's Biggest Market

Stuart Hart met Coimbatore Krishnarao Prahalad, known universally as C.K., in 1985. Hart was then a newly minted PhD hired by the University of Michigan. Prahalad was already a full professor at its Ross School of Business and a growing legend. His ideas about "core competencies" and "cocreation" sparked a revolution in the management world, and his 1994 book *Competing for the Future,* coauthored with Gary Hamel, became a classic. Moreover, in his consulting work, Prahalad had a reputation for unorthodoxy and a significant track record for doing the impossible: convincing multinational corporations that nimble and collaborative was a better approach than staid and defensive.

Over the next few years, Hart and CK got to know each other. They taught classes together and became friends. In the late eighties, when most of Hart's professional colleagues were telling him to abandon his interest in the environment and stay focused on business, Prahalad was one of the few who encouraged his passion. "In fact," says Hart, "were it not for CK, I never would have made the conscious decision (which I did in 1990) to devote the rest of my professional life to sustainable enterprise. That was the best decision I ever made."

During their time at Michigan, the duo never collaborated. Hart left to run the Center for Sustainable Enterprise at the University of North Carolina. (Now he's the chair of the Cornell Center for Sustainable Global Enterprise.) From that post, in 1997, he wrote his now-seminal "Beyond

Greening: Strategies for a Sustainable World," which helped launch the sustainability movement. But that article, published in the *Harvard Business Review,* raised a number of follow-up questions that peaked Prahalad's interest, and the following year, the pair teamed up to answer them.

The result was another article, this one just sixteen pages long, that was destined to change the world—although, as Hart points out, that didn't happen overnight. "It took us four years before anyone would publish it. The paper went through literally dozens of revisions before coming out in 2002 as 'The Fortune at the Bottom of the Pyramid' [in the journal *Strategy + Business*]. That paper became an underground hit before it was ever published and spawned a whole new field: BoP business. For me, this was a life-changing experience. For C.K., it was another day at the office."

Their article made a simple point: the four billion people occupying the lowest strata of the economic pyramid, the so-called bottom billion, had lately become a viable economic market. They didn't claim that the bottom of the pyramid (BoP) was an ordinary market, rather that it was extraordinary. While the majority of BoP consumers lived on less than $2 a day, it was their aggregate purchasing power that made for extremely profitable possibilities. Of course, this radically different business environment demanded radically different strategies, but for those companies that could adapt to business unusual, both Hart and Prahalad felt that the opportunities were immense.

Backing up this claim was a quick survey of a dozen big-name companies that had all enjoyed considerable success in BoP markets after adopting business practices that were a little outside their comfort zone. Arvind Mills, for example, the world's fifth-largest denim manufacturer, had a history of struggling in India. At $40 to $60 a pair, its jeans weren't affordable for the masses, and its distribution system had almost zero penetration into rural markets. "So Arvind introduced Ruf & Tuf jeans," Hart and Prahalad wrote in *The Fortune at the Bottom of the Pyramid,* "a ready-to-make kit of jean components—denim, zipper, rivets, and a patch—priced at about six dollars. Kits were distributed through a network of thousands of local tailors, many in rural towns and villages, whose self-interest motivated them to market the kits extensively. Ruf & Tuf jeans are now the largest-selling jeans in India, easily surpassing Levi's and other brands from the US and Europe."

In 2004 these ideas were expanded into Prahalad's book *The Fortune at the Bottom of the Pyramid.* He opened with a strong statement of purpose: "If we stop thinking of the poor as victims or as a burden and start recognizing them as resilient and creative entrepreneurs and value-conscious consumers, a whole new world of opportunity will open up," and an even stronger statement of possibility: "The BoP market potential is huge: 4 to 5 billion underserved people and an economy of more than $13 trillion PPP (purchasing power parity)." While Prahalad's book presented twelve case studies of BoP business success, its biggest selling point was social rather than fiscal: finding cocreative ways to serve this market was a developmental activity, one that could pull the poor out of poverty.

One of the best examples is the telecom Grameenphone, which started in Bangladesh in 1997, and, as of February 2011, had thirty million subscribers in that country. Along the way, Grameenphone invested $1.6 billion in network infrastructure—which means that money made in Bangladesh actually stayed in Bangladesh. But the even bigger impact has been on poverty reduction. Economists at the London School of Business and Finance figured out that adding ten phones per one hundred people adds 0.6 percent to the GDP of a developing country. Nicholas Sullivan, in his book about the rise of microloans and cellular technology, *You Can Hear Me Now: How Microloans and Cell Phones Are Connecting the World's Poor to the Global Economy,* explains what this really means:

"Extrapolating from UN figures on poverty reduction (1 percent of GDP growth results in a 2 percent poverty reduction), that 0.6 percent growth would cut poverty by roughly 1.2 percent. Given 4 billion people in poverty, that means that with every 10 new phones per 100 people, 48 million *graduate* from poverty, to borrow a phrase from Mohammad Yunus."

Critics have pointed out that this approach can take us only so far, but they fail to mention that may actually be far enough. Hart and Prahalad's BoP argument is essentially one of commodification: take existing goods and services and make them orders of magnitude cheaper, then sell them on a massive scale. But there are two additional features. First, the methodology required to open these markets is based on cocreating products with the BoP consumer. Second, the products and services being commodified—soaps, clothes, home-building supplies, solar energy, microscopes, prosthetic limbs, heart surgery, eye surgery, neonatal baby care, cell phones, bank accounts, pumps, and irrigation systems, to name only the more famous success stories—may seem a random lot, but they share exactly what's needed to move massive numbers of people up the abundance pyramid.

When Hindustan Unilever, a subsidiary of Unilever, developed a hygiene-based marketing campaign for BoP markets in India, its goal was to sell more soap (which the company did, with sales increasing 20 percent). But for our purposes, more important was the fact that 200 million people learned that diarrheal disease—which kills 660,000 people in India each year—can be prevented simply by washing one's hands. This form of improvement quickly becomes empowerment, since the better health that results from hand washing adds income (fewer sick days from work) and keeps kids in school, and thus becomes a self-reinforcing cycle.

But the benefits don't just flow toward the consumer. As Hart explains in his (also now classic) 1995 book *Capitalism at a Crossroads: The Unlimited Business Opportunities in Solving the World's Most Difficult Problems,* "[I]t is very difficult to remove cost from a business model aimed at higher-income customers without affecting quality or integrity." To compete in BoP markets, a new wave of disruptive technology is required. Take Honda's motorcycles. In the 1950s, Honda began selling very stripped-down and inexpensive motorized bicycles in Japan's jam-packed, poverty-stricken cities. When these bikes entered the American market in the 1960s, they reached a considerably larger population than those who could afford Harley-Davidsons. Hart explains: "Honda's base in impoverished Japan gave it a huge competitive advantage in disrupting American motorcycle makers because it could make money at prices that were unattractive to established leaders."

Ratan Tata, the CEO of the gargantuan multinational Tata Industries, offers another great example. In 2008 he created the Nano, the world's first \$2,500 automobile. In 2008 the *Financial Times* reported, "If ever there were a symbol of India's ambitions to become a modern nation, it would surely be the Nano, the tiny car with the even tinier price tag. A triumph of homegrown engineering, the Nano encapsulates the dream of millions of Indians groping for a shot at urban prosperity." Besides benefiting India, Tata's efforts jump-started an innovation trend. A dozen plus companies, including Ford, Honda, GM, Renault, and BMW, are now developing cars for emerging markets, a development that will introduce a level of choice in transportation into BoP communities that was unimaginable just ten years ago.

Choice was the missing ingredient. Suddenly the rising billion—all four billion of them—have a way and a reason to participate in the global conversation. "This new generation growing up with freedom of communication," says Tata, "are plugged into an information and entertainment world that didn't exist before. They have needs and wants that exceed those of the older generation. And they're going to be demanding in terms of the quality of their life."

For the first time, not only are their voices being heard, their ideas—ideas that we've never had access to before—are joining the global conversation. And if for no other reason than the law of large numbers and the power of these ideas, this puts the rising billion in the same category as exponential technology, the DIY-ers, and the technophilanthropists: as a potent force for abundance.

Quadir's Bet

In 1993 Iqbal Quadir was working as a venture capitalist in New York when a temporary power outage shut down his computer. The inconvenience reminded him of his childhood in Bangladesh, when he once spent an entire day walking to buy medicine for his brother, only to arrive and find the pharmacy closed. Then, like now, poor communications led to wasted time and lowered productivity. In fact, by comparison, the power outage was just a minor inconvenience. So Quadir quit his job and moved back to Bangladesh to tackle this communication problem. Cell phones, he thought, were an obvious solution, but this was 1993. Back then, the cheapest cell phone available ran about $400 and had an operating cost of about fifty-two cents per minute, while the average yearly income in Bangladesh was $286, so how to pull this off was anybody's guess.

"When I first proposed the idea," says Quadir, "I was told I was crazy. I was thrown out of offices. Once, in New York, I was pitching the idea to a cell phone company, and they said, 'We're not the Red Cross; we don't want to go to Bangladesh.' But I knew what was happening in the Western world. I knew that cell phones were analog, and they were about to become digital, and that meant their core components would be subject to Moore's law—so they would continue to get exponentially smaller and cheaper. I also knew that connectivity equals productivity, so if we could get cell phones into the hands of BoP consumers, it would translate into their ability to pay for the phones."

Quadir won his bet. Cell phones followed an exponential price-performance curve, and Grameenphone transformed life in Bangladesh. By 2006, sixty million people had access to a cell phone, and the technology had added $650 million to Bangladesh's GDP. Other companies filled the gaps in other countries. In India, by 2010, fifteen million new cell phone users were being added *each month.* As of early 2011, over 50 percent of the world had cellular connectivity. And it's this technology that's transforming the "bottom billion" into the rising billion. "We snuck powerful computers into the hands of the people," explains Quadir. "They crept in through the killer app of voice communication." As a result, over the next few decades, these devices bring with them the potential to completely reshape the world.

We're already seeing this happen in banking. There are 2.7 billion people in the developing world without access to financial services. Impediments to change are considerable. In Tanzania, for example, less than 5 percent of the population have bank accounts. In Ethiopia, there's one bank for every 100,000 people. In Uganda (circa 2005), there were 100 ATM machines for 27 million people. Opening an account in Cameroon costs $700—more than most people make in a year—and a woman in Swaziland can manage that feat only with the consent of a father, brother, or husband.

Enter mobile banking. Allowing the world's poor to set up digital bank accounts accessible via cell phones has a significant impact on quality of life and poverty reduction. M-banking allows people to check their balances, pay bills, receive payments, and send money home without giant transfer fees, as well as avoid the increased personal security risks that come from carrying cash. In Kenya, where many poor people work very far away from home, workers would frequently disappear for three to four days after getting paid—the amount of time it took to get that money to their families—so being able to transfer cash wirelessly saves them incredible amounts of time.

For all of these reasons, mobile banking has seen exponential growth in a few short years. M-PESA, launched in Kenya in 2007 by Safaricom, had 20,000 customers its first month. Four months later, it was 150,000; four years after that, 13 million. A market that did not exist as of 2007—the mobile payment market (making payments via mobile phones)—exploded into a $16 billion industry by 2011, with analysts predicting that it would grow an additional 68 percent by 2014. And the benefits appear to be considerable. According to the *Economist,* over the past five years, incomes of Kenyan households using M-PESA have increased by 5 percent to 30 percent.

Beyond banking, cell phones are now enabling improvement at every level of our abundance pyramid. For water, there's already SMS-delivered information available on everything from hand washing to conservation techniques and technology is now being pioneered that turns a smart phone into a testing device for water quality. In food, fishermen can check in advance which ports are paying top dollar before hauling their catch into shore, and farmers can do the same before bringing fruits and vegetables to market, in both cases maximizing their time and revenue. The impacts of mobile telephony on health stretch from being able to quickly locate the nearest doctor to a smart phone app invented by Peter Bentley, a researcher from University College London, that turns an iPhone into a stethoscope and has since been downloaded by over 3 million doctors. And it is only one of *6,000* health care apps now available through Apple.

These examples go on and on, but what they all have in common is that they empower the individual like never before. Most of these services used to require tremendous amounts of infrastructure, resources, and well-trained professionals, making them accessible primarily in the developed world. If one of the definitions of abundance is the widespread availability of goods and services—such as stethoscopes and water-quality testing—then the now-networked rising billion are rapidly gaining access to many of the fundamental mechanisms of first world prosperity.

The Resource Curse

The majority of mobile phones at work in BoP markets are on 2G networks, which provide voice and text-messaging capabilities. As should be clear by now, just these features alone have enabled incredible progress at every level of our pyramid, but they've also done what many considered impossible: help the rising billion break out of the "resource curse."

Over the past fifty years, researchers have spent a lot of time trying to figure out what was keeping the bottom billion pinned to the bottom. As economist William Easterly has frequently pointed out, "The West spent

$2.3 trillion in foreign aid over the past five decades and still has not managed to get twelve-cent medicines to children to prevent half of all malarial deaths." The issue comes down to so-called poverty traps. Being a landlocked nation without access to shipping ports is one kind of poverty trap; being stuck in a cycle of civil war is another. One of the most insidious of these is the resource curse, which goes like this:

When a developing nation discovers a new natural resource, this causes its currency to rise against other currencies and has the downstream effect of making other exportable commodities uncompetitive. The discovery of oil reserves in Nigeria in the 1970s destroyed the country's peanut and cocoa industries. Then, in 1986, the world price of oil crashed, and, as Oxford University economist Paul Collier writes in *The Bottom Billion: Why the Poorest Countries Are Failing and What Can Be Done About It,* "the Nigerian gravy train came to an end. Not only was oil revenue drastically reduced, but the banks were not willing to continue lending: they actually wanted to be paid back. This swing from big oil and borrowing to little oil and repayment approximately halved Nigerian living standards."

There is no easy way to break the resource curse, but two of the more effective measures are the development of diversified markets and the emergence of a free press (and the transparency it brings). Thirty years of aid failures have taught us that neither is easy to jump-start, but both are now a part of the wireless landscape. Microcredit gives people outside the natural resource game access to money, thus encouraging the creation of small businesses not linked to the boom-and-bust cycle. The crowdsourcing of tiny jobs—known as microtasking—gives the poor access to novel revenue streams that further break this cycle. According to the *New York Times,* freelancers the world over are "increasingly taking on assignments like customer service, data entry, writing, accounting, human resources, payroll—and virtually any 'knowledge process' that can be performed remotely." This is a huge step forward. By helping disperse productivity, communication technology helps disperse power, which, as Quadir once wrote, "makes it harder for individuals or groups to corner resources or advance state policies that favor narrow interests." Furthermore, the free flow of information enabled by cell phones replaces the need for a free press and, as recent events in the Middle East bear out, can have serious impacts on the spread of democracy.

What's more incredible is that all this was possible with yesterday's technology. However, smart phones relying on 3G and 4G networks are arriving in the developing world, and that makes tomorrow's potential exponentially greater. Former Harvard business professor Jeffrey Rayport, now CEO of the consulting firm MarketShare, writes in *Technology Review*: "Today's mobile device is the new personal computer. The average smart phone is as powerful as a high-end Mac or PC of less than a decade ago . . . With over five billion individuals currently armed with mobile phones, we're talking about unprecedented levels of access and insight into the psyches of over two-thirds of the world's population."

The World Is My Coffee Shop

In his excellent book *Where Good Ideas Come From: The Natural History of Innovation,* author Steven Johnson explores the impact of coffeehouses on the Enlightenment culture of the eighteenth century. "It's no accident," he says, "that the age of reason accompanies the rise of caffeinated beverages." There are two main drivers at work here. The first is that before the discovery of coffee, much of the world was intoxicated much of the day. This was mostly a health issue. Water was too polluted to drink, so beer was the beverage of choice. In his *New Yorker* essay "Java Man," Malcolm Gladwell explains it this way: "Until the eighteenth century, it must be remembered, many Westerners drank beer almost continuously, even beginning their day with something called 'beer soup.' Now they begin each day with a strong cup of coffee. One way to explain the industrial revolution is as the inevitable consequence of a world where people suddenly preferred being jittery to being drunk."

But equally important to the Enlightenment was the coffeehouse as a hub for information sharing. These new establishments drew people from all walks of life. Suddenly the rabble could party alongside the royals, and this allowed all sorts of novel notions to begin to meet and mingle and, as Matt Ridley says, "have sex." In his book *London Coffee Houses,* Bryant Lillywhite explains it this way:

> The London coffee-houses provided a gathering place where, for a penny admission charge, any man who was reasonably dressed could smoke his long, clay pipe, sip a dish of coffee, read the newsletters of the day, or enter

> into conversation with other patrons. At the period when journalism was in its infancy and the postal system was unorganized and irregular, the coffee-house provided a centre of communication for news and information . . . Naturally, this dissemination of news led to the dissemination of ideas, and the coffee-house served as a forum for their discussion.

But researchers in recent years have recognized that the coffee-shop phenomenon is actually just a mirror of what occurs within cities. Two thirds of all growth takes place in cities because, by simple fact of population density, our urban spaces are perfect innovation labs. The modern metropolis is jam-packed. People are living atop one another; their ideas are as well. So notions bump into hunches bump into offhanded comments bump into concrete theories bump into absolute madness, and the results pave the way forward. And the more complicated, multilingual, multicultural, wildly diverse the city, the greater its output of new ideas. "What drives a city's innovation engine, then—and thus its wealth engine—is its multitude of differences," says Stewart Brand. In fact, Santa Fe Institute physicist Geoffrey West found that when a city's population doubles, there is a 15 percent increase in income, wealth, and innovation. (He measured innovation by counting the number of new patents.)

But just as the coffeehouse is a pale comparison to the city, the city is a pale comparison to the World Wide Web. The net is allowing us to turn ourselves into a giant, collective meta-intelligence. And this meta-intelligence continues to grow as more and more people come online. Think about this for a moment: by 2020, nearly 3 billion people will be added to the Internet's community. That's 3 billion new minds about to join the global brain. The world is going to gain access to intelligence, wisdom, creativity, insight, and experiences that have, until very recently, been permanently out of reach.

The upside of this surge is immeasurable. Never before in history has the global marketplace touched so many consumers and provided access to so many producers. The opportunities for collaborative thinking are also growing exponentially, and since progress is cumulative, the resulting innovations are going to grow exponentially as well. For the first time ever, the rising billion will have the remarkable power to identify, solve, and implement their own abundance solutions. And thanks to the net, those solutions aren't going to stay balkanized in the developing world.

Perhaps most importantly, the developing world is the perfect incubator for the technologies that are the keys to sustainable growth. "Indeed," writes Stuart Hart, "new technologies—including renewable energy, distributed generation, biomaterials, point-of-use water purification, wireless information technologies, sustainable agriculture, and nanotechnology—could hold the keys to addressing environmental challenges from the top to the base of the economic pyramid."

However, he adds, "Because green technologies are frequently 'disruptive' in character (that is, they threaten incumbents in existing markets), the BoP may be the most appropriate socioeconomic segment upon which to focus initial commercialization attention . . . If such a strategy were widely embraced, the developing economies of the world become the breeding ground for tomorrow's sustainable industries and companies, with the benefits—both economic and environmental—ultimately 'trickling up' to the wealthy at the top of the pyramid."

Thus this influx of intellect from the rising billion may turn out to be the saving grace of the entire planet. Please, please, please, let the bootstrapping begin.

Dematerialization and Demonetization

So let's return to where we began: with One Planet Living. Jay Witherspoon explained that if everyone on Earth wants to live like a North American, then we're going to need five planets' worth of resources to do so—but is this really the case anymore? Bill Joy, cofounder of Sun Microsystems turned venture capitalist, feels that one of the advantages of contemporary technology is "dematerialization," which he describes as one of the benefits of miniaturization: a radical decrease in footprint size for a great many of the items we use in our lives. "Right now," says Joy, "we're fixated on having too much of everything: thousands of friends, vacation homes, cars, all this crazy stuff. But we're also seeing the tip of the dematerialization wave, like when a phone dematerializes a camera. It just disappears."

Just think of all the consumer goods and services that are now available with the average smart phone: cameras, radios, televisions, web browsers, recording studios, editing suites, movie theaters, GPS navigators, word processors, spreadsheets, stereos, flashlights, board games, card games, video games, a whole range of medical devices, maps, atlases, encyclopedias, dictionaries, translators, textbooks, world class educations (more on this in chapter 14), and the ever-growing smorgasbord known as the app store. Ten years ago, most of these goods and services were available only in the developed world; now just about anyone anywhere can have them. How many goods and services? In summer 2011 the Android and Apple App stores boasted 250,000 and 425,000 applications, respectively, with a staggering 20 billion downloads combined.

Moreover, all of these now dematerialized goods and services used to require significant natural resources to produce, a physical distribution system to disperse, and a cadre of highly trained professionals to make sure that everything ran smoothly. None of these elements remain in the picture. And the list of those items no longer necessary keeps growing. When you also consider that robotics and AI will soon be replacing material possessions such as the automobile (think time-shared, on-demand access to the robo car of your choice), the potential for sustainably increasing standards of living becomes much more apparent. "It used to be that you were considered healthy and wealthy if you were fat," says Joy. "Now it's not. So now we think it's healthy and wealthy if we have all these things; well, what if it's actually the opposite? What if healthy and wealthy means you don't need all those things because, instead, you've got these really simple devices that are low maintenance and encapsulate everything you need?"

Furthermore, for most of the twentieth century, pulling oneself out of poverty demanded having a job that—one way or another—relied on these same natural resources, but today's greatest commodities aren't physical objects, they're ideas. Economists use the terms *rival goods* and *nonrival goods* to explain the difference. "Picture a house that is under construction," says Stanford economist Paul Romer. "The land on which it sits, capital in the form of a measuring tape, and the human capital of the carpenter are all rival goods. They can be used to build the house, but not another simultaneously. Contrast this with the Pythagorean theorem, which the carpenter uses implicitly by constructing a triangle with sides in the proportion of three, four, and five. This idea is nonrival: every carpenter in the world can use it at the same time to create a right angle."

Today the fastest-growing job category is the "knowledge worker." Since knowledge is nonrival, most of the jobs in the future will produce nonrival goods, and this removes another constraint on abundance: it allows the rising billion to earn a living in a way that does not require burning through our ever-diminishing supply of natural resources. And this trend, as Stuart Hart explains, will only continue as we move forward:

> Bio- and nanotechnology create products and services at the molecular level, holding the potential to completely eliminate waste and pollution. Biomimicry emulates nature's processes to create novel products and services without relying on brute force to hammer goods from large stocks of virgin raw materials. Wireless information technology and renewable energy are distributed in character, meaning they can be applied in the most remote and small-scale settings imaginable, eliminating the need for centralized infrastructure and wire-line distribution, both of which are environmentally destructive. Such technologies thus hold the potential to meet the needs of the billions of rural poor (who have thus far been largely ignored by global business) in a way that dramatically reduces environmental impact.

Alongside dematerialization, there's also the demonetization exemplified by Chris Anderson's drones to consider. In the past decade, this force has been steadily reshaping markets across the globe. eBay demonetized transactions, putting local stores out of business, yet increasing the availability of goods while simultaneously reducing their cost. Then there's Craigslist, which demonetized advertising, taking 99 percent of the profits out of the newspaper industry and putting them back into the pockets of the consumer. Or iTunes, which tanked the record store and liberated audiophiles. And the list of similar examples runs long. While short-term job loss is the inevitable and often painful result of demonetization and dematerialization, the long-term payoff is undeniable: goods and services once reserved for the wealthy few are now available to anyone equipped with a smart phone—which, these days, thankfully, includes the rising billion.

It's here, then, with the rising billion rising, that we conclude part 4 of this book. We'll continue working our way up the pyramid in part 5, then, in part 6, we'll return to one of our basic premises: this transformation is not inevitable. To go where we need to go also requires accelerating the rate of innovation, increasing global collaboration, and—perhaps most importantly—expanding our notions of the possible. But first, our world of abundance is going to need a lot of energy, so let's look at how we can power our planet in the decades to come.

PART FIVE

PEAK OF THE PYRAMID

CHAPTER THIRTEEN

ENERGY

Energy Poverty

Archaeologists differ on when humanity first tamed fire. Some believe that it was only 125,000 years ago; others point to evidence dating back some 790,000 years. Either way, once our ancestors learned the benefits of rubbing two sticks together, they never looked back. Fire provided a reliable source of heat, warmth, and light that forever altered our history. Unfortunately, for roughly one out of three people alive today, very little has changed in the past 100,000 years.

The United Nations estimates that one and a half billion people live without electricity and three and a half billion still rely on primitive fuels such as wood or charcoal for cooking and heating. In sub-Saharan Africa, the numbers are even higher, with more than 70 percent of the population living without access to electricity. This bottleneck brings with it a collection of consequences. Energy is arguably the most important lynchpin for abundance. With enough of it, we solve the issue of water scarcity, which also helps address a majority of our current health problems. Energy also brings light, which facilitates education, which, in turn, reduces poverty. The interdependencies are so profound that the United Nations Development Programme warned that none of the Millennium Development Goals aimed at reducing poverty by half can be met without major improvements in developing countries' energy services.

For Mercy Njima, a Kenyan doctoral student, about 85 percent of her nation is still ravaged by energy poverty. Mercy spent the summer of 2010 at Singularity University, where she painted me a picture of the complex problems she observed in her youth:

> Imagine being forced to rely on burning poor-grade wood, dung, or crop waste to cook, suffering the effects of the potentially fatal toxic fumes given off by this fuel. Imagine being desperately ill and turned away from a clinic because it has no electricity and can't offer even the simplest treatment. Imagine your friends living under the shadow of life-threatening disease because there's no vital vaccine, due to a lack of refrigeration. Imagine if you or your partner were pregnant and went into labor at night and had no light, no pain relief and no way of saving you or the baby if there were complications.

Mercy describes herself as part of the new-breed "cheetah generation" of Africans who are fast-moving, entrepreneurial leaders working to snatch back the continent from the jaws of poverty, corruption, and poor governance—three issues she believes could be changed significantly with more access to energy. "Consider the women and children who spend hours every day searching for increasingly scarce energy resources. They are at risk from wild animals and sometimes rape. And once they start burning biomass, the acrid smoke causes serious lung disease and turns kitchens into death traps. Children and their mothers are most at risk, choking, retching, and gasping. More people die from smoke inhalation than from malaria. Indoor air pollution is linked to respiratory diseases such as pneumonia, bronchitis, and lung cancer. Women and children who spend long periods every day around traditional open fires inhale the equivalent of two packs of cigarettes a day."

She also points out that because children have to help collect fuel during school hours, time spent on their education is severely reduced. This problem compounds at night, when students need to do their homework but have no light for studying. Kerosene can help matters, but it's both expensive and dangerous. In addition, Mercy says, teachers don't want to work in communities with no lights and little equipment. But the consequences of energy poverty extend further than homes and schools. "Lack of energy also means people struggle to start simple businesses," she explains. "This shortage impacts every aspect of Kenyan life, and it's mostly the same across the continent. This is the stark reality for most Africans living in energy poverty."

However, it doesn't have to be a permanent reality, maintains Emem Andrews, a former senior program manager for Shell Nigeria and now a Silicon Valley energy entrepreneur. "Without question," she says, "Africa could become energy independent. Nigeria alone has enough oil for the entire continent. Ultimately, though, the biggest opportunity is the sun. It's decentralized, fully democratic, and available to all. Africa is endowed with underutilized deserts and lies within latitudes with high solar isolation levels. Sunlight is plentiful and essentially free. We just lack the technology to access it."

According to the Trans-Mediterranean Renewable Energy Cooperation, an international network of scientists and experts founded by the Club of Rome, enough solar power hits one square kilometer of Africa's deserts to produce the equivalent of one and a half million barrels of oil or three hundred thousand tons of coal. The German Aerospace Center estimates that the solar power in the deserts of North Africa is enough to supply forty times the present world electricity demand. Furthermore, David Wheeler, a research fellow at the Center for Global Development, found that Africa has nine times the solar potential of Europe and an annual equivalent to one hundred million tons of oil. When coupled to its vast reserves of wind, geothermal, and hydroelectric, the continent has enough energy to meet its own needs *and* export the surplus to Europe. Perhaps Africa's greatest asset in exploiting this vast potential for renewables lies in the paradoxical fact that it has a complete and total absence of existing energy infrastructure.

Just as Africa's lack of copper landlines allowed for the explosive deployment of wireless systems, its lack of large-scale, centralized coal and petroleum power plants could pave the way for decentralized, renewable-power generation architectures. While wealthier early adopters, primarily in first world nations, will likely pay for and develop these technologies (ideally, in cocreative ways with the rising billion), once they do find their way to Africa, these systems have an immediate advantage over existing options. Many forget that there is a significant price paid for hauling and safeguarding kerosene and generators to remote locations. In most places, this raises the cost of electricity to 35 cents per kilowatt-hour. So even today, with existing solar options at 20 cents per kilowatt-hour (and including the cost of the batteries required for storage), solar would total out around 25 cents per kilowatt-hour—a 30 percent savings over existing technologies.

And existing solar technologies; well, they're far from the end of this story.

A Bright Future

Like many who survived the dot-com bust, Andrew Beebe got out just in time. In 2002 he sold his Internet company, Bigstep, and went looking for greener pastures. Inspired by visionary physicist Freeman Dyson's ideas about "hacking photosynthesis," Beebe sought those pastures in the field of renewable energy. Initially he teamed up with Bill Gross, CEO of Idealab, to launch Energy Innovations (EI), a high-concentration photovoltaic (PV) business. They soon split into two companies, with Beebe taking the systems-installation end of the enterprise, EI Solutions. Over the next few years, he grew EI Solutions into a $25 million company, installing PV panels at the headquarters of corporations such as Google, Sony, and Disney, then selling the operation to Suntech, the largest PV manufacturer in the world. He ran global product management there, then took over global sales and marketing—a position he still holds. As the person in charge of selling the most PV in the world, Beebe has his finger of the pulse of solar. According to him, that pulse is strong:

> The solar market is a great econ-101 story. PV production and installation have grown at 45 percent to 50 percent per year for the last decade. That is epic, as the remainder of global energy growth is only increasing at 1 percent annually. In 2002, when I got started in this industry, total capacity sold was something like 10 megawatts per year. This year, it'll probably be eighteen gigawatts. That's nearly a 2,000-fold increase in less than a decade. At the same time, cost has been plummeting. Four years ago, when I was buying solar panels for Google, it was $3.20 per watt using extremely mature technology. Today the global average price per installed watt is below $1.30. I'm on calls night and day coming up with even more radical price reductions. It's weird to be in a business where one of the major goals is to find a way to sell our product for less money, but that's exactly what's happening.

And the bottom is nowhere in sight. Over the past thirty years, the data show that for every cumulative doubling of global PV production, costs have dropped by 20 percent. This is another of those exponential price-performance curves, now known as Swanson's law (after Dick Swanson, cofounder of SunPower). According to Swanson, the cost improvement is essentially a learning curve for manufacturing techniques and production efficiencies.

"The expensive crystalline silicon has been the biggest cost in the panel," he says, "and we have been steadily making wafers thinner and thinner. We use half the amount of silicon to produce a watt of power than we did five years ago." Lowering the cost of silicon wafers another tenfold is the mission of 1366 Technologies, a solar start-up launched by MIT professor of mechanical engineering Emanuel Sachs. (The name refers to the average number of watts of solar energy that hit each square meter of Earth per year.) Having found a way to make thin sheets of silicon without having to first slice them from solid chunks of the element, 1366 dramatically reduces the most expensive part of any PV system.

This type of discovery shouldn't surprise anyone. Solar's potential marketplace and benefit to humanity are so vast that reducing the cost of PVs, increasing the ease of installation, and stepping up global production are the objectives of hundreds, if not thousands, of entrepreneurs, large corporations, and university labs. In the United States, the number of clean-tech patents hit a record high of 379 during the first quarter of 2010, while the number of solar-related patents nearly tripled between mid-2008 and the start of 2010.

And since then, the pace of discovery has only continued to accelerate. Scientists at IBM recently announced that they've found a way to replace expensive, rare Earth elements such as indium and gallium, with less expensive elements like copper, tin, zinc, sulfur, and selenium. Engineers at MIT, meanwhile, using carbon nanotubes to concentrate solar energy, have made PV panels one hundred times more efficient than traditional models. "Instead of needing to turn your whole roof into a photovoltaic cell," says Dr. Michael Strano, leader of the research team, "you could have tiny PV spots with antennas that would drive photons onto them."

But why have rooftop panels at all? The Maryland-based New Energy Technologies has discovered a way to turn ordinary windows into PV panels. Its technology uses the world's smallest organic solar cell, which, unlike conventional systems, can generate electricity from both natural and artificial light sources, outperforming today's commercial solar and thin-film technologies by as much as tenfold.

All this work could soon be eclipsed by far more revolutionary breakthroughs. At the University of Michigan, physicist Stephen Rand recently discovered that light, traveling at the right intensity through a nonconductive material such as glass, can create magnetic fields 100 million times stronger than previously believed possible. "You could stare at the equations of motion all day and not see this possibility," says Rand. "We've all been taught this doesn't happen." But in his experiments, the fields are strong enough to allow for energy extraction. The result would be a way to make PV panels without using semiconductors, reducing their cost by orders of magnitude.

Beebe, though, doesn't think that these sorts of radical breakthroughs are required. "I'm happy with the glide slope we're on," he says. "Italy and the US will achieve grid parity [the point when renewables become as cheap as traditional sources] in two and five years, respectively. In California today, home owners with good credit can install PV solar with no money down and pay less for energy in their first month on PV than they did in the previous month buying it from the grid. Of course, this works because of a thirty percent California tax credit, but once solar costs decrease by another thirty percent, which is expected in the next four years, we won't need the tax credit anymore. Once solar hits subsidy-free grid parity, it will go crazy. When you fly into LAX, you look down and see miles and miles of flat roofs. Why don't they all have solar on them? Eventually, with grid parity, those buildings will be covered with the stuff."

Making solar cheap enough to cover our roofs and compete with coal is also the goal of US Energy Secretary Stephen Chu's recently announced SunShot Initiative, an ambitious effort modeled on President John F. Kennedy's 1961 "moonshot" speech, wherein he challenged the nation to land a man on the Moon before the end of the decade. Dunshot's aim is to spur American innovation and reduce the total cost of solar energy systems another 75 percent by 2020. This reduction would put costs around $1 per watt, or six cents per kilowatt-hour—a price capable of undercutting even coal.

Lest we focus only on solar, wind power is also approaching grid parity. According to a 2011 report by *Bloomberg New Energy Finance,* in parts of Brazil, Mexico, Sweden, and the United States, onshore wind power is down to $68 per megawatt (MW), while coal in those same regions is about $67 per MW. Demand is growing too. Between 2009 and 2010, Vestas, one of the world's largest wind energy firms, reported orders rising by 182 percent. In 2011, worldwide turbine installations climbed 20 percent and are projected to double by 2015.

Yet despite these considerable gains, other forms of energy innovation are also required. Solar and wind are sources of electricity, but they represent only 40 percent of America's energy needs. The remainder is split between transportation (29 percent) and home and office heating/cooling (31 percent). Of the fuel used for transportation, 95 percent is petroleum based, while our buildings rely on both petroleum and natural gas. To end our oil addiction, we're going to need to displace this remaining 60 percent. Many believe this won't be easy. "The oil and gas industries are very well funded and very entrenched," says Beebe. "The question is: How do we change that? These industries don't want to let go, and they have enough money to hold on for a very long time."

Synthetic Life to the Rescue

But what if the change was coming from within these same entrenched petroleum giants? In 2010 Emil Jacobs, ExxonMobil's vice president of research and development, announced an unprecedented $600 million six-year commitment to develop a new generation of biofuels. Of course, the older generation of biofuels, primarily corn-based ethanol, was a disaster. These fuels have caused considerable environmental damage and displaced millions of acres of crops, thus helping to drive food prices sky-high. But Exxon's biofuel isn't based on food crops, nor does it have the considerable land requirements of first-generation technology. Instead Exxon plans to grow its biofuel from algae.

The US Department of Energy says that algae can produce thirty times more energy per acre than conventional biofuels. Moreover, because pond scum grows in almost any enclosed space, it's now being tested at several major power plants as a carbon dioxide absorber. Smokestacks feed into ponds and algae consumes the CO_2. It's a delicious possibility, but to make it more of a reality, Exxon has partnered with biology's bad boy, Craig Venter, and his most recent company, Synthetic Genomics Inc. (SGI).

To study algae-growing methods and oil extraction techniques, Exxon and SGI built a new test facility in San Diego. Venter calls it "an algae halfway house." On a sunny afternoon in February 2011, I was given a tour. From the outside, the facility looks like a high-tech greenhouse: clear plastic panes, white struts, and a set of airlock doors. As we step through those doors, Paul Roessler, who heads the project, explains the basics: "Our biofuel has three requirements: sunlight, CO_2, and seawater. The rationale for using seawater is that we don't want to compete for agricultural land or agricultural water. CO_2 is the bigger issue. That's why CO_2 sequestration would be great: it both slows global warming and provides a concentrated source."

We walk through another door, and we're inside the main room, a football-field-sized area with not much by way of decoration save for a half dozen vats of green algae and a large "Life of the Cell" poster on the wall. Roessler points to the poster: "I don't know how much you remember from school, but photosynthesis is how plants convert light energy into chemical energy. During the day, plants use sunlight to split water into hydrogen and oxygen, then combine it with carbon dioxide and turn the result into a hydrocarbon fuel called 'bio oil,' which they typically use at night for repair. Our goal is to reliably mass-produce these bio oils."

Venter, who has also joined the tour, jumps into the conversation. "Paul's being modest. He actually found a way to cause algae cells to voluntarily secrete their collected lipids, turning them into micromanufacturing plants." Roessler picks up the explanation. "In theory, once perfected, we could run this process continuously and just harvest the oil. The cells just keep cranking it out. This way you don't have to harvest all the cells; instead just scoop up the oils they excrete."

The efficiencies are considerable. "When compared to conventional biofuels," says Venter, "corn produces 18 gallons per acre per year and palm oil about 625 gallons per acre per year. With these modified algae, our goal is to get to 10,000 gallons per acre per year, and to get it to work robustly, at the level of a two-square-mile facility."

To understand how ambitious Venter's goals are, let's do the math: two square miles is 1,280 acres. At 10,000 gallons of fuel per acre, that's 12.8

million gallons of fuel per year. Using today's average of twenty-five miles per gallon and twelve thousand miles driven per year, two square miles of algae farms produce enough fuel to power around 26,000 cars. So how many acres does it take to power America's entire fleet? With roughly 250 million automobiles in the United States today, that translates to about 18,750 square miles, or about 0.49 percent of the US land area (or about 17 percent of Nevada). Not bad. Just think what can happen when our cars start getting 100 miles per gallon or when more of us make the switch to electric automobiles.

Even if SGI falls short of this goal, Exxon isn't the only player in the race. The Bay Area energy company LS9 has partnered with Chevron (and Procter & Gamble) to develop its own biofuel, while not far away in Emeryville, California, Amyris Biotechnologies has done the same with Shell. The Boeing Company and Air New Zealand are starting to develop an algae-based jet fuel, and other companies are even further along. Virgin Airlines is already using a partial biofuels mix (coconut and babassu oil) to move 747s around the sky, and in July 2010 the San Francisco–based Solazyme delivered 1,500 gallons of algae-based biofuels to the US Navy, thus winning a contract for another 150,000 gallons. Meanwhile, the DOE is funding three different biofuel institutes, and Clean Edge, which tracks the growth of renewable energy markets, reports in its tenth annual industry overview that global production and wholsesale pricing of biofuels reached $56.4 billion in 2010—and is projected to grow to $112.8 billion by 2020.

Clearly, interest in carbon-neutral, low-cost fuels is at an all-time high, but problems remain. None of the aforementioned companies (or any of their unmentioned competitors) have figured out how to bring this technology to scale. To really meet our needs, Secretary Chu says, production has to be increased a millionfold, maybe even ten millionfold, although he also points out that the same scientists working on biofuels have already scaled up products such as antimalaria drugs. "So it's a possibility," he says, "and with the quality of scientists involved, maybe—I'd like to believe—a likelihood."

But the DOE isn't betting only on biofuels to meet this need. The agency is also interested in hacking photosynthesis. Chu's SunShot Initiative has now funded the Joint Center for Artificial Photosynthesis, a $122 million multi-institution project being led by Caltech, Berkeley, and Lawrence Livermore National Laboratory. JCAP's goal is to develop light absorbers, catalysts, molecular linkers, and separation membranes—all the necessary components for faux photosynthesis. "We're designing an artificial photosynthetic process," says Dr. Harry Atwater, director of the Caltech Center for Sustainable Energy Research and one of the project's lead scientists. "By 'artificial,' I mean there's no living or organic component in the whole system. We're basically turning sunlight, water, and CO_2 into storable, transportable fuels—we call 'solar fuels'—to address the other two-thirds of our energy consumption needs that normal photovoltaics miss."

Not only will these solar fuels be able to power our cars and heat our buildings, Atwater believes that he can increase the efficiency of photosynthesis tenfold, perhaps a hundredfold—meaning solar fuels could completely replace fossil fuels. "We're approaching a critical tipping point," he says. "It is very likely that, in thirty years, people will be saying to each other, 'Goodness gracious, why did we ever set fire to hydrocarbons to create heat and energy?' "

The Holy Grail of Storage

In addition to their energy density and on-demand nature, another reason that we've relied so heavily on hydrocarbons is because they're easy to store. Coal sits in a pile, oil in a drum. But solar works only when the sun shines, and wind works only when the wind blows. These limits remain the largest impasse toward widespread renewable adoption. Until solar and wind can provide reliable 7x24 baseload power, neither will provide a significant portion of our energy supply. Decades ago, Buckminster Fuller proposed a global energy grid that could bring power collected on the sunny side of our planet to the dark side. But most people pin their hopes on the creation of large amounts of local, grid-level storage capable of "firming" or "time shifting" energy—that is, collecting energy during the day and releasing it at night. This, then, has become the holy grail of the green energy movement.

Ultimately, it doesn't matter how cheap solar gets unless we can store that energy, and storage on this scale has never been achieved before. Grid-level storage requires colossal batteries. Today's lithium-ion batteries are woefully inadequate. Their storage capacity would need to be improved ten- to twentyfold, and—if we really want them to be scalable—they have to be built from Earth-abundant elements. Otherwise we're just exchanging an economy built on the importation of petroleum for one built on the importation of lithium.

Thankfully, progress is being made. Recently, the market for grid-level storage has seen enough improvement that venture capitalists have gotten interested. Lead among them is Kleiner Perkins Caufield & Byers (KPCB). With over 425 investments, including AOL, Amazon, Sun, Electronic Arts, Genentech, and Google, Kleiner has a habit of picking winners. And since John Doerr, Kleiner's lead partner, is passionate about the environment and fighting global warming, many of those winners have been in the energy space.

During the winter of 2011, I caught up with Bill Joy, formerly of Sun Microsystems and now KPCB's lead green energy partner, to get a progress report on storage. He told me of two recent investments aimed at transforming the marketplace. Primus Power, the first, builds rechargeable "flow" batteries, in which electrolytes flow through an electrochemical cell that converts chemical energy directly to electricity. These devices are already firming wind energy in a new $47 million, 25-megawatt, 75-megawatt-hour energy storage system in Modesto, California.

Kleiner's second bet, Aquion Energy, builds a battery similar to today's lithium-ion designs, but with a serious twist. Rather than relying on lithium, a rare and toxic element, its battery uses sodium and water, two cheap and ubiquitous ingredients with the added advantage of being neither lethal nor flammable. The result is a battery that releases energy evenly, doesn't corrode, is based on Earth-abundant elements, and, literally, is safe enough to eat.

"Using these technologies," says Joy, "I think we're going to be able to store and retrieve a kilowatt-hour for a total cost of one cent. So I can put the intermittent flow of wind energy through my Aquion system and firm it for about one cent more per kilowatt-hour. And that's all up and all in. In a few years, you'll see these products in the marketplace. After that, there's no reason that we can't have reliable, grid-level renewables."

MIT professor Donald Sadoway, one of the world's foremost authorities on solid-state chemistry, is also optimistic about the future of grid-level storage. Backed by funds from the Advanced Projects Research Agency-Energy (ARPA-E) and Bill Gates, he's developed and demonstrated a Liquid Metal Battery (LMB) originally inspired by the high current density and enormous scale of aluminum smelters. Inside an LMB, the temperature is hot enough to keep two different metals liquid. One is high density, like antimony, and sinks to the bottom. The other is low density, such as magnesium, and rises to the top. Between them, a molten salt electrolyte helps the exchange of electrical charge. The result is a battery with currents ten times higher than present-day high-end batteries and a simple, cheap design that prices at $250 a kilowatt-hour fully installed—less than one-tenth the cost of current lithium-ion batteries. And Sadoway's design scales.

"Today's working LMB prototypes are the size of a hockey puck and capable of storing twenty watt-hours," says Sadoway, "but larger units are in the works. Imagine a device the size of a deep freezer that's able to store thirty kilowatt-hours of energy, enough to run your home for a day. We've designed them to be 'install and forget'—that is, able to operate for fifteen to twenty years without need of human intervention. It's cheap, quiet, requires no maintenance, produces no greenhouse gases, and is made of Earth-abundant elements." At $250 per kilowatt-hour, a home unit would go for about $7,500. Spread over fifteen years, adding the cost of capital and installation, one of these home LMBs would run a home owner under $75 per month.

But the real beauty of these systems is their ability to scale up. An LMB the size of a shipping container can power a neighborhood; one the size of a Walmart Supercenter could power a small city. "Within the next decade, we plan to deploy the shipping-container-sized LMB, soon followed by the family-sized unit," says Sadoway. "There's a clear line of sight to get there, and no miraculous breakthroughs needed."

Of course, when we do solve the storage problem, this would give solar and wind a major boost, so what to do with those dirty coal plants becomes a real question. Here too, Bill Joy has an idea. "It's hard to believe power companies would shut down a completely amortized asset that's still cranking out money every day. What we ought to do is flip the model and make coal plants into emergency backup plants. We can employ one hundred percent renewables for our baseload, and only turn on the coal plants when the weather forecast says we're going to have a real problem. We just pay the utilities to maintain them and run them occasionally, like you would run your emergency generator."

Nathan Myhrvold and the Fourth Generation

Nathan Myhrvold likes a good challenge, perhaps more than most. He started college at age fourteen and finished—with three masters degrees and a PhD from Princeton University—at twenty-three. Afterward, he spent a year with physicist Stephen Hawking, studying cosmology, later becoming a world-renowned paleontologist, prize-winning photographer, and gourmet chef—all in his spare time. In his work life, Myhrvold was Microsoft's chief technology officer, retired with a sum that, as *Fortune* once said, "runs well into nine figures," then cofounded the innovation accelerator Intellectual Ventures. But all this was just the warm-up round. "To me, the problem to solve this century is how do we supply US levels of carbon-free energy to everyone in the world?" he says. "It's a massive energy challenge."

Myhrvold is not wrong. Civilization currently runs on sixteen terawatts of power—mostly from CO_2-generating sources. If we're serious about fighting energy poverty and raising global living standards, then we'll need to triple—perhaps even quadruple—that figure over the next twenty-five years. Concurrently, if we want to stabilize the amount of CO_2 in the atmosphere at 450 parts per million (the agreed-upon number for staving off dramatic climate change), we've got to replace thirteen of those sixteen terawatts with clean energy. To put it another way: every year, we humans dump nearly 26 billion tons of CO_2 into the atmosphere, or about five tons for every person on the planet. We have little more than two decades to bring that number close to zero, while at the same time increasing global energy production to meet the needs of the rising billion.

Certainly there are plenty who believe that solar will scale and storage will materialize, and meeting those needs with renewables is entirely feasible. But there are plenty of others, Myhrvold included, who believe that the only other option is nuclear power. In fact, widespread belief in this option has never been stronger.

Both the George W. Bush administration and the current Obama administration back the proposal, as do serious greens such as Stewart Brand, James Lovelock, and Bill McKibben. This much overwhelming support of a previously dismissed technology is confusing to people, but that's mainly because they're basing their opinions on facts that are now forty years out of date. "When most people argue about nuclear energy," says Tom Blees, author of *Prescription for the Planet: The Painless Remedy for Our Energy and Environmental Crises,* "they're arguing about Three Mile Island and 1970s technology—which is about when the US nuclear industry ground to a halt. But research didn't die off, just new construction. We're two generations beyond that earlier tech, and the changes have been massive."

Scientists denote nuclear power by generations. Generation I reactors were built in the 1950s and 1960s; generation II refers to all the reactors supplying power in the United States today. Generation III is considerably cheaper and safer than previous iterations, but it's generation IV that explains the recent outpouring of support. The reason is simple: this fourth-generation technology was developed to solve all the problems long associated with nuclear power—safety, cost, efficiency, waste, uranium scarcity, and even the threat of terrorism—without creating any new ones.

Generation IV technologies come in two main flavors. The first are fast reactors, which burn at higher temperatures because the neutrons inside bounce around at a faster rate than in traditional light-water reactors. This extra heat gives fast reactors the ability to turn nuclear waste and surplus weapons-grade uranium and plutonium into electricity. The second category are liquid fluoride thorium reactors. These burn the element thorium, which is four times more plentiful than uranium, and don't create any long-lived nuclear waste in the process.

As a general rule, all generation IV technologies are "passively safe"—meaning that in case of trouble, they're able to shut themselves down without human intervention. Most fast reactors, for example, burn liquid metal fuels. When a liquid metal fuel overheats, it expands, so its density decreases, and the reaction slows down. According to retired Argonne National Laboratory nuclear physicist George Stanford, the reactors can't melt down. "We know this for certain," he says, "because in public demonstrations, Argonne duplicated the exact conditions that led to both the Three Mile Island and Chernobyl disasters, and nothing happened."

But what has people most excited are so-called backyard nukes. These self-contained small-scale modular generation IV nuclear reactors (SMRs) are built in factories (for cheaper construction), sealed completely, and designed to run for decades without maintenance. A number of familiar faces such as Toshiba and Westinghouse, and a number of nuclear newcomers such as Nathan Myhrvold's company TerraPower, have gone into this area because of SMRs' tremendous potential for providing the entire world with carbon-free energy.

With coinvestments from Bill Gates and venture capitalist Vinod Khosla, Myhrvold founded TerraPower to develop the traveling wave reactor (TWR), a generation-IV variation that he calls the "the world's most simplified passive fast breeder reactor." The TWR has no moving parts, can't melt down, and can run safely for fifty-plus years, literally without human intervention. It can do all this while requiring no more enrichment operations, zero spent-fuel handling, and no reprocessing or waste storage facilities. What's more, the reactor vessel serves as the unit's (robust) burial cask. Essentially, TWRs are a "build, bury, and forget" power supply for a region or city, making them ideal for the developing world.

Of course, powering the developing world would require tens of thousands of nuclear power plants. Myhrvold recognizes the size of this challenge, but he correctly points out that "if we're going to reach our goal of energy abundance, places like Africa and India are where the massive increase will be needed most. This is exactly why we've designed these reactors with safe, easy-to-maintain, and proliferation-proof features. We have to make them appropriate for use in the developing world." He is also quick to point out the environmental upside his system brings: "We could power the world for the next one thousand years just burning and disposing of the depleted uranium and spent fuel rods in today's stockpiles."

So when might we see one of these reactors? Myhrvold wants a demonstration unit up and running by 2020. If this timetable is accurate, then TerraPower has a real advantage. Outside of a handful of projects, most generation-IV reactors won't make it to market until 2030. More importantly, Myhrvold believes that the power provided by TWRs can be priced to undercut coal—which is exactly what it would take to spread them around the globe.

Perfect Power

Where we source our power is only one part of this issue; how we distribute it is equally important. Imagine an intelligent network of power lines, switches, and sensors able to monitor and control energy down to the level of a single lightbulb. This is the dream of today's smart grid engineers. Currently the only network this extensive is the Internet, which is why Bob Metcalfe is constantly comparing today's electric "dumb grid" to the early days of telephony. Metcalfe, founder of 3Com Corporation and today a general partner at Polaris Venture Partners, is an expert in energy-related investments. He began his career as one of the creators of both Arpanet and the Ethernet, and knows what it takes to build something as vast as the World Wide Web. "In the early days, everything was stovepiped," he says. "Computing was done by IBM; communications was done by AT&T. Voice, video, and data were distinct services: voice was synonymous with telephone, video with television, and data with a teletype machine plugged into a time-sharing computer system. These were three different worlds with different networks and regulatory agencies. The Internet has dissolved these distinctions and boundaries."

Today we see similar balkanization in energy, but Metcalfe believes that the distinctions among production, distribution, sensing, control, storage, and consumption will ultimately disappear. "When the traffic on Arpanet began to explode," he says, "our first reaction was to try and squeeze it through the old AT&T infrastructure by focusing on compression efficiency. We conserved data in the same way we're trying to conserve energy today. Then, like now, the problem was a centralized grid not robust enough to handle our needs. But forty years after Arpanet, it's not about conservation at all; in fact, it's about a world of data abundance. The Internet's architecture has ultimately allowed a millionfold increase in data flow. So if the Internet is any guide, once we're able to build the next generation energy network—what I call the Enernet—I believe we'll be awash in energy. In fact, once we have the Enernet, I believe we'll have a squanderable abundance of energy."

So what are the features for such a smart grid? Metcalfe envisages a distributed mesh network, not unlike the Internet, which would allow the exchange of power between a multitude of producers and consumers over local and wide-area networks. "It must also be desynchronized," he adds, "so anyone can put power in or take power out, as easily as computers, phones, or modems plug into the Internet today."

Perhaps the biggest change that Metcalfe predicts is the massive addition of storage. "The old telecom network had absolutely no storage and looked very much like today's power grid," he says. "Your analog voice entered the network on one end and went flying out the other. But this has changed dramatically. Today's Internet is filled with all kinds of storage at every possible location—at the switch, on the server, in your building, on your phone. Tomorrow's smart grid will also have storage everywhere: storage at your appliances, your home, your car, your building, the community, and at every point of energy production."

Cisco, one of the world's largest networking companies, has made a huge commitment to build the smart grid. Laura Ipsen, senior vice president in charge of Cisco's energy business, explains the opportunity: "Today we have more than one and a half billion connections to the Internet. But this is small in comparison to the number of connections to the electric grid, which is at least tenfold larger. Just think of the number of electric appliances you have plugged in at home, compared to the number of IP addressable devices. This is a huge opportunity."

Ipsen feels that we're moving rapidly toward a world where every device that consumes power has an IP address and is part of a distributed intelligence. "These connected devices," she says, "no matter how small, will communicate their energy usage and turn themselves off when not needed. Ultimately, we should be able to double or triple efficiency of a building or a community."

Cisco has an aggressive time line for this vision. "In the near term," Ipsen says, "the next seven years, the smart grid will be dominated by 'sensing and response.' IP-connected sensors will monitor energy use and manage demand, time shifting noncritical applications like delaying the start of your dishwasher to the middle of the night, when energy is cheaper. Starting in 2012 and for the next dozen years, we envision that solar and wind will rapidly be integrated, enabling commercial and residential property owners to go off grid for the majority of their needs." Ultimately, the goal is integrated distributed generation, coupled with smart IP-enabled appliances, and ubiquitous distributed storage allowing for what Ipsen calls "perfect power."

So What Does Energy Abundance Really Mean?

In this chapter, we've focused principally on solar, biofuels, and nuclear. There are certainly plenty of other technologies to consider. I've not spoken about natural gas, which, given the large US supplies, is currently all the rage. Nor have I discussed geothermal energy, which is reasonably reliable and clean, but can lack easy geographic access.

Yet there are reasons this chapter places an emphasis on solar power. It is pollution, carbon, and stigma free. Should we be able to crack the storage infrastructure challenges ahead, sunlight is ubiquitous and democratic. There is more energy in the sunlight that strikes the Earth's surface in an hour than all the fossil energy consumed in one year. More importantly, if we want to achieve energy abundance, we need to choose technologies that scale—ideally, on exponential curves. Solar fits all of those criteria.

According to Travis Bradford, chief operating officer of the Carbon War Room and president of the Prometheus Institute for Sustainable Development, solar prices are falling 5 percent to 6 percent annually, and capacity is growing at a rate of 30 percent per year. So when critics point out that solar currently accounts for 1 percent of our energy, that's linear thinking in an exponential world. Expanding today's 1 percent penetration at an annual growth of 30 percent puts us eighteen years away from meeting 100 percent of our energy needs with solar.

And growth doesn't end there, but it certainly gets interesting. Ten years later—twenty-eight years from now—at this rate we'd be producing 1,550 percent of today's global energy needs via solar. And, even better, at the same time that production is going up, technology is making every electron go even further. Whether it's the smart grid making energy use two- or threefold more efficient, or innovations like the LED lightbulb dropping the energy needed to light a room from one hundred watts to five watts, there is dramatic change ahead. With efficiencies lowering our usage and innovation increasing our supply, the combination really could produce a squanderable abundance of energy.

So what do we do with a squanderable abundance of energy? Of course, Metcalfe's been thinking about this for some time. "First," he proposes, "why not drop the price of energy by an order of magnitude, driving the planet's economic growth through the roof? Second, we could truly open the space frontier, using that energy to send millions of people to the Moon or Mars. Third, with that amount of energy, you can supply every person on the Earth with the American standard of fresh, clean water every day. And fourth, how about using that energy to actually remove CO_2 from the Earth's atmosphere. I know a professor at the University of Calgary,

Dr. David Keith, who has developed such a machine. Back it up with cheap energy, and we might even solve global warming. I'm sure there's a much longer list of great examples."

To see how much longer that list might be, I tweeted Metcalfe's question. My favorite answer came from a Twitter handle BckRogers, who wrote: "All struggles are effectively conflicts over the energy potential of resources. So end war." I'm not entirely sure it's that simple, but considering everything we've discussed in this chapter, one thing seems certain: we are going to find out.

CHAPTER FOURTEEN
EDUCATION

The Hole-in-the-Wall

In 1999 the Indian physicist Sugata Mitra got interested in education. He knew there were places in the world without schools and places in the world where good teachers didn't want to teach. What could be done for kids living in those spots was his question. Self-directed learning was one possible solution, but were kids living in slums capable of all that much self-direction?

At the time, Mitra was head of research and development for NIIT Technologies, a top computer software and development company in New Delhi, India. His posh twenty-first-century office abutted an urban slum but was kept separate by a tall brick wall. So Mitra designed a simple experiment. He cut a hole in the wall and installed a computer and a track pad, with the screen and the pad facing into the slum. He did it in such a way that theft was not a problem, then connected the computer to the Internet, added a web browser, and walked away.

The kids who lived in the slums could not speak English, did not know how to use a computer, and had no knowledge of the Internet, but they were curious. Within minutes, they'd figured out how to point and click. By the end of the first day, they were surfing the web and—even more importantly—teaching one another how to surf the web. These results raised more questions than they answered. Were they real? Did these kids really teach themselves how to use this computer, or did someone, perhaps out of sight of Mitra's hidden video camera, explain the technology to them?

So Mitra moved the experiment to the slums of Shivpuri, where, as he says, "I'd been assured no one had ever taught anybody anything." He got similar results. Then he moved it to a rural village and found the same thing. Since then, this experiment has been replicated all over India, and all over the world, and always with the same outcome: kids, working in small, unsupervised groups, and without any formal training, could learn to use computers very quickly and with a great degree of proficiency.

This led Mitra to an ever-expanding series of experiments about what else kids could learn on their own. One of the more ambitious of these was conducted in the small village of Kalikkuppam in southern India. This time Mitra decided to see if a bunch of impoverished Tamil-speaking, twelve-year-olds could learn to use the Internet, which they'd never seen before; to teach themselves biotechnology, a subject they'd never heard of; in English, a language none of them spoke. "All I did was tell them that there was some very difficult information on this computer, they probably wouldn't understand any of it, and I'll be back to test them on it in a few months."

Two months later, he returned and asked the students if they'd understood the material. A young girl raised her hand. "Other than the fact that improper replication of the DNA molecule causes genetic disease," she said, "we've understood nothing." In fact, this was not quite the case. When Mitra tested them, scores averaged around 30 percent. From 0 percent to 30 percent in two months with no formal instruction was a fairly remarkable result, but still not good enough to pass a standard exam. So Mitra brought in help. He recruited a slightly older girl from the village to serve as a tutor. She didn't know any biotechnology, but was told to use the "grandmother method": just stand behind the kids and provide encouragement. "Wow, that's cool, that's fantastic, show me something else!" Two months later, Mitra came back. This time, when tested, average scores had jumped to 50 percent, which was the same average as high-school kids studying biotech at the best schools in New Delhi.

Next Mitra started refining the method. He began installing computer terminals in schools. Rather than giving students a broad subject to learn—for example, biotechnology—he started asking directed questions such as "Was World War II good or bad?" The students could use every available resource to answer the question, but schools were asked to restrict the number of Internet portals to one per every four students because, as Matt Ridley wrote in the *Wall Street Journal,* "one child in front of a computer learns little; four discussing and debating learn a lot." When they were tested on the subject matter afterward (without use of the computer), the mean score was 76 percent. That's pretty impressive on its own, but the question arose as to the real depth of learning. So Mitra came back two months later, retested the students, and got the exact same results. This wasn't just deep learning, this was an unprecedented retention of information.

Mitra has since taken a job as a professor of education technology at the University of Newcastle in England, where he's developing a new model of primary school education he calls "minimally invasive education." To this end, he's created "self-organized learning environments" (SOLES) in countries around the world. These SOLES are really just computer workstations with benches in front of them. The benches seat four. Because SOLES are also installed in places where good teachers cannot be found, these machines are hooked up to what Mitra calls the "granny cloud"—literally groups of grandmothers recruited from all over the United Kingdom who have agreed to donate one hour a week of their time to tutor these kids via Skype. On average, he's discovered, the granny cloud can increase test scores by 25 percent.

Taken together, this work reverses a bevy of educational practices. Instead of top-down instruction, SOLES are bottom up. Instead of making students learn on their own, this work is collaborative. Instead of a formal in-school setting for instruction, the Hole-in-the-Wall method relies on a playground-like environment. Most importantly, minimally invasive education doesn't require teachers. Currently there's a projected global shortage of 18 million teachers over the next decade. India needs another 1.2 million. America needs 2.3 million. Sub-Saharan Africa needs a miracle. As Peter Smith, the United Nations' assistant director-general for education, explained recently, "This is the Darfur of children's future in terms of literacy. We have to invent new solutions, or we are as good as writing off this generation."

But Mitra discovered that solutions already exist. If what's really needed are students with no special training, grandmothers with no special training, and a computer with an Internet connection for every fourth student, then the Darfur of literacy need not be feared. Clearly, both kids and grandmothers are plentiful. Wireless connectivity already exists for over 50 percent of the world and is rapidly extending to the rest. And affordable computers? Well, that's exactly where the work of Nicholas Negroponte comes in.

One Tablet Per Child

One of the first people to recognize the educational potential of computers was Seymour Papert. Originally trained as a mathematician, Papert spent many years working with famed child psychologist Jean Piaget before moving to MIT, where he and Marvin Minsky cofounded the Artificial Intelligence Lab. From that perch, in 1970 Papert delivered a now-famous paper, "Teaching Children Thinking," in which he argued that the best way for children to learn was not through "instruction," but rather through "construction"—that is, learning through doing, especially when that doing involved a computer.

As this was five years before the Homebrew Computer Club had its first meeting, a lot of people laughed at Papert's ideas. Computers were gigantic and expensive. How exactly were they going to get into the hands of children? But an architect named Nicholas Negroponte took him seriously. Now known as one of the founding fathers of the Information Age, the founder of MIT's Architecture Machine Group, and the cofounder of MIT's Media Lab, Negroponte too felt that computers might be a way to bring a quality education to the 23 percent of the world's children currently not in school.

To this end, in 1982 Papert and Negroponte brought Apple II computers to schoolchildren in Dakar, Senegal, confirming what Mitra had confirmed previously: that poverty-stricken rural children take to computers just as quickly as all other children. A few years later, at the Media Lab, the duo created the "School of the Future," which moved computers into the classroom, and served as a test bed for ideas. In 1999 Negroponte took those ideas abroad and began setting up schools in Cambodia. Each student was provided with a laptop and an Internet connection. They also learned their first word in English: *Google.*

The experience was powerful. Negroponte left Cambodia with two firm beliefs. One, that children everywhere loved the Internet. Two, that the market wasn't particularly interested in making low-cost computers, especially ones cheap enough for the developing world, where annual educational budgets could be as low as $20 per child. In 2005 he started working on a solution, One Laptop Per Child (OLPC), an initiative aimed at providing every child on the planet with a rugged, low-cost, low-power, connected laptop.

While the computer's fabled $100 price tag has yet to materialize (it's roughly $180 today), OLPC has delivered laptops to three million children around the world. Because the initiative is based on a learning-by-doing education model, rote-memorization-based tests and other traditional measures of success do not apply. But there are metrics available. "[T]he most compelling piece of evidence that I have found that this program is working," says Negroponte, "is that everywhere we go, truancy drops to zero. And we go into some place where it's as high as thirty percent of the kids, and suddenly it's zero."

Truancy isn't exclusive to the Third World. On average, only two-thirds of American public school students finish high school—the lowest graduation rate in the industrialized world. In some areas, the dropout rate is over 50 percent; in Native American communities, it's higher than 80 percent. Many assumed that these students leave school because they're unable to do the requisite work, but research conducted by the Gates Foundation found that this isn't the case. "In a national survey of nearly 500 dropouts from around the country," writes Tony Wagner, codirector of Harvard's Change Leadership Group, in his book *The Global Achievement Gap: Why Even Our Best Schools Don't Teach the New Survival Skills Our Children Need—And What We Can Do About It,* "about half of these people said they left school because their classes were boring and not relevant to their lives or career aspirations. A majority also said that schools did not motivate them to work hard. More than half dropped out with just two years or less remaining to earn a high school diploma, and 88 percent had passing grades at the time that they dropped out. Nearly three-quarters of the interviewees said they could have graduated if they wanted to."

Whether OLPC will have these same effects in the United States is an open question (the North American version didn't launch until 2008), but its global impact continues to grow. Uruguay has made OLPC the backbone of primary school education, and other countries are starting to follow suit. In April 2010, the organization partnered with the East African community to deliver fifteen million laptops to children in Kenya, Uganda, Tanzania, Rwanda, and Burundi.

Helping fulfill Negroponte's vision is OLPC's recent switch from a $100 laptop to a $75 tablet. Of course, as Nokia is currently developing a $50 smart phone—which will most likely spread organically instead of requiring significant governmental investment—this does raise the question "Why bother?" But Negroponte feels that the smart phone is the wrong device to deliver an education, arguing that tablets provide what he calls "the book experience," which he believes is fundamental to learning. Considering the Media Lab's track record with machine-human interfaces, we'd be foolish not to consider his opinion. And even if smart phones do end up as tomorrow's favorite platform, who cares, as long as every kid get access to an education?

Another Brick in the Wall

Our current education system was forged in the heat of the industrial revolution, a fact that not only influenced what subjects were taught but also how they were taught. Standardization was the rule, conformity the desired outcome. Students of the same age were presented with the same material and assessed against the same scales of achievement. Schools were organized like factories: the day broken into evenly marked periods, bells signaling the beginning and the end of each period. Even teaching, as Sir Ken Robinson put it in his excellent book *Out of Our Minds: Learning to Be Creative,* was subject to the division of labor: "Like an assembly line, students progressed from room to room to be taught by different teachers specializing in separate disciplines."

In their defense, the transition from education as a rare treat reserved for the clergy and aristocracy to one where everyone was entitled to free schooling was nothing if not radical. But it has been over 150 years since then, and our education system has not kept up. Robinson himself has become one of the loudest voices calling for reform, arguing that today's schools—with their emphasis on extreme conformity—are killing creativity and squelching talent. "As humans, we all have immense potential," he says, "but most people pass through their entire lives with that potential untapped. Human culture, and school is a fundamental component of how we pass along that culture, is really a set of permissions. Permission to be different, permission to be creative. Our education systems rarely give people permission to be themselves. But if you can't be yourself, it's hard to know yourself, and if you don't know yourself, how can you ever tap into your true potential?"

So if our current system isn't doing the job it was designed to do, what exactly is it doing? This is not an easy question to answer for any number of reasons, not the least of which is that we no longer agree on what comprises success. In America, for example, after the passage of the No Child Left Behind Act of 2001, we now have the stated goal of 100 percent proficiency in reading and math by 2014. Most consider this a serious long shot, but even if we pull it off, does it really get us where we want to go?

Harvard's Tony Wagner isn't so sure:

> So-called advanced math is perhaps the clearest example of the mismatch between what is being taught and tested in high school versus what's needed for college and in life. It turns out that knowledge of algebra is required to pass state tests . . . because it is a near-universal requirement for college admissions. But why is that? If you are not a math major, you usually don't have to take any advanced math in college, and most of what you need for other courses is knowledge of statistics, probability, and basic computational skills. This is even more evident after college. Graduates from the Massachusetts Institute of Technology were recently surveyed regarding the math that this very technically trained group used most frequently in their work. The assumption was that if any adults use higher-level math, it would be MIT grads. And while a few did, the overwhelming majority reported using nothing more than arithmetic, statistics, and probability.

Taken together, Wagner and Robinson are pointing out that we're teaching the wrong stuff, but just as alarming is the fact that the stuff we're teaching isn't sticking. Two-fifths of all high school students need remedial courses upon entering college. In the state of Michigan, alone, the Mackinac Center for Public Policy estimates that remediation costs college and businesses about $600 million a year. A 2006 report on the subject by the think tank the Heritage Foundation observed: "If the other 49 states and the District of Columbia are anything like Michigan, the country spends tens of billions of dollars each year making up for public schools' shortcomings." A few years back, the National Governors Association interviewed 300 college professors about their freshman classes. The results: 70 percent said students couldn't understand complex reading materials; 66 percent said students couldn't think analytically; 62 percent said students wrote poorly; 59 percent said students don't know how to do research; 55 percent said students couldn't apply their knowledge. No surprise then that 50 percent of all students entering college do not graduate.

Even for those that do graduate, if the goal of college is to prepare students for the workforce, here too we are failing. In 2006, executives from four hundred major corporations were asked a simple question: "Are students graduating from school ready to work?" Their answer: "Not really." And that's right now. This year's kindergarten class will be retiring around 2070 (provided that we don't change the retirement age). So what will the world look like in 2070? What skills will our kids need to thrive then? No one has a clue.

What we do know is that the industrialized model of education, with its emphasis on the rote memorization of facts, is no longer necessary. Facts are what Google does best. But creativity, collaboration, critical thinking, and problem solving—that's a different story. These skills have been repeatedly stressed by everyone from corporate executives to education experts as the fundamentals required by today's jobs. They have become the new version of the three *R*'s (reading, writing, and arithmetic); the basics of what's recently been dubbed "twenty-first-century learning."

Twenty-first-century learning has dozens of moving parts, but at the center of them is a simple idea. "Over and over again," says Wagner, "in hundreds of interviews with business leaders and college professors, they stressed the ability to ask the right questions." As Ellen Kumata, managing partner of the Fortune 200 consultancy Cambria Consulting, explains:

> When I talk to my clients, the challenge is this: How do you do things that haven't been done before, where you have to rethink or think anew, or break set in a fundamental way. It's not incremental improvement anymore. That just won't cut it. The markets are changing too fast, the environments are changing too fast . . . You have to spend the time to ask the next question. There is something about understanding what the right questions are, and there is something about asking the nonlinear, counterintuitive question. These are the ones that take you to the next level.

If educational abundance is our goal, these facts leave us with serious quality and quantity concerns. For quality, what kind of learning system teaches kids to ask the right questions? That system needs to be able to teach the three *R*'s (because, yes, even in this digital age, these basics are still critical) and the twenty-first-century skills kids need to succeed. The quantity issue is equally important. We're already short millions of teachers. Forget about infrastructure. Schools in America are falling apart; schools in Africa don't even exist. So even if we do figure out what to teach our children, how to do this at scale remains equally perplexing.

But overshadowing both of these is a third problem. The twenty-first century is a media-rich environment. Between the Internet, video games, and those five hundred channels of cable, the competition for our children's attention has become ruthless. If boredom is the number one cause of truancy, then our new education system needs to be effective, scalable, and wildly entertaining. In fact, wildly entertaining might not be enough. If we really want to prepare our children for the future, then learning needs to become addictive.

James Gee Meets Pajama Sam

About ten years ago, Dr. James Gee sat down to play Pajama Sam for the first time. Gee is a linguist at Arizona State University. His early work examined syntactic theory, his more recent research delves into discursive analysis. Pajama Sam falls into neither of those categories. It's a problem-solving video game aimed at young children. But Gee had a six-year-old son, and he wanted to help him develop better problem-solving skills.

The game surprised Gee. The problems, as it turned out, were a little harder than expected. More stunning was how well the game held his son's attention. This piqued Gee's curiosity. He started to wonder about adult video games, so he picked up a copy of The New Adventures of the Time Machine—mostly because he liked the H. G. Wells reference in the title. "When I sat down to play, it wasn't anything like I expected," he recalls. "I had this idea that video games were relaxing, like television is relaxing. Time Machine was hard, long, and complex. All of my normal ways of thinking didn't apply. I had to relearn how to learn. I couldn't believe people would pay fifty bucks to be this frustrated."

But then it clicked: lots of young people were paying lots of money to engage in activities this frustrating. "As an educator, I realized this was the same problem our schools face: how do you get students to learn things that are long, hard, and complex?" Gee became intrigued by the implications. He also became intrigued by the games. Gee may be the only linguist in the world whose recent academic research includes the phrase: "The Legend of Zelda: The Windwalker," but that research has helped turn upside down much of what people believed about video games.

For example, the idea that games are a waste of time holds up only if you consider serious, deep learning a waste of time. "Take young kids playing Pokémon," says Gee. "Pokémon is a game for five-year-olds, but it requires a lot of reading to play. And the text isn't written for five-year-olds, it's written at about a twelfth-grade level. In the beginning, Mom has to play with her child, reading the text aloud. This is great, of course, because this is just how kids learn to read—by reading aloud with their parents. But then something funny happens. The kid realizes that Mom might be good at reading, but she's not very good at playing. So the kid starts reading, just so he can kick Mom out of the game and play with his friends."

This is just the beginning. Studies have shown that games outperform textbooks in helping students learn fact-based subjects such as geography, history, physics, and anatomy, while also improving visual coordination, cognitive speed, and manual dexterity. For example, surgeons and pilots trained on video games perform better than those who were not. But the real advantage is an ability to do what today's schools cannot: teach twenty-first-century skills. World-building games like SimCity and RollerCoaster Tycoon develop planning skills and strategic thinking. Interactive games are great teachers of collaborative skills; customizable games do the same for creativity and innovation. "Some educators compare game play to the scientific method," a recent *Christian Science Monitor* article on the subject reported. "Players encounter a phenomenon that doesn't make sense, observe problems, form hypotheses, and test them while being mindful of cause and effect." Considering all of this, many experts have come to the obvious conclusion: we need to find ways to make learning a lot more like video games and a lot less like school.

There are many different ways to do this. Jeremiah McCall, a history teacher at the Cincinnati County Day School, makes his students compare the battle depictions in Rome: Total War against the historical evidence. Lee Sheldon, meanwhile, a professor at the University of Indiana, has thrown out the traditional grading systems, where one bad grade can slide students backward. "This is demotivating," said Carnegie Mellon University professor of entertainment technology Jesse Schell in a recent talk on the subject. "A game designer would never put it in a game because people hate that." Instead Sheldon has implemented an "experience points" game-based design. Students begin the semester as a level zero avatar (equivalent to an F), and strive toward a level 12 (an A). This means that anything you do in the class produces forward motion, and students always know exactly where they stand—two conditions that serve to motivate.

Taking things even further are new schools like Quest2Learn. Founded by Katie Salen, a former associate professor of design and technology at Parsons the New School for Design, Q2L is a New York public school with a curriculum based on game design and digital culture. What does that look like in real life? *Popular Science* explains it this way: "In one sample curriculum, students create a graphic novel based on the epic Babylonian poem 'Gilgamesh,' record their understanding of ancient Mesopotamian culture through geography and anthropology journals, and play the strategic board game Settlers of Catan."

There are plenty of other examples as well, with many more to come. At the earlier mentioned X PRIZE Visioneering meeting, US Chief Technology Officer Aneesh Chopra and Scott Pearson of the Department of Education headed up a conversation on the use of incentive prizes to spark a brand-new generation of "effective, engaging and viral" educational games to be released on the net. A few months later, President Obama said, "I'm calling for investments in educational technology that will help create . . . educational software as compelling as the best video game." This revolution is upon us. Soon we're going to be able to create gamed-based learning that is so deep, immersive, and totally addictive that we're going to look back on the hundred-year hegemony of the industrial model and wonder why it ever hung around for so long.

The Wrath of Khan

In 2006 Salman Khan was a successful hedge fund analyst living in Boston, with younger cousins living in New Orleans whom he'd agreed to help in school. Khan began tutoring them remotely by making simple digital videos. Usually no more than ten minutes long, these self-narrated videos consisted of an animated digital chalkboard on which he would draw equations, chemical reactions, and the like. Kahn taught the basic subjects covered in school. Because he saw no reason not to make the tutorials public, he began posting them on YouTube. Surprisingly, his cousins preferred Khan on YouTube to him tutoring them in person.

"Once you get over the backhand nature of that," Khan told audiences at TED 2011, "there's actually something profound. They were saying they preferred the automated version of their cousin to their cousin . . . [F]rom their point of view, this makes a ton of sense. You have this situation where they can pause and repeat their cousin. If they have to review something they learned a couple of weeks ago or a couple of years ago, they don't have to be embarrassed and ask their cousin, they can just watch those videos. If they're bored, they can skip ahead. They can watch on their own time and at their own pace."

The tutorials struck a nerve. Very quickly, the Khan Academy, as it is now known, became an underground Internet sensation. By 2009, over fifty thousand people a month were watching the videos. A year later, the number had risen to two hundred thousand a month. A year after that, it had grown to a million. As of summer 2011, the Khan Academy was pulling in over two million visitors a month—exponential growth driven almost entirely by word of mouth.

As users have grown, so have the subjects covered. The academy now has 2,200 videos on topics ranging from molecular biology, to American history, to quadratic equations. They are adding three lessons a day—roughly 1,000 a year—and have plans to open up the site and begin crowdsourcing content. "Our vision is a free virtual school," says President and COO Shantanu Sinha. "We want to get enough content up that anyone in the world can start at one plus one equals two and go all the way through quantum mechanics. We also want to translate the site into the ten most common languages [Google is now driving this effort] and then crowdsource further translation into hundreds of languages. We think, at that level, the site is scalable into billions of visitors a month."

And for those who prefer their education in a physical setting, the Khan Academy has recently partnered with the Los Altos School District in Northern California. Together they are taking an approach that inverts the two-hundred-year-old schoolhouse model. Instead of teachers using classroom time to deliver lectures, students are assigned to watch Khan Academy videos as homework, so that class time can be spent solving problems (also provided by Khan) and getting points along the way (ten correct answers earns them a merit badge). This lets teachers personalize education, trading their sage-on-a-stage role for that of a coach. Students now work at their own pace and advance to the next topic only once they have thoroughly learned the last. "This is called mastery-based learning," says Sinha, "and there's research going back to the seventies that shows it produces greater student engagement and better results."

And better results are exactly what Los Altos is seeing. In the first twelve weeks of the project, students doubled their scores on exams. "It's like a game," John Martinez, a thirteen-year-old from Los Altos, told *Fast Company.* "It's kind of an addiction—you want a ton of badges." And it's because of responses like this that Bill Gates, after Khan's TED talk, told attendees they "just got a glimpse of the future of education."

This Time It's Personal

Gates is partially right. For some, the Khan Academy really is the future of education, but it's not the only future available. Of the lessons to be learned from industrial education, foremost among them is the fact that not every student is the same. There are those who enjoy the head-on collision with knowledge that is a Khan video; others prefer it presented tangentially, which is how information usually arrives in video games. Whatever the case, digitally delivered content means that it's no longer one size fits all. Students are now able to learn what they want, how they want, and when they want. And with the exponential expansion of IT technologies such as Negroponte's tablets and Nokia's smart phones, personalized learning will soon be available to just about anyone who wants it, no matter where in the world he or she lives.

But for digitally delivered universal education to be truly effective, we also need to change the way progress is measured. "We can't get deeper learning until we change the tests," says Gee, "because the tests drive the system." Here too video games offer a solution. "A video game is just an assessment," continues Gee. "All you do is get assessed, every moment, as you try to solve problems. And if you don't solve a problem, the game says you failed, try again. And you do. Why? Because games take testing, the most ludicrous, painful part of school, and make it fun." Even better is the data-capturing ability of video games, which can collect fine-grain feedback about student progress moment by moment, literally measuring growth every step of the way. As this technology develops, games will be able to record massive amounts of data about every aspect of each student's development—a far superior metric for progress than the one-size-fits-all testing method we currently favor.

We should not assume that all of these developments mean an end to teachers. Study after study show that students perform better when coached by someone who cares about their progress. This means that in places where teachers are in short supply, we'll need to expand the reach of Mitra's granny cloud. Even more potential exists for peer-to-peer tutoring networks; the John D. and Catherine T. MacArthur Foundation is currently beta testing one model. Most critically, since these newer models of education turn teachers into coaches, we'll need to expand our research into ways to make these coaches more effective. Right now the majority of education research is based around classroom management techniques that are no longer necessary with digital delivery. Instead there is a great need for new data about how to make the best use of the one-on-one attention that is now becoming possible.

Lastly, for those who prefer their instruction machine based, with the increasing development of artificial intelligence, an always-available, always-on AI tutor will soon be in the offing. Early versions of such systems, such as Apangea Learning's math tutor, has raised scores a staggering amount. For example, the Bill Arnold Middle School in Grand Prairie, Texas, used Apangea Math to help at-risk students prepare for their final exams, increasing pass rates from 20 percent to 91 percent. But such systems merely scratch the surface. In his novel *The Diamond Age: Or, A Young Lady's Illustrated Primer,* author Neal Stephenson gives readers a glimpse of what AI experts call a "lifelong learning companion": an agent that tracks learning over the course of one's lifetime, both insuring a mastery-level education and making exquisitely personalized recommendations about what exactly a student should learn next.

"The mobility and ubiquity of future AI tutors will enable one teacher per adult or child learner, anywhere and anytime," explains Singularity

University cochair for AI and Robotics Neil Jacobstein. "Learning will become real time, embedded into the fabric of everyday life and available on demand as needed. Children will still gather together with each other and with human teachers to collaborate in teams and learn social skills, but, fundamentally, the paradigm for education will shift dramatically."

The benefits to this shift are profound. Recent research into the relationship between health and education found that better-educated people live longer and healthier lives. They have fewer heart attacks and are less likely to become obese and develop diabetes. We also know that there's a direct correlation between a well-educated population and a stable, free society: the more well educated the population, the more durable its democracy. But these advances pale before what's possible if we start educating the women of tomorrow alongside the men.

Right now, of the 130 million children who are not in school, two-thirds of them are girls. According to the United Nations Educational, Scientific, and Cultural Organization (UNESCO), providing these girls with an education is "the key to health and nutrition; to overall improvements in the standard of living; to better agricultural and environmental practices; to higher gross national product; and to greater involvement and gender balance in decision making at all levels of society." In short, educating girls is the greatest poverty-reduction strategy around.

And if educating girls can have this much impact, imagine what educating everyone can do. With the convergence of infinite computing, artificial intelligence, ubiquitous broadband coverage, and low-cost tablets, we can provide a nearly free and personalized education to anyone, anywhere, at any time. This is an incredible force for abundance. Imagine billions of newly invigorated minds, thrilled by the voyage of discovery, using their newly gained knowledge and skills to improve their lives.

CHAPTER FIFTEEN
HEALTH CARE

Life Span

It's hard to measure how much our health has improved over the course of history, though life span is a fairly good indicator. Evolutionary pressures shaped Homo sapiens to have an average life expectancy of roughly thirty years. The logic is easily understood. "Natural selection favors the genes of those with the most descendants," explains MIT's Marvin Minsky. "Those numbers tend to grow exponentially with the number of generations, and so natural selection prefers the genes of those who reproduce at earlier ages. Evolution does not usually preserve genes that lengthen lives beyond the amount adults need to care for their young." Thus, for most of human evolution, men and women would enter puberty in their early teens and have children quickly thereafter. Parents would raise their kids until *they* reached the child-bearing years, at which point the parents—now thirty-year-old grandparents—became an expensive luxury. In early hominid societies, where life was difficult and food in short supply, an extra pair of grandparents' mouths to feed meant less food for the children. Thus, evolution built in a fail-safe: a three-decade life span.

Historically, though, as our living conditions improved, this number has improved. In the Neolithic period, life was a nasty, brutish, and short twenty years. This jumped to twenty-six in the Bronze and Iron Ages; and up to twenty-eight in ancient Greece and Rome, making Socrates a seventy-year-old anomaly when he died in 399 BCE. By the early Middle Ages, we pushed into the forties, but our ascendancy was still limited by appallingly high infant mortality rates. During the early sixteen hundreds in England, two-thirds of all children died before the age of four, and the resulting life expectancy was only thirty-five years.

It was the industrial revolution that started us on our trend toward longevity. A more robust food supply coupled with simple public health measures such as building sewers, collecting garbage, providing clean water, and draining mosquito-infested swamps made a huge difference. By the early twentieth century, we had added fifteen years to our historical average, with numbers rising into the low forties. With the creation of modern medicine and hospitals, that mean age skyrocketed into the middle seventies. While centenarians and supercentenarians are becoming increasingly common in the developed world (the verified age record stands at 122 years old), a combination of killers such as lower respiratory infections, AIDs, diarrheal infections, malaria, and tuberculosis, coupled with war and poverty, have played havoc with sub-Saharan Africa, where a large chunk of the population still doesn't make it much past forty.

Creating a world of health care abundance means addressing the needs at both ends of this spectrum and plenty in the middle. We'll need to provide clean water, ample nutrition, and smoke-free air. We'll also need to extinguish already curable ailments such as malaria and learn to detect and prevent those pesky pandemics that seem to threaten our survival with ever-increasing frequency. In the developed world, we'll need to find new ways to improve quality of life for an increasingly long-lived population. In total, creating a world of health care abundance appears to be a very tall order—except that almost every component of medicine is now an information technology and therefore on an exponential trajectory. And this, my friends, makes it a whole new ball game.

The Limits of Being Human

"Code blue, Baker five!" was the urgent page over the loudspeakers, snapping me out my brief slumber. It was four o'clock in the morning, I was catnapping on a stretcher in the hallway of Massachusetts General Hospital. As a third-year medical student, sleep was a rare commodity, and I'd learned to take it whenever and wherever possible. But code blue meant cardiac arrest; Baker five meant the fifth floor of the Baker building. I was just upstairs, on Baker six, now wide awake, adrenaline pumping, already racing down the stairwell. I was the second person to arrive in the room of a sixty-year-old man, less than twenty-four hours post-op from a triple coronary bypass. The resident giving him CPR barked an order my way, and I found myself taking over the chest compressions. It's the sound I remember most: the cracking of his surgically split sternum under the force of my repeated compressions. It was then that I realized it didn't matter what I learned in the classroom; none of it prepared me for the reality of this situation and the frailties of the human body.

That classroom learning had started two years earlier at Harvard Medical School. The first year was standard stuff: the basics of normal anatomy and physiology; how it all fits together and is meant to work. The second year was all about pathophysiology: where and how it all goes wrong. And with ten trillion cells in our body, there's plenty of opportunity for havoc. It was a dizzying amount of information. I remember a single moment when, while studying for my national board exams at the end of that second year, I felt like I had successfully stuffed all the concepts, systems, and terminology into my brain. But that moment was fleeting, especially in the hospital wards, where reality met flesh and blood as it did that early morning on Baker five. In that situation, I realized quickly how much I still had to learn and, even more, how much *we* really didn't know.

And that's our first problem. Learning takes time. It takes practice. Our brains process information at a limited pace, but medicine is growing exponentially, and there's no way we can keep up. Our second problem is a common refrain heard in medical school: five years after graduation, half of what one learns will probably be wrong—but no one knows which half. Regardless of how much medical progress has been made over these past centuries, our third problem is that we're never really satisfied with our health care. We continually have higher and higher standards, but with humans as the conduits of such care, there will always be limitations on how much information any doctor can know, let alone master.

A recent RAND Corporation report illustrates these points precisely, finding that preventable medical errors in hospitals result in tens of thousands of deaths per year; preventable medication errors occur at least one and a half million times annually; and, on average, adults receive only 55 percent of recommended care, meaning that 45 percent of the time, our doctors get it wrong.

In spite of these dismal numbers, having inaccurate doctors is much better than having no doctor at all. Fifty-seven countries currently don't have enough health care workers, a deficit of 2.4 million doctors and nurses. Africa has 2.3 health care workers per 1,000 people, compared with the Americas, which have 24.8 per 1,000. Put another way, Africa has 1.3 percent of the world's health workers caring for 25 percent of the global disease burden.

But things aren't rosy in the developed world either. The Association of American Medical Colleges warned recently that if training and graduation rates don't change, the United States could be short 150,000 doctors by 2025. And if America can't produce enough staff to cover its own medicinal needs, where are we possibly going to find the tenfold increase in health care workers needed to care for the rising billion?

Watson Goes to Medical School

"IBM Watson Vanquishes Human *Jeopardy!* Foes," read *PCWorld* magazine on February 16, 2011. Nearly fourteen years after Deep Blue had beaten world chess champion Garry Kasparov, IBM's silicon progeny challenged humanity to another battle. This time combat took place on the quiz show *Jeopardy!* There was $1.5 million in prize money at stake. Meet Watson, a supercomputer named after IBM's first president, Thomas Watson Sr. Over the course of three days, Watson bested both Brad Rutter, *Jeopardy!*'s biggest all-time money winner, and Ken Jennings, the show's record holder for the longest championship streak—meaning two men enter, one computer comes out.

It was something of an inevitable defeat. During the competition, Watson had access to 200 million pages of content, including the full text of Wikipedia. To be fair, the machine didn't have access to the Internet and could use only what was stored in its 16-terabyte brain. That said, Watson's brain is a massively parallel system composed of a cluster of ninety IBM Power 750 servers. The end product could handle 500 gigabytes of data per second, or the equivalent of 3.6 billion books per hour.

And that's only the hardware. The bigger breakthrough was the DeepQA software, which allows Watson to "understand" natural language—for example, the kinds of questions and answers found on *Jeopardy!* To make this possible, Watson had to not only comprehend context, slang, metaphors, and puns but also gather evidence, analyze data, and generate hypotheses.

Of course, not all good things come in small packages. Right now it takes a medium-sized room to hold Watson. But that's soon to change. If Moore's law and exponential thinking have taught us anything, it's that what fills a room today will soon require no more than a pocket. Moreover, this much computational power could soon be ubiquitous—hosted on one of the many clouds being developed—at little or no cost.

So what can we do with a computer like that? Well, a company called Nuance Communications (which used to be Kurzweil Computer Products, Kurzweil's first start-up) has teamed with IBM, the University of Maryland Medical School, and Columbia University to send Watson to medical school.

"Watson has the potential to help doctors reduce the time needed to evaluate and determine the correct diagnosis for a patient," says Dr. Herbert Chase, professor of clinical medicine at Columbia. He also has the ability to develop personalized treatment options for every patient, a capability that Dr. Eliot Siegel, professor and vice chair at Maryland's Department of Diagnostic Radiology, explains this way: "Imagine a supercomputer that can not only store and collate patient data but also interpret records in a matter of seconds, analyze additional information and research from medical journals, and deliver possible diagnoses and treatments, with the probability of each outcome precisely calculated."

But delivering correct diagnoses depends on having accurate data, which sometimes don't come from a conversation with the patient. Even the most brilliant diagnostician needs X-rays, CT scans, and blood chemistries to make the right call. But most of today's high-tech hospital equipment is large, expensive, and power hungry—unfit for the cost-conscious consumer, let alone the developing world. But now ask yourself that fabled DIY question: What would MacGyver do?

Well, MacGyver would empty his pockets and get the job done with a roll of Scotch tape, a piece of paper napkin, and a ball of spit—which, as it turns out, is exactly the solution we need.

Zero-Cost Diagnostics

Scotch tape? Really? When Carlos Camara entered his doctoral program at the University of California at Los Angeles, to study high energy-density physics, the last thing he imagined was that he'd soon find himself in a dark room experimenting with Scotch tape, or that this tape could drastically lower health care costs around the world. All he knew at first was that certain materials, when crushed together, create light—which is why, when you crunch a wintergreen Life Saver in your mouth, there's a little flash. It's called triboluminescence. Camara was experimenting with triboluminescence in a moderate vacuum and discovered that some materials don't just release visible light, they also release X-rays. So the question became, which materials? He started working his way through a wide range. Then it happened. Camara unrolled some Scotch tape in the dark. "I was shocked," he says. "Not only was it one of the brightest materials I had tested, but it also generated X-rays."

This was big news. It made the cover of *Nature,* then popped up on an episode of *Bones*. Soon after its television premiere, Camara teamed up with serial start-up entrepreneur Dale Fox to found Tribogenics, a company aiming to build the world's smallest and cheapest X-ray machine. Instead of a quarter-million-dollar dishwasher-sized device relying on eighteenth-century technology—basically, vacuum tubes connected to a power supply—the key component of the Tribogenics version (what Camara calls an "X-ray pixel") costs less than $1, is half the size of a thumb drive, and uses triboluminescence to create X-rays. Groups of these pixels can be arranged into any size or shape. A fourteen-by-seventeen-inch array takes a chest X-ray; a long curve gives you a CT scan. As these pixels require very little energy—less than 1/100th of a traditional X-ray machine—a solar panel or a hand crank can power one. "Imagine an entire radiological suite in a briefcase," continues Fox. "Something powered with batteries or solar, easily transportable, and capable of diagnosing anything from a broken arm to an abdominal obstruction. It will bring a whole new level of care to field medicine and the developing world."

Fox sees additional possibilities in mammography. "Today mammography requires an expensive, large, stationary machine that takes a crude, two-dimensional picture. But imagine a 'bra' that has tiny X-ray pixel emitters on the top and X-ray sensors on the bottom. It's self-contained, self-powered, has a 3G or Wi-Fi-enabled network, and can be shipped to a patient in a FedEx box. The patient puts on the bra, pushes a button, and the doctor comes online and starts talking: 'Hi. All set to take your mammogram? Hold still.' The X-ray pixels fire, the detectors assemble and transmit the image, and the doctor reads it on the spot. The patient ships back the package, and she's done. With little time and little money.

The X-ray pixel array is the first step toward what Harvard chemistry professor turned uber-entrepreneur George Whitesides calls "zero-cost diagnostics." Exactly as it sounds, Whitesides wants to drop the cost of diagnosing disease as low as possible, which, here in *MacGyver*-land, is pretty low indeed. Toward that end, Whitesides recently turned his attention to the diseases plaguing the rising billion. The only way to develop the vaccines needed to fight HIV, malaria, and tuberculosis is to find a method for accurately and inexpensively diagnosing and monitoring large numbers of patients. You can't do this with today's technology.

So Whitesides took a page out of C. K. Prahalad's BoP development model. Instead of starting with a $100,000 machine and trying to lower its cost by orders of magnitude, he started with the cheapest materials available: a piece of paper about one centimeter on a side, able to wick fluid. Place a pinprick of blood or a drop of urine on the edge of Whitesides's paper, and the liquid soaks in, migrating through the fibers. A hydrophobic polymer printed on this paper guides the fluid along prescribed channels, toward a set of testing wells, wherein the sample interacts with specific reagents, turning the paper different colors. One chamber tests urine for glucose, turning brown in the presence of sugar. Another turns blue in the presence of protein. Since paper isn't very expensive, Whitesides's goal of zero-cost diagnostics isn't that far fetched. "The major cost is the wax printer," he says. "These printers are around eight hundred dollars apiece. If you run them twenty-four hours a day, each of them can make some ten million tests per year, so it's really a solved problem."

The final stop in our MacGyver triad—the spit sample—holds even more promise. This is the input necessary for the aforementioned Lab-on-a-Chip array developed by Dr. Anita Goel at her company, Nanobiosym. Place a drop of saliva (or blood) on Goel's nanotechnology platforms, and the DNA and RNA signature of any pathogen in your system will get detected, named, and reported to a central supercomputer—aka

Dr. Watson. These chips are a serious step toward zero-cost diagnostics, and a critical component in helping to solve a trio of major health care challenges: arresting pandemics, decreasing the threat of bioterrorism, and treating widespread diseases like AIDS. Already mChip, a technology out of Columbia University, is demonetizing and dematerializing the HIV testing process. What once required long doctor visits, a vial of blood, and days or weeks of anxious waiting now needs no visit, a single drop of blood, and a fifteen-minute read, all for under $1 using a microfluidic optical chip smaller than a credit card.

Since Dr. Watson will soon be accessible through a mobile device, and that mobile device has a GPS, the computer can both diagnose your infection *and* detect an unusually high incidence of, say, flu symptoms in Nairobi—thus alerting WHO to a possible pandemic. Better yet, because the incremental cost of Watson's diagnosis is simply the expense of computing power (which is really just the cost of electricity), the price comes to pennies. To help accelerate this process, on May 10, 2011, the wireless provider Qualcomm teamed up with the X PRIZE Foundation and announced plans to develop the Qualcomm Tricorder X PRIZE—named after *Star Trek*'s medical scanning technology. This competition will offer $10 million to the first team able to demonstrate a consumer-friendly low-cost mobile device able to diagnose a patient better than a group of board-certified doctors.

But even this much MacGyver thinking still falls short of our ultimate health care goals, since knowing what's wrong with a patient is only half the battle. We still need to be able to treat and cure that patient. We've already addressed many of the "preventable" ailments, which are obviated through clean water, clean energy, basic nutrition, and indoor plumbing, but there's also another category to consider: easily treatable and/or curable diseases. Many of these are managed with simple medicines, but others require surgical intervention. In the same way that technology has revolutionized diagnostics, what if it were possible to do the same with surgery?

Paging Dr. da Vinci to the Operating Room

According to the World Health Organization, age-related cataracts are the world's largest cause of blindness, accounting for eighteen million cases, primarily in Africa, Asia, and China. Cataracts are a clouding of the eye's normally transparent lens. Although they can be easily removed, and this form of blindness completely cured, surgical services in many developing countries are inadequate, inaccessible, and too expensive for much of the affected population.

The best chance many have is a nonprofit humanitarian organization called ORBIS International, which teaches cataract surgery in developing countries and operates a Flying Eye Hospital. ORBIS's refurbished DC-10 swoops into a region with doctors, nurses, and technicians aboard. Once there, they provide treatment to a limited number of cases, and train local physicians. But only so many doctors can be trained this way. Physician and robotics expert Catherine Mohr sees a future without these limitations. "Imagine," she says, "specialized robots able to conduct this type of simple and repeatable surgery with complete accuracy, at little to no cost."

The earliest versions of this type of surgical robot, called the da Vinci Surgical System, were built by Mohr's company, Intuitive Surgical. Da Vinci actually came out of the DARPA's desire to get surgeons off the front line of the battlefield while still treating the wounded during the first "golden hour" after injury. The best way to do this is with a robot tending the injured soldier, and a telepresent physician running the show from a remote location. In recent years, this technology has evolved rapidly, moving from the battlefield to the surgical suite, initially at the behest of cardiac surgeons looking for ways to operate without splitting the sternum. It was next taken up by surgeons seeking to conduct rapid and repeatable prostatectomies and gastric bypasses. Current iterations, like the MAKO surgical robot, are skilled enough to assist orthopedists with delicate procedures such as knee replacements.

Today's technology doesn't completely replace surgeons; instead it enhances their abilities and allows them to operate remotely. "By fully digitizing an image of the injured site being repaired," explains Mohr, "you put a digital layer between the tissue and the surgeon's eyes, which can then be augmented with overlaid information or magnification. Also, by digitizing hand movements and placing a digital layer between the surgeon and the robotic instruments, you can take out jitter, make motions more precise, and even transmit the surgical incisions over a long distance, allowing an expert in Los Angeles to conduct surgery in Algiers during their free time without spending twenty hours on airplanes."

Over the next five to ten years, Mohr predicts a proliferation of smaller, special purpose robots, extending far beyond cataract removal. One might handle glaucoma surgery, another a gastric bypass, while a third performs dental repairs. Mohr thinks the fifteen- to twenty-year horizon is even more exciting. "In the future, we'll be able detect cancers by monitoring blood, urine, or breath and, once detected, remove them robotically. The robot will find the tiny cancerous lesion, insert a needle, and obliterate it, just like you do a cancerous mole today."

Robo Nurse

Cancer is only one of the problems that our aging population will have to face. In fact, when it comes to health care costs and quality of life, caring for the aged is a multitrillion-dollar expense that we'd better get used to. The oldest baby boomers turned sixty-five in 2011. When the trend peaks in 2030, in the United States alone, the number of people over age sixty-five will have soared to 71.5 million. In developed countries, the centenarian population is doubling every decade, pushing the 2009 total of 455,000 to 4.1 million by 2050. And the average annual growth rate of those over eighty is twice as high as the growth rate for those over sixty. In 2050 we'll have 311 million octogenarians in the world. As the elderly lose the ability to care for themselves, many, according to the National Center for Health Statistics, are sent to nursing homes at an annual cost per person of between $40,000 and $85,000. Bottom line: with hundreds of millions of people soon heading down this road, how will we ever afford it?

For Dr. Dan Barry, the answer is easy: let the robots do the work. Barry brings an eclectic background to this problem, including an MD, a PhD, three Space Shuttle flights, a robotics company, and a starring role as a contestant on the reality TV show *Survivor.* Barry is also the cochair of the AI and robotics track at SU, where he spends considerable time thinking about how robots can be applied to the future of health care. "The biggest contribution that robots will make to health care is taking care of an aging population: people who have lost spouses or lost the ability to take care of themselves," he says. "These robots will extend the time they are able to live independently by providing emotional support, social interaction, and assisting them with the basic functional tasks like answering the door, helping them if they fall, or assisting them in the bathroom. They will be willing to listen to the same story twenty-five times and respond appropriately every time. And for some with sexual dysfunction or need, these robots will also play a huge role."

When will these robots become available, and what will they cost? "Within five years," continues Barry, "robots will hit the market that can recognize you uniquely, react to your movements and facial expressions with recognizable emotive responses, and perform useful tasks around the home, like clean up while you sleep. Fast-forward fifteen to twenty years, and we'll be delivering robotic companions that will have real, nuanced conversations, making them able to serve as your friend, your nurse, perhaps even your psychologist."

The anticipated cost is almost as shocking as their capabilities. "I expect the initial robots will cost on the order of a thousand dollars," says Barry. He goes on to explain that the cost of a three-dimensional laser range finder has plummeted from $5,000 to about $150 because of new technology and the massive scale of production for Microsoft's Xbox Kinect. "A five-thousand-dollar laser range finder was the typical way for a robot to navigate a cluttered environment," he says. "It's mind boggling how powerful and cheap they've become. The result is a tsunami of new code and applications and an explosion in the number of people developing DIY robots. As soon as the price dropped low enough, an army of graduate students began playing, experimenting, and coming up with amazing new applications."

Just like laser range finders, all other robo-nurse components are on similar price-performance reduction curves. Pretty soon, the requisite sensors and computing power will be nearly free. All that's left to buy is the mechanical body, which is why Barry believes that $1,000 is the ballpark figure for these bots. So here's your comparison: if we assume that the majority of the octogenarians in our future will need some form of assisted living care, we can either spend (at today's costs) trillions of dollars on nursing homes or we can, as Barry suggests, let robots do the work.

The Mighty Stem Cell

In the early 1990s, accomplished neurotrauma surgeon Robert Hariri was growing frustrated with his field, especially with the limitations of the scalpel. "We could do some limited repairs and keep people alive after an accident," he says, "but surgery couldn't return them to normal." So Hariri went looking for ways to restore the natural developmental processes that allow the brain to regrow and rewire itself. In the late 1990s, he realized that he might be able to inject stem cells into patients to treat and potentially cure diseases in the same way that one now injects drugs. Hariri believed that to harness the true potential of cellular medicine he had to ensure a steady source of stem cells for future procedures, so he created his first company to bank both placenta-derived stem cells and cord blood from newborns. Four years later, LifeBank/Anthrogenesis merged with $30 billion pharmaceutical giant Celgene Corporation, which saw the technology's potential to reinvent medicine.

But it's not just Celgene that wants in on this action. "We all start out as a single fertilized egg that develops into a complex organism of ten trillion cells, made up of over two hundred tissue types, each working twenty-four/seven at specialized functions," says Dr. Daniel Kraft, a specialist in bone marrow transplantation (a form of stem cell therapy) and chair of the medicine track at SU. "Stem cells drive this incredible process of differentiation, growth, and repair. They have the ability to revolutionize many aspects of health care like almost nothing else in the pipeline."

Dr. Hariri agrees:

> The potential for this technology is immense. In the next five to ten years, we're going to be able to use stem cells to correct chronic autoimmune diseases such as rheumatoid arthritis, multiple sclerosis, ulcerative colitis, Crohn's disease, and scleroderma. After that, I think neurodegenerative diseases will be the next big frontier; this is when we'll reverse the effects of Parkinson's disease, Alzheimer's disease, even stroke. And it'll be affordable too. Cell manufacturing technology has seen vast improvements over the past decade. To give you an idea, we've gone from thinking that cell therapy would cost over $100,000, to believing that we can do it for about $10,000. Over the next decade, I think we can lower costs significantly more. So we're speaking about the potential for "curing" chronic diseases and revitalizing key organs for less than the price of a new laptop.

And should your liver or kidney fail before you have a chance to revitalize it, fear not, there is another solution. One of Hariri's issued patents,

"The renovation and repopulation of cadaveric organs and tissue matrices by stem cells," is the basis for growing new and transplantable organs in the lab, which is an approach that tissue-engineering pioneer Anthony Atala of Wake Forest University Medical Center has demonstrated successfully.

"There is a huge need for organs worldwide," says Atala. "In the past decade, the number of patients on the organ transplant waiting list has doubled, while the number of actual transplants has remained flat. Thus far, we've been able to grow human ears, fingers, urethras, heart valves, and whole bladders in the lab."

Atala's next major challenge is to grow one of the most intricate organs in the human body: the kidney. About 80 percent of patients on the transplant list are waiting for a kidney. In 2008 there were over sixteen thousand kidney transplants in the United States alone. To accomplish this feat, he and his team have moved beyond the use of cadaveric organs and tissue matrices and are literally "3-D printing" early versions of the organ. "We started by using a normal desktop ink-jet printer that we rigged to print layers of cells one at a time," he explains. "We've been able to print an actual minikidney in a few hours." While the full organ may need another decade of work, Atala is cautiously optimistic, given that sections of his printed kidney tissue are already excreting a urine-like substance.

"Whether it's organ regeneration or repairing tissues affected by aging, trauma, or disease," says Dr. Kraft, "this fast-moving field will impact almost every clinical arena. The recent invention of induced pluripotent stem cells, which can be generated by reprogramming a patient's own skin cells, gives us controversy-free access to this powerful technology. And with the coming convergence of stem cells, tissue engineering, and 3-D printing, we'll soon have an incredibly potent arsenal for achieving health care abundance."

Predictive, Personalized, Preventive, and Participatory

While many believe that stem cells will soon give us the ability to repair and replace failed organs, if P4 medicine does its job, the situation might never get that desperate. P4 stands for "predictive, personalized, preventative, and participatory," and it's where health care is heading. Combine cheap, ultrafast, medical-grade genome sequencing with massive computing power, and we're en route to the first two categories: predictive and personalized medicine.

During the past decade, sequencing costs have dropped from Craig Venter's historic $100 million genome in 2001 to an anticipated $1,000 version of equal accuracy. Companies such as Illumina, Life Technologies, and Halcyon Molecular are vying for the trillion-dollar sequencing market. Soon every newborn will have his or her genome sequenced. Genetic profiles will be part of standard patient care. Cancer victims will have their tumors DNA analyzed, with the results linked to a massive data correlation effort. If done properly, all three efforts will yield a myriad of useful predictions, changing medicine from passive and generic to predictive and personalized. In short, each of us will know what diseases our genes have in store for us, what to do to prevent their onset, and, should we become ill, which drugs are most effective for our unique inheritance.

But rapid DNA sequencing is only the beginning of today's biotech renaissance. We are also unraveling the molecular basis for disease and taking control of our body's gene expression, which together can create an era of personalized and preventative medicine. One example is the potential to cure what the WHO now recognizes as a global epidemic: obesity. The genetic culprit here is the fat insulin receptor gene that instructs our body to hold on to every calorie we consume. This was a helpful gene in the era before the invention of Whole Foods and McDonald's, when early hominids could never be certain about their next harvest or even their next meal. But in our fast-food nation, this genetic edict has become a death sentence.

However, a new technology called RNA interference (RNAi) turns off specific genes by blocking the messenger RNA they produce. When Harvard researchers used RNAi to shut off the fat insulin receptor in mice, the animals consumed plenty of calories but remained thin and healthy. As an added bonus, they lived almost 20 percent longer, obtaining the same benefit as caloric restriction, without the painful sacrifice of an extreme diet.

Participatory medicine is the fourth category of our health care future. Powered by technology, each of us is becoming the CEO of our own health. The mobile phone is being transformed into a mission control center where our body's real-time data can be captured, displayed, and analyzed, empowering each of us to make important health decisions day by day, moment by moment. Personal genomics companies such as 23andMe and Navigenics, meanwhile, allow users to gain a deeper understanding of their genetic makeup and its health implications. But equally important is the effect of our environment and daily choices—which is where a new generation of sensing technology comes into play.

"Sensors have plummeted in cost, size, and power consumption," explains Thomas Goetz, executive editor of *Wired* and author of *The Decision Tree: Taking Control of Your Health in the New Era of Personalized Medicine.* "An ICBM guidance sensor from the 1960s used to cost one hundred thousand dollars and weigh many kilograms. Now that same capability fits on a chip and costs less than a dollar." Taking advantage of these breakthroughs, members of movements such as Quantified Self are increasing self-knowledge through self-tracking. Today they're tracking everything from sleep cycles, to calories burned, to real-time electrocardiogram signals. Very soon, should one choose to go this route, we'll have the ability to measure, record, and evaluate every aspect of our lives: from our blood chemistries, to our exercise regimen, to what we eat, drink, and breathe. Never again will ignorance be a valid excuse for not taking care of ourselves.

An Age of Health Care Abundance

As should be apparent, the field of health care is entering a period of explosive transformation. However, the major drivers here are not just technological. As the baby boomers age, there is no amount of money that the richest among them won't spend for a little more quality time with their loved ones. Thus, every new technology inevitably finds its way into the service of health, driven by an older, wealthier, and more motivated population.

In the same way that Wall Street tycoons talking on briefcase-sized mobile phones in the 1970s underwrote the development of the hundreds of millions of Nokia handsets now scattered through sub-Saharan Africa, so too will the billions of health care research dollars and entrepreneurial inventions described in this chapter soon benefit all nine billion of us. And given the rigorous, somewhat calcified, nature of the first-world health care regulatory process, there's every reason to believe that more than a few of these groundbreaking technologies will first make their way to less bureaucratic regions of the developing world before being legally allowed onto Main Street, USA.

While the developing world will certainly benefit from these high-tech cures, the truth of the matter is that the majority of their needs are still basic: bed nets and antimalarial drugs; antibiotics to combat bronchitis and diarrhea; education about the realities of HIV and the necessity of contraception. In many cases, the remedies exist, but the necessary infrastructure does not. However, there are now a host of mobile-phone-enabled education programs that can help. Project Masiluleke, for example, in South Africa, uses text messages to broadcast an HIV-awareness bulletin. Johnson & Johnson's Text4Baby effort has served more than twenty million pregnant women and new parents in China, India, Mexico, Bangladesh, South Africa, and Nigeria. This is also where technophilanthropists like Bill Gates and his war on malaria can make a huge difference. Ultimately, however, meeting the medical needs of the entire world means empowering the rising billion with the basic resources—food, water, energy, and education—while at the same time driving forward the breakthroughs outlined in this chapter. If we can do this, we can create an age of health care abundance.

CHAPTER SIXTEEN
FREEDOM

Power to the People

Freedom, the subject of this chapter, is both the peak of our pyramid and the place where this book must get a little philosophical. In other sections, we've explored how the combination of collaboration and exponential technology can conspire to better lives over the next few decades. But the deliverables in those chapters are goods and services: food, water, education, health care, and energy. Freedom falls into a different category. It's both an idea and access to ideas. It's a state of being, a state of consciousness, and a way of life. On top of that, it's a catch-all term with meanings stretching from the right to gather a few people around a coffee table to the right to carry a fully automatic weapon down a city street—which is to say that freedom is also a number of things beyond the scope of this book.

What's within our scope are economic freedom, human rights, political liberty, transparency, the free flow of information, freedom of speech, and, empowerment of the individual. These are all categories impacted directly by the forces of change discussed in this book, all liberties liberated on the road to abundance. We'll take them one at a time.

Not having enough to eat and drink, having no way to obtain remedies for treatable illnesses, lacking access to clothing or shelter or affordable health care or education or sanitary facilities—all are, to quote Nobel laureate Amartya Sen, "major sources of unfreedom." As the previous chapters have made clear, exponentials are already making an impact here. Whether it's the Khan Academy's algebraic offerings or Dean Kamen's Slingshot water purifier, these tools of prosperity do double duty as crusaders for liberation: freeing up time and money, improving quality of life, and creating even greater opportunity for opportunity. This trend will continue. With each tiny step taken toward clean water or cheap energy or any other level of our pyramid, these basic freedoms are the direct beneficiaries of progress.

Human rights, too, have been aided by exponentials. The website Ushahidi was created to chart outbreaks of violence in Kenya, but its success has lead to a flurry of "activist mapping." This crowdsourced combination of social activism, citizen journalism, and geospatial mapping has been used in countries all over the world to defend human rights. Activist mapping protects sexual minorities in Namibia, ethnic minorities in Kenya, and potential victims of military abuse in Colombia. Sites like World Is Witness document stories of genocide, while sites such as WikiLeaks blow the whistle on human rights violations of all sorts.

WikiLeaks is also an example of how information and communications technology promote political liberty and greater transparency—although it's not the only one. In 2009 a version of Ushahidi was modified to let Mexican citizens self-police their elections, while the $130,000 Enough Is Enough Nigeria grant from the Omidyar Network to activists in Nigeria utilizes Twitter, Facebook, and local social media tools to provide a nonpartisan one-stop online portal designed to aid voter registration, supply candidate information, and monitor elections. Arguably, ICT's biggest impact has been at the intersection of transparency and sociopolitical liberty. Before the advent of the Internet, a shy gay man living in Pakistan was in for a rough ride. These days, while the ride is still plenty bumpy, at least that man is a couple of mouse clicks away from the advice and companionship of several million other people in similar situations.

That the free flow of information has benefited most from the rise of mobile communications and the Internet is obvious. As mentioned earlier, the majority of humanity, even those in the poorest of developing nations, now have access to better mobile phone systems than the president of the United States did twenty-five years ago, and if they're hooked up to the Internet, they have access to more knowledge than the president did fifteen years ago. The free flow of information has become so important to all of us that in 2011 the United Nations declared "access to the Internet" a fundamental human right.

Free speech and freedom of expression, too, have found plenty of allies in the Information Age. "Think of it this way," says Google executive chairman Eric Schmidt. "We've gone from a hierarchical messaging structure where people are broadcast to, and information usually had local context; to a model where everyone's an organizer, a broadcaster, a blogger, a communicator." Sure, there are difficult issues concerning censorship to deal with (the so-called Great Firewall of China, for starters), but the fact remains that never before in history has the ordinary citizen had both the power to make himself heard and the access to a global audience. Nor is this access in jeopardy. "[The] Internet tends to shift power from centralized institutions to many leaders representing different communities," Ben Scott, Secretary of State Hillary Clinton's policy advisor on innovation, recently told the *Christian Science Monitor.* "Governments who want to censor are fighting a battle against the nature of the technology."

But of all the categories in question, self-empowerment has been and will continue to be the one most significantly affected by the rising tide of abundance. So important is this change and—for good or for ill—so far reaching are its effects, that we'll spend the next few sections examining it in depth.

One Million Voices

In 2004, while doing graduate work as a Rhodes scholar at Oxford University, Jared Cohen decided that he wanted to visit Iran. Since Iran's stance against the United States is based partially on US support of Israel, Cohen—a Jewish American—didn't think he stood much chance of getting a visa. His friends told him not to bother applying. Experts told him he was wasting his time. But after four months and sixteen trips to the Iranian Embassy in London, he received permission to travel to, as Cohen later recounted in his book *Children of Jihad: A Young American's Travels Among the Youth of the Middle East,* "a country that President Bush had less than two years ago labeled as one of the three members of the 'axis of evil.' "

The purpose of Cohen's trip was to expand his knowledge of international relations. He wanted to interview opposition leaders, government officials, and other reformers, but after successful conversations with the Iranian vice president and several members of the opposition, the government's Revolutionary Guard sauntered into his hotel room late one night, found his potential interview list, and foiled his plans. But rather than leaving Iran and flying back to England defeated, Cohen decided to explore the country and see what kinds of friends he made along the way.

He made plenty of friends, most of them young. Two-thirds of Iran is under the age of thirty. Cohen dubbed them "the real opposition," a massive, not-especially-dogmatic youth movement hungry for Western culture and suffocating under the current regime. He also discovered that technology was allowing this movement to flourish—a lesson that crystallized for him at a busy intersection in downtown Shiraz, where Cohen noticed a half dozen teens and twentysomethings leaning up against the sides of buildings, staring at their cell phones.

He asked one boy what was going on and was told this was the spot everyone came to use Bluetooth to connect to the Internet.

"Aren't you worried?" asked Cohen. "You're doing this out in the open. Aren't you worried you might get caught?"

The boy shook his head no. "Nobody over thirty knows what Bluetooth is."

That was when it hit him: the digital divide had become the generation gap, and this, Cohen realized, opened a window of opportunity. In countries where free speech was wishful thinking, folks with basic technological savvy suddenly had access to a private communication network. As people under thirty constitute a majority in the Muslim world, Cohen came to believe that technology could help them nurture an identity not based on radical violence.

These ideas found a welcome home in the US State Department. When Cohen was twenty-four years old, then Secretary of State Condoleezza Rice hired him as the youngest member of her policy planning staff. He was still on her staff a few years later when strange reports about massive anti-FARC protests started trickling in. The FARC, or Revolutionary Armed Force of Colombia, a forty-year-old Colombia-based Marxist-Leninist insurgency group, had long made its living on terrorism, drugs, arms dealing, and kidnapping. Bridges were blown up, planes were blown up, towns were shot to hell. Between 1999 and 2007, the FARC controlled 40 percent of Colombia. Hostage taking had become so common that by early 2008, seven hundred people were being held, including Colombian presidential candidate Íngrid Betancourt—who'd been kidnapped during the 2002 campaign. But suddenly, and seemingly out of nowhere, on February 5, 2008, in cities all over the world, 12 million people poured into the streets, protesting the rebels and demanding the release of hostages.

Nobody at State quite understood what was going on. The protestors appeared spontaneously. They appeared to be leaderless. But the gathering seemed to have been somehow coordinated through the Internet. Since Cohen was the youngest guy around—the one who supposedly "spoke" technology—he was told to figure it out. In trying to do that, Cohen discovered that a Colombian computer engineer named Oscar Morales might have been responsible. "So I cold-called the guy," recounts Cohen. "Hi. How are you? Can you tell me how you did this?"

What had Morales done to bring millions of people into the streets in a country where, for decades, anyone who said anything against the FARC wound up kidnapped or dead or worse? He'd created a Facebook group. He called it A Million Voices Against FARC. Across the page, he typed, in all capital letters, four simple pleas: "NO MORE KIDNAPPING, NO MORE LIES, NO MORE DEATH, NO MORE FARC."

"At the time, I didn't care if only five people joined me," said Morales. "What I really wanted to do was stand up and create a precedent: we young people are no longer tolerant of terrorism and kidnapping."

Morales finished building his Facebook page around three in the morning on January 4, 2008, then went to bed. When he woke up twelve hours later, the group had 1,500 members. A day later it was 4,000. By day three, 8,000. Then things got really exponential. At the end of the first week, he was up to 100,000 members. This was about the time that Morales and his friends decided that it was time to step out of the virtual world and into the real one.

Only one month later, with the help of 400,000 volunteers, A Million Voices mobilized some 12 million people in two hundred cities in forty countries, with 1.5 million taking to the streets of Bogotá alone. So much publicity was generated by these protests that news of them penetrated deep into FARC-held territory, where news didn't often penetrate. "When FARC soldiers heard about how many people were against them," says Cohen, "they realized the war had turned. As a result, there was a massive wave of demilitarization."

Cohen was fascinated. He flew down to Colombia to meet with Morales.

What surprised him most was the structure of the organization. "Everything I saw had the structure of a real nongovernmental organization—but there was no NGO. There was the Internet. You had followers instead of members, volunteers instead of paid staff. But this guy and his Facebook friends helped take down the FARC." For Cohen and the rest of the State Department, it was something of a watershed moment. "It was the first time we grasped the importance of social platforms like Facebook and the impact they could have on youth empowerment."

This was also about the time that Cohen decided technology needed to be a fundamental part of US foreign policy. He found willing allies in the Obama administration. Secretary of State Clinton had made the strategic use of technology, which she termed "twenty-first-century statecraft," a top priority. "We find ourselves living in a moment in human history when we have the potential to engage in these new and innovative forms of diplomacy," said Secretary Clinton, "and to also use them to help individuals empower their development."

Toward this end, Cohen had become increasingly concerned about the gap between local challenges in developing nations and the people who made the high-tech tools of the twenty-first century. So, wearing his State Department hat, he started bringing technology executives to the Middle East, primarily to Iraq. Among those invited were Twitter founder Jack Dorsey. Six months after that trip, when Iranian postelection protestors overran the streets of Tehran, and a government news blackout threatened all traditional lines of communication, Cohen called Dorsey and asked him to postpone a routine maintenance shutdown of the Twitter site. And the rest, as they say, is history.

Twitter, of course, soon became the only available pipeline to the outside world, and while the Twitter revolution didn't topple the Iranian government, in combination with Morales's efforts and other Internet-based activism campaigns, all of these events paved the way for what we would soon be calling the Arab Spring (more on this later).

"It didn't happen intentionally," says Cohen. "Bluetooth was a technology invented so people could talk and drive—nobody who built it expected their peer-to-peer network would be used to get around an oppressive regime. But the message of the events of the past few years are clear: modern information and communication technologies are the greatest tools for self-empowerment we've ever seen."

Bits Not Bombs

In 2009, when Eric Schmidt was still the CEO of Google (before he became executive chairman), he went to Iraq at the behest of Jared Cohen and the State Department. During that trip, Schmidt and Cohen became friendly. They had long conversations about the reconstruction of the country and how technology should have played a much earlier role in the effort. Iraq, under dictator Saddam Hussein, had no cell phone structure. The United States had spent over $800 billion on regime change, but, as Schmidt says, "What we should have done is laid down fiber-optic cable and built out a wireless infrastructure to empower the Iraqi citizens."

This idea led the duo to an interesting realization: technology, at least in its current form, seems to have a bias toward individual empowerment. Schmidt explains further: "The individual gets to decide what to do, as opposed to the traditional systems, but this has a whole bunch of implications. Technology doesn't just empower the good people, it also empowers the bad. Everyone can be a saint or everyone can be a terrorist."

This is no small matter. The Internet has proved to be a fantastic recruiting tool for Hamas, Hezbollah, and Al Qaeda. In 2011, the terrorists who sailed from Karachi to Mumbai used GPS devices to navigate, satellite phones to communicate, and Google maps to locate their targets. In Kenya, hateful text messages were used to direct waves of ethnic violence after the disputed 2007 election. But it was also in Kenya where the aforementioned Ushahidi was created. Schmidt feels that sites like this are a critical counterforce. "We have greater safety when the majority of people are empowered," he says. "Technologically empowered people can tell you things, they can report things, they can take pictures."

In November 2010, a few months before Cohen left the State Department to join Google as director of ideas, he teamed up with Schmidt to write "The Digital Disruption," an article for the magazine *Foreign Affairs* that examined the impact ICT will have on international relations over the next decade or so. As a basis for prognostication, the duo used a combination of a nation's current political system and its current state of communications technology. Strong countries like the United States and the European and Asian giants appear able to regulate what Cohen and Schmidt call "the interconnected estate" in ways that reflect national values. Partially connected autocratic, corrupt, or unstable governments, though, could prove volatile. "In many cases," they write, "the only thing holding the opposition back is the lack of organizational and communication tools, which connection technologies threaten to provide cheaply and widely."

This is just what we saw in the Arab Spring. One of the defining features of the revolutions that swept the Middle East in early 2011 was their use of communication technologies. During the protests in Cairo, Egypt, that brought down President Hosni Mubarak, one activist summed this up nicely in a tweet: "We use Facebook to schedule the protests, Twitter to coordinate, and YouTube to tell the world."

Yet this blade too cuts both ways. In Egypt, the government shut down the Internet to quell revolt. In the Sudan, protestors were arrested and tortured into revealing Facebook passwords. In Syria, progovernment messages popped up on dissidents' Facebook pages, and the Twitter #Syria hashtag—which had carried accounts of the protests—was flooded with sports scores and other nonsense. "In the same way that, a few years ago, it became commonplace to talk about Web 2.0, we're now seeing Repression 2.0," Daniel B. Baer, a deputy assistant secretary of state for democracy, human rights, and labor, told the *Washington Post.* And repression 2.0 may soon give way to repression 3.0, as authoritarian governments become better acquainted with the technology now at their disposal. In *The Net Delusion: The Dark Side of Internet Freedom,* Evgeny Morozov, a contributing editor at *Foreign Policy* and a Schwartz Fellow at the New America Foundation, writes:

> Google already bases the ads it shows us on our searches and the text of our emails; Facebook aspires to make its ads much more fine-grained, taking into account what kind of content we have previously "liked" on other sites and what our friends are "liking" and buying online. Imagine building censorship systems that are as detailed and fine-tuned to the information needs of their users as the behavioral advertising we encounter every day. The only difference between the two is that one system learns everything about us to show us more relevant advertisements, while the other learns everything about us to ban us from accessing relevant pages. Dictators have been somewhat slow to realize that the customization mechanisms underpinning so much of Web 2.0 can easily be turned to purposes that are much more nefarious than behavioral advertising, but they are fast learners.

So while ICT is clearly the greatest tool for self-empowerment we've yet seen, it's still only a tool, and, like all tools, is fundamentally neutral. A hammer can build bridges or bash brains. Connection technologies are not much different. While their bias toward self-empowerment is clear, there's no guarantee that a safer, freer world will be the result. What ICT does guarantee is an exceptionally broad platform for cooperation. Nations can partner with corporations, which can partner with citizens, who can partner with one another to use these tools to promote positive self-empowerment, democracy, equality, and human rights. In fact, with the complexity of today's world, this sort of cooperation appears to be mandatory. As Schmidt and Cohen point out, "In a new age of shared power, no one can make progress alone."

But we can all make progress together—which is, after all, the point.

PART SIX
STEERING FASTER

CHAPTER SEVENTEEN

DRIVING INNOVATION AND BREAKTHROUGHS

Fear, Curiosity, Greed, and Significance

Now that we've finished exploring the upper levels of our abundance pyramid, it should be clear that the rate of technology innovation has never been greater and the tools at our disposal have never been more powerful. But will this be enough? While abundance is a very real possibility, we're also in a race against time. Can some version of today's world handle a population of nine billion? Can we feed, shelter, and educate everyone without the radical changes discussed in this book? What happens if somewhere along the way, the prophets of peak oil or peak water or peak whatever turn out to be right before some breakthrough technology can prove them wrong? Until the innovations of abundance bear fruit, scarcity remains a real concern. And nearly as bad as scarcity is the threat of scarcity and the devastating violence it can often incite.

In many cases, we know where we want to go but not how to get there. In others, we know how to get there but want to get there faster. This chapter focuses on how we can steer innovation and step on the gas. When bottlenecks arise, when breakthroughs are needed, when acceleration is the core commandment, how can we win this race?

There are four major motivators that drive innovation. The first, and weakest of the bunch, is curiosity: the desire to find out why, to open the black box, to see around the next bend. Curiosity is a powerful jones. It fuels much of science, but it's nothing compared to fear, our next motivator. Extraordinary fear enables extraordinary risk-taking. John F. Kennedy's

Apollo program was executed at significant peril and tremendous expense in response to the early Soviet space successes. (You can ballpark the ratio of fear to curiosity as a driver for human innovation: it's the ratio of the defense budget to the science budget, which in 2011 was roughly $700 billion compared to $30 billion.) The desire to create wealth is the next major motivator, best exemplified by the venture capital industry's backing of ten ideas, expecting nine to fail and hoping for one grand-slam winner. The fourth and final motivator is the desire for significance: the need for one's life to matter, the need to make a difference in the world.

One tool that harnesses all four of these motivators is called the incentive prize. If you need to accelerate change in specific areas, especially when the goals are clear and measurable, incentive competitions have a biological advantage. Humans are wired to compete. We're wired to hit hard targets. Incentive prizes are a proven way to entice the smartest people in the world, no matter where they live or where they're employed, to work on your particular problem. As Raymond Orteig discovered in the early portion of the last century, such competitions can change the world.

The New Spirit of St. Louis

Raymond Orteig grew up a shepherd in France, on the slopes of the Pyrenees. By age twelve, he'd followed in his uncle's footsteps and immigrated to America. With little money, he took the only job he could find, working as a busboy at the Hotel Martin in Midtown Manhattan. Over the course of a decade, he rose to café manager, then hotel manager, and then, with monies saved, eventually purchased the establishment. He changed its name to the Hotel Lafayette, and a few months later bought the nearby Hotel Brevoort.

In the years after World War I, French airmen often stayed at these hotels. Orteig loved listening to their combat stories. He developed a serious passion for aviation, dreaming of the good that air travel could do and wanting to find a way to help progress along. Then two British pilots, John Alcock and Arthur Whitten Brown, made the first nonstop flight from Newfoundland to Ireland in 1919, and Orteig had an idea. On May 22, 1919, he laid out his plan in a short letter to Alan Hawley, president of the Aero Club of America in New York City:

"Gentlemen, as a stimulus to courageous aviators, I desire to offer, through the auspices and regulations of the Aero Club of America, a prize of $25,000 to the first aviator of any Allied country crossing the Atlantic in one flight from Paris to New York or New York to Paris, all other details in your care."

The prize would be offered for a period of five years, but the 3,600 miles between Paris and New York was almost twice the previous record for nonstop flight, and those years passed without anyone claiming victory. Orteig was unfazed: he renewed the offer for another five. This next round of competition brought casualties. In the summer of 1926, Charles W. Clavier and Jacob Islamoff died when their plane, grossly overloaded, ripped apart on takeoff. In the spring of 1927, it was Commander Noel Davis and Lieutenant Stanton H. Wooster who perished during their final test flight. Weeks later, on May 8, 1927, French aviators Charles Nungesser and François Coli flew westward into the dawn over Le Bourget, France, and were never seen again. Then came Charles A. Lindbergh.

Out of everyone who entered Orteig's competition, Lindbergh was, by far, the least experienced pilot. No aircraft manufacturer, in fact, even wanted to sell him an airframe or an engine, fearing that his death would give their product a bad reputation. The media dubbed him the "flying fool," then promptly dismissed him. But this is an aspect of incentive competitions: they're open to all comers—and all comers often show up, including the underdog. Sometimes the underdog wins. On May 20, 1927, eight years after the original challenge, Lindbergh did just that: departing Roosevelt Field in New York and flying solo and nonstop for thirty-three hours and thirty minutes before landing safely at Le Bourget Airdrome outside of Paris.

The impact of Lindbergh's flight cannot be overemphasized. The Orteig Prize captured the world's attention and ushered in an era of change. A landscape of daredevils and barnstormers was transformed into one of pilots and passengers. In eighteen months, the number of paying US passengers grew thirtyfold, from about 6,000 to 180,000. The number of pilots in the United States tripled. The number of airplanes quadrupled. Gregg Maryniak, a pilot and the aforementioned director of the McDonnell Planetarium, says, "Lindbergh's flight was so dramatic that it changed how the world thought about flight. He made it popular with consumers and investors. We can draw a direct connection between his winning of the Orteig Prize and today's three-hundred-billion-dollar aviation industry."

In 1993 it was also Maryniak who gave me a copy of Lindbergh's 1954 Pulitzer Prize–winning book *The Spirit of St. Louis.* He was hoping to inspire me to finish my pilot's license—which he did, but the inspiration didn't stop there. Before I read the book, I'd always believed that Lindbergh woke up one day and decided to head east, crossing the Atlantic as a stunt. I had no idea that he made the flight to win a prize. Nor did I know what extraordinary leverage such competitions could provide. Nine teams cumulatively spent $400,000 to try to win Orteig's $25,000 purse. That's sixteenfold leverage. And Orteig didn't pay one cent to the losers: instead his incentive-based mechanism automatically backed the winner. Even better, the resulting media frenzy created so much public excitement that an industry was launched.

I wanted to launch another. Since early childhood, I'd been dreaming of the day when the public could routinely buy tickets to space. I waited patiently, expecting that NASA would eventually make this happen. But thirty years later, I realized this wasn't the agency's goal—and not even their responsibility. Getting the public into space was our job, possibly *my* job, and by the time I finished reading *The Spirit of St. Louis,* the concept of an incentive prize for the "demonstration of a suborbital, private, fully reusable spaceship" had formed in my mind.

Not knowing who my "Orteig" would be, I called it the X PRIZE. The letter *X* was a variable, a place holder, to be replaced with the name of the person or company who put up the $10 million purse. I thought raising the money would be easy. Over the course of the next five years, I pitched the project to over two hundred philanthropists and CEOs. Everyone said the same three things: "Can anyone really do this? Why isn't NASA doing it? And isn't someone going to die trying?" All of them turned me down. Finally, in 2001 I met our ultimate purse benefactors: Anousheh, Hamid, and Amir Ansari. They didn't care about the risks involved and said yes on the spot. By then, the *X* had stuck around for so long that we'd grown attached. As a result, we ended up calling the competition the Ansari X PRIZE.

The Power of Incentive Competitions

Orteig didn't invent incentive prizes. Three centuries before Lindbergh crossed the Atlantic by plane, the British Parliament wanted some help crossing the Atlantic by ship. In 1714 it offered £20,000 to the first person to figure out how to accurately measure longitude at sea. This was called the Longitude Prize, and not only did it help Parliament solve its navigation problem, its success launched a long series of incentive competitions. In 1795 Napoléon I offered a 12,000-franc prize for a method of food preservation to help feed his army on its long march into Russia. The winner, Nicolas Appert, a French candy maker, established the basic method of canning, still in use today. In 1823 the French government once again offered a prize, this time 6,000 francs for the development of a large-scale commercial hydraulic turbine. The winning design helped to power the burgeoning textile industry. Other prizes have driven breakthroughs in transportation, chemistry, and health care. As a recent McKinsey & Company report on the subject said, "Prizes can be the spur that produces a revolutionary solution . . . For centuries, they were a core instrument of sovereigns, royal societies and private benefactors alike who sought to solve pressing societal problems and idiosyncratic technical challenges."

The success of these competitions can be boiled down to a few underlying principles. First and foremost, large incentive prizes raise the visibility of a particular challenge while helping to create a mind-set that this challenge is solvable. Considering what we know about cognitive biases, that is no small detail. Before the Ansari X PRIZE, few investors seriously considered the market for commercial human spaceflight; it was assumed to be the sole province of governments. But after the prize was won, a half dozen companies were formed, nearly $1 billion has been invested, and hundreds of millions of dollars' worth of tickets for carriage into space have been sold.

Secondly, in areas where market failures have hindered investment or entrenched incumbents have prevented progress, prizes break bottlenecks. In the spring of 2010, the failure of the BP Deepwater Horizon oil platform created a disaster in the Gulf of Mexico. A lot of people wanted to make sure nothing like this ever happened again, myself included. Through a sequence of conversations among Francis Béland, vice president of prize development at the X PRIZE Foundation, David Gallo of the Woods Hole

Oceanographic Institution, and the foundation's newest trustee, filmmaker James Cameron, it was decided that we should develop a "flash prize" to deal with the emergency.

The focus of the prize was clear. The technology used to clean up the BP spill in 2010 was the same technology used to clean up the *Exxon Valdez* spill in 1989. In fact, it was not only the same technology but also the same equipment. It was clearly time for an upgrade. A prize for a better way to clean oil off the surface of the ocean seemed like the way to go. Philanthropist Wendy Schmidt, head of the Schmidt Family Foundation and the 11th Hour Project, agreed. Within twenty-four hours of our announcement, she stepped forward to underwrite the competition. "When I watched what was happening last year in the Gulf," she said, "I felt a sense of disbelief—a horror at the scale of the disaster and its impact on the lives of people, wildlife, and natural systems. I knew we could do something to lessen the impact of this kind of manmade disaster in the future. Incentive prizes seemed like the fastest path I could imagine to finding a solution." And it worked. The results of the competition were spectacular. The winning team quadrupled the performance of the industry's existing technology.

Besides being a way to raise the profile of key issues and rapidly address logjams, another key attribute of incentive prizes is their ability to cast a wide net. Everyone from novices to professionals, from sole proprietors to massive corporations, gets involved. Experts in one field jump to another, bringing with them an influx of nontraditional ideas. Outliers can become central players. At the time of England's Longitude Prize, there was considerable certainty that the purse would go to an astronomer, but it was won by a self-educated clock maker, John Harrison, for his invention of the marine chronometer. Along similar lines, in the first two months of the Wendy Schmidt Oil Cleanup X CHALLENGE, some 350 potential teams from over twenty nations preregistered for the competition.

The benefits of incentive prizes don't stop here. Because of the competitive framework, people's appetite for risk increases, which—as we'll explore in depth a little later—further drives innovation. Since many of these competitions require significant capital to field a team (in other words: no bucks, no Buck Rogers), it's fortunate that the sporting atmosphere lures legacy-craving wealthy benefactors and corporations looking to distinguish themselves in a media-cluttered environment. Finally, competitions inspire hundreds of different technical approaches, which means that they don't just give birth to a single-point solution but rather to an entire industry.

The Power of Small Groups (Part II)

The American anthropologist Margaret Mead once said, "Never doubt that a small group of thoughtful, committed citizens can change the world. Indeed, it is the only thing that ever has." There are, as it turns out, pretty good reasons for this. Large or even medium-sized groups—corporations, movements, whatever—aren't built to be nimble; nor are they willing to take large risks. Such organizations are designed to make steady progress and have considerably too much to lose to place the big bets that certain breakthroughs require.

Fortunately, this is not the case with small groups. With no bureaucracy, little to lose, and a passion to prove themselves, small teams consistently outperform larger organizations when it comes to innovation. Incentive prizes are perfectly designed to harness this energy. A great example was the 2009 Northrop Grumman Lunar Lander X CHALLENGE. This was a $2 million purse put up by NASA and managed by the X PRIZE Foundation as part of the NASA Centennial Challenges program. The competition asked teams to build a rocket-powered vehicle capable of vertical takeoffs and landings, for going back to the Moon's surface. Not since the Defense Department's DC-X program fifteen years earlier had the government possessed this capability—and that vehicle, which ultimately crashed during testing, had cost taxpayers some $80 million.

Neither of the two teams that ultimately split this purse (meeting all of NASA's requirements) looked anything like a traditional aerospace contractor. Both were small, started by software entrepreneurs, and staffed by a few part-time engineers with no experience in the space industry. Engineer John Carmack, creator of the video games Quake and Doom, who founded and funded Armadillo Aerospace (which placed second in the competition), summed up this point nicely: "I think the biggest benefit that NASA can possibly get out of this is to witness an operation like ours go from concept to (almost) successful flight in under six months with a team of eight part-time people for a total cost of only $200,000. That should shame some of their current contractors who are going to be spending tens of billions of dollars doing different things."

A similar outcome was reached in 2007, when, in partnership with the Progressive Insurance Company, the X PRIZE Foundation launched a competition for the world's first fast, affordable, production-ready car able to achieve over one hundred miles per gallon equivalent (MPGe). Over 130 teams from twenty nations entered the competition. Three winners split the $10 million purse (achieving mileage figures ranging from 102.5 to 187.5 MPGe), and none of them had more than a few dozen employees.

"Right now the foundation has two more active X PRIZEs," says its president and vice chairman, Robert K. Weiss. "There's the thirty-million-dollar Google Lunar X PRIZE and the ten-million-dollar Archon Genomics X PRIZE presented by Medco. To win the first, all you have to do is build a robot, land it on the surface of the Moon, send back photos and videos, then rove or hop five hundred meters, and send back more photos and videos. To win the second, teams have to sequence the genomes of one hundred healthy centenarians in ten days." Not much more than a decade ago, both of these missions would have required billions of dollars and thousands of people. I don't know who will win either competition, but whoever it turns out to be, I can all but guarantee that it's going to be a small group of thoughtful, committed citizens—because, as Mead pointed out and incentive prizes validate, this is exactly what it takes to change the world.

The Power of Constraints

Creativity, we are often told, is a kind of free-flowing, wide-ranging, "anything goes" kind of thinking. Ideas must be allowed to flourish unhindered. There's an entire literature of "think-outside-the-box" business strategies to go along with these notions, but, if innovation is truly the goal, as brothers Dan and Chip Heath, the best-selling authors of *Made to Stick: Why Some Ideas Survive and Others Die,* point out in the pages of *Fast Company,* "[D]on't think outside the box. Go box shopping. Keep trying on one after another until you find the one that catalyzes your thinking. A good box is like a lane marker on the highway. It's a constraint that liberates."

In a world without constraints, most people take their time on projects, assume fewer risks, spend money wastefully, and try to reach their goals in comfortable and traditional ways—which, of course, leads nowhere new. But this is another reason why incentive prizes are such effective change agents: by their very nature, they are nothing more than a focusing mechanism and a list of constraints.

For starters, the prize money defines spending parameters. The Ansari X PRIZE was $10 million. Most teams, perhaps optimistically (and who would pursue a space prize without being an optimist?), told their backers that they could win for less than purse value. In reality, teams go over budget, spending considerably more than the prize money in solving the problem (because, by design, there's a back-end business model in place to help them recoup their investment). But this perceived upper limit tends to keep out risk-adverse traditional players. In the case of the X PRIZE, my goal was to dissuade the likes of Boeing, Lockheed Martin, and Airbus from entering the competition. Instead I wanted a new generation of entrepreneurs reinventing space flight for the masses—which is exactly what happened.

The time limit of a prize competition serves as another liberating constraint. In the pressure cooker of a race, with an ever-looming deadline, teams must quickly come to terms with the fact that "the same old way" won't work. So they're forced to try something new, pick a path, right or wrong, and see what happens. Most teams fail, but with dozens or hundreds competing, does it really matter? If one team succeeds, within the constraints, they've created a true breakthrough.

Having a clear, bold target for the competition is the next important restriction. After Venter sequenced the human genome, many companies started selling whole genome-sequencing services. But none of their products had sufficient fidelity to be medically relevant. So the Archon Genomic X PRIZE was created. It challenges teams to sequence one hundred human genomes accurately (one error in one million base pairs), completely (98 percent of the human genome), rapidly (within ten days) and cheaply (and a cost of less than $1,000 per genome)—a quadruple combination that's a 365 millionfold price-time-performance improvement over Venter's original 2001 work. Moreover, as the genomes to be sequenced belong to a hundred healthy centenarians, this competition's results will further unlock the secrets of longevity and drive us to our goal of health care abundance.

Fixed-Price Solutions

Incentive prizes are not a panacea; they can't fix all that ails us. But on the road toward abundance, when a key technology is missing, or a specific end goal has been identified but not yet achieved, incentive prizes can be an efficient and highly leveraged way to get from A to B. Of course, this is what we're doing at the X PRIZE Foundation. We've launched six competitions, awarded four of them, and conceived of another eighty-plus that are awaiting funding. Ultimately, though, this chapter isn't about the X PRIZE—that isn't the point. The point is that incentive prizes have a three-hundred-year track record of driving progress and accelerating change. They are a great way to steer toward the future we really want. So start your own. Help with ours. Whatever.

In areas like chronic disease, where governments spend billions of dollars, the offer of a massive incentive prize seems like a no-brainer. AIDS costs the US government over $20 billion a year; that's more than $100 billion during a five-year period. Imagine, for example, a $1 billion purse offered for the first team to demonstrate a cure or vaccine. Sure, the marketplace is vast and the corporation that develops this cure will reap huge rewards, but what if the government's $1 billion was paid directly to the scientists who made the discovery? How many more brilliant minds might be turned on to this problem? How many graduate students might start daydreaming about solutions?

Now apply this thinking to Alzheimer's, Parkinson's, or your cancer of choice. Whatever you like. The advantage here is an army of brilliant people around the world thinking about your problem and working on their own nickel to solve it. Properly executed, this mechanism offers the potential for fixed-cost science, fixed-cost engineering, and fixed-cost solutions. I've always believed (to paraphrase computer scientist Alan Kay) that the best way to predict the future is to create it yourself, and in my five decades of experience, there is no better way to do just that than with incentive prizes.

CHAPTER EIGHTEEN

RISK AND FAILURE

The Evolution of a Great Idea

Sir Arthur C. Clarke, inventor of the geostationary communication satellite and author of dozens of best-selling science-fiction books, knew something about the evolution of great ideas. He described three stages to their development. "In the beginning," says Clarke, "people tell you that's a crazy idea, and it'll never work. Next, people say your idea might work, but it's not worth doing. Finally, eventually, people say, I told you that it was a great idea all along!"

When Tony Spear was given the job of landing an unmanned rover on the Martian surface, he had no inkling that Clarke's three stages would be precisely his experience. A jovial, white-haired cross between Albert Einstein and Archie Bunker, Spear started his career at NASA's Jet Propulsion Laboratory in 1962. Over the next four decades, he worked on missions from Mariner to Viking, but it was his final assignment, project manager on the Mars Pathfinder, that he describes as his "greatest mission challenge ever."

The year was 1997, and the United States had not landed a probe on Mars since July 1976. That was Viking, a complex and expensive mission, costing some $3.5 billion (in 1997 dollars). Spear's assignment was to find a way to do everything that the previous mission had done, just "faster, better, cheaper." And when I say cheaper, I mean a *whole lot* cheaper: fifteen times cheaper, to be exact, for a fixed and total development cost of only $150 million. Out the window went the expensive stuff, the traditional stuff, and the proven stuff, including the types of retro-rockets for landing that got the job done on Viking.

"To pull this off under these impossible constraints, we had to do everything differently," reflects Spear, "from how I managed, to how we landed. That really scared people. At NASA headquarters, I was assigned six different managers in rapid sequence—each of the first five found a different excuse to get off the project. Finally I was assigned someone about to retire who didn't mind sticking with me at the end of his career. Even the NASA administrator, Daniel Golden, nearly flipped out when he received his initial mission briefing—he couldn't get past how many new things we were trying out."

Among the many things Spear tried out, nothing struck people as zanier than using air bags to cushion the initial impact, helping the craft bounce around like a beach ball on the Martian surface, before settling down into a safe landing spot. But air bags were cheap, they wouldn't contaminate their landing site with foreign chemicals, and Spear was pretty certain that they would work. The early tests, however, were a disaster, so the experts were summoned.

The experts had a pair of opinions. The first was: Don't use airbags. The second was: No, we're totally serious, don't even *consider* using air bags. "Two of them," recounts Spear, "told me flat out that I was wasting government money and should cancel the project. Finally, when they realized I wasn't going to give up, they decided to dig in and help me."

Together they tested more than a dozen designs, skidding them along a faux rocky Martian surface to see which would survive without shredding to pieces. Finally, just eight months prior to launch, Spear and his team completed qualification testing of a design composed of twenty-four interconnected spheres, loaded it aboard Pathfinder, and launched it into space. But the anxiety didn't end there. The trip to Mars took eight months, during which there was plenty of time to worry about the fate of the mission. "In the weeks just prior to landing," Spear recalls, "everyone was very nervous, speculating whether we'd have a big splat when we arrived. Golden himself was wondering what to do: should he come to the JPL control room for the landing or not? Just a few days before our July 4 descent to the surface, the administrator took a bold tack, holding a press conference and proclaiming, 'The Pathfinder mission demonstrates a new way of doing business at NASA, and is a success whether or not we survive the landing.' "

The landing, though, went exactly as planned. They had spent one-fifteenth the cost of Viking, and everything worked perfectly—especially the air bags. Spear was a hero. Golden was so impressed, he insisted that air bags be used to land the next few Mars missions and was quoted as saying, "Tony Spear was a legendary project manager at JPL and helped make Mars Pathfinder the riveting success that it was."

The point here, of course, is that Clarke was right. Demonstrating great ideas involves a considerable amount of risk. There will always be naysayers. People will resist breakthrough ideas until the moment they're accepted as the new norm. Since the road to abundance requires significant innovation, it also requires significant tolerance for risk, for failure, and for ideas that strike most as absolute nonsense. As Burt Rutan puts it, "Revolutionary ideas come from nonsense. If an idea is truly a breakthrough, then the day before it was discovered, it must have been considered crazy or nonsense or both—otherwise it wouldn't be a breakthrough."

The Upside of Failure

Rutan is spot on, but he's leaving something out—sometimes crazy ideas are just that, crazy. Some are plain bad. Others are ahead of their time, or miss their market, or are financially impractical. Whatever the case, these notions are doomed. But failure is not necessarily the disaster that everyone assumes. In an article for *Stanford Business School News,* Professor Baba Shiv explains it this way: "Failure is a dreaded concept for most business people. But failure can actually be a huge engine of innovation. The trick lies in approaching it with the right attitude and harnessing it as a blessing, not a curse."

Shiv studies the role that the brain's liking and wanting systems play in shaping our decisions, a field now known as neuroeconomics. When it comes to risk, he divides the world into two mind-sets: type 1 people are fearful of making mistakes. For them, failure is shameful and disastrous. As a result, they are risk averse, and whatever progress they make is incremental at best. On the other hand, type 2 people are fearful of losing out on opportunities. Places like Silicon Valley are full of type 2 entrepreneurs. "What is shameful to these people," says Shiv, "is sitting on the sidelines while someone else runs away with a great idea. Failure is not bad; it can actually be exciting. From so-called failures emerge those valuable gold nuggets—the 'ah-ha!' moments of insight that guide you toward your next innovation."

One of the most famous cases of this was Thomas Edison's invention of the lightbulb, which took him a thousand tries to get right. When asked by a reporter what it felt like to fail a thousand times in a row, Edison responded, "I have not failed. I've just found a thousand ways that don't work." Or take the Newton, considered one of Apple's few fiascos. The world's first personal digital assistant (PDA) was ahead of its time, rushed to market, buggy, and seriously overpriced. The handwriting recognition software, its core feature, never worked quite right. Apple spent $1.5 billion (in 2010 dollars) on development and recouped less than a quarter of that. Critics panned the project. But a decade after the device's cancellation, the concepts that drove the Newton were rejiggered into the epic success known as the iPhone—which sold 1.4 million units in its first ninety days and was *Time* magazine's 2007 Invention of the Year.

Arianna Huffington, CEO and founder of the Huffington Post website, agrees:

> You'll never be able to achieve big-time success without risking big-time failure. If you want to succeed big, there is no substitute for simply sticking your neck out. Of course, nobody likes to fail, but when the fear of failure translates into taking fewer risks and not reaching for our dreams, it often means never moving ahead. Fearlessness is like a muscle: the more we use it, the stronger it becomes. The more we are willing to risk failure and act on our dreams and our desires, the more fearless we become and the easier it is the next time. Bottom line, taking risks is an indispensable part of any creative act.

Tony Spear never would have achieved his breakthrough by taking incremental steps. He did it by facing his fears and facing down the parade of experts who discouraged him along the way. So if you're interested in solving grand challenges, driving breakthroughs, and changing the world, you'll need to get ready. Go to the gym, start working out your fearlessness muscles and thickening your skin against the rain of criticism to come. Most importantly, do not seek to change the world unless you seek it, to paraphrase the nineteenth-century Indian mystic Sri Ramakrishna, "as a man whose hair is on fire seeks a pond." Ultimately, one must have passion and purpose in order to convince the world of anything—which is, of course, the first step to changing it.

Born Above the Line of Supercredibility

If your goal is to reshape the world, then how the world learns about your plan is every bit as important as the plan itself. In May 1996, my challenge was getting the world to believe that the X PRIZE was a viable way to open the space frontier, even though I had no prize money and no competing teams. Four months earlier, inspired by Charles Lindbergh's autobiography, I'd found a group of visionary Saint Louisians who convinced me that the arched city was the right place to base my efforts. Our next goal was to convince local philanthropists that a $10 million competition could birth a private space industry and simultaneously return St. Louis to its 1927 glory. Ultimately, we collected about $500,000—not nearly enough to run the competition, but more than enough to announce it in a bold and convincing fashion, above what I later came to call the "line of supercredibility."

Each of us has an internal "line of credibility." When we hear of an idea that is introduced below this line, we dismiss it out of hand. If the teenager next door declares his intent to fly to Mars, you smirk and move on. We also have an internal line of supercredibility. Should it be announced that Jeff Bezos, Elon Musk, and Larry Page have committed to fund a private mission to Mars, "When is it going to land?" becomes a much more reasonable question. When we hear an idea presented above the supercredible line, we immediately give it credence and use it to anchor future actions.

On May 18, 1996, my goal was nothing less than supercredibility. On stage with me were Erik and Morgan Lindbergh, Charles's grandsons, and twenty veteran NASA astronauts. Directly to my right was Patti Grace Smith, the associate administrator for spaceflight at the Federal Aviation Administration (FAA); on my left, NASA Administrator Daniel Golden. It was a collection of many of the world's leading space experts. Sure, I was just a guy with a crazy idea. But with this crew backing me up, did it sound that crazy after all?

Obviously, the greatest benefit to having these people on stage was the halo effect they brought to the announcement. But equally important were the countless hours I spent speaking to each of them, presenting the X PRIZE concept, honing the ideas, and addressing their concerns.

And it worked. After the ceremony, front pages around the world announced, "$10 Million prize created to spur private spaceships." Hundreds of articles followed—none bothering to mention that we had no prize money, no teams, and no remaining funds. Yet because we'd launched above the line of supercredibility, other people jumped in to share our dream. Funding began to arrive; teams began to step forward. While we didn't raise the $10 million purse—that would have to wait five more years, until I met the Ansari family—we did pull in enough to keep both the organization and the competition alive.

That day, I learned how a powerful first impression (in other words, announcing your idea in a supercredible way) is fundamental to launching a breakthrough concept. But I also saw the importance of mind-set. *My* mind-set. Sure, I had wanted to open up space since my childhood, but was I really sure this approach would work? In getting to supercredibility, I had to lay out my ideas before the aerospace industry's best and brightest, testing my premises and answering uncomfortable questions. In doing so, whatever doubts I'd had vanished along the way. By the time I was on stage with my dignitaries, the idea that the X PRIZE could work wasn't a hopeful fantasy, it was the tomorrow I was certain would soon arrive. This is the second thing I learned that day: the awesome power of the right mind-set.

Think Different

In 1997 Apple introduced its "Think Different" advertising campaign with the now-famous declaration: "Here's to the crazy ones":

> Here's to the crazy ones, the misfits, the rebels, the troublemakers, the round pegs in the square holes . . . the ones who see things differently—they're not fond of rules . . . You can quote them, disagree with them, glorify or vilify them, but the only thing you can't do is ignore them because they change things . . . they push the human race forward, and while some may see them as the crazy ones, we see genius, because the ones who are crazy enough to think that they can change the world are the ones who do.

If you were to just hear these words, they'd seem like bravado—marketingspeak from a company not known for marketingspeak. But Apple coupled sight to sound. Accompanying those words were images: Bob Dylan as a misfit; Dr. Martin Luther King Jr. as a troublemaker; Thomas Edison as the one without respect for the status quo. Suddenly everything changes. Turns out this campaign is not all bluster. In fact, it seems to be a fairly accurate retelling of historical events.

The point, however obvious, is pretty fundamental: you need to be a little crazy to change the world, and you can't really fake it. If you don't believe in the possibility, then you'll never give it the 200 percent effort required. This can put experts in a tricky situation. Many have built their careers buttressing the status quo, reinforcing what they've already accomplished, and resisting the radical thinking that can topple their legacy—not exactly the attitude you want when trying to drive innovation forward.

Henry Ford agreed: "None of our men are 'experts.' We have most unfortunately found it necessary to get rid of a man as soon as he thinks himself an expert because no one ever considers himself expert if he really knows his job . . . Thinking always ahead, thinking always of trying to do more, brings a state of mind in which nothing is impossible." So if you're going after grand challenges, experts may not be your best coconspirators.

Instead, if you need a group of people who thrive on risk, are overflowing with crazy ideas, and don't have a clue that there's a "wrong way" to do things, there's one particular place to look. In the early 1960s, when President Kennedy launched the Apollo program, very few of the necessary technologies existed at the time. We had to invent almost everything. And we did, with one of the main reasons being that those engineers involved didn't know they were trying to do the impossible, because they were too young to know. The engineers who got us to the Moon were in their mid- to late twenties. Fast-forward thirty years, and once again it was a group of twenty-somethings driving a revolution, this time in the dot-com world. This is not a coincidence: youth (and youthful attitudes) drives innovation—always has and always will. So if we're serious about creating an age of abundance, then we're going to have to learn to think differently, think young, roll the dice, and perhaps most importantly, get comfortable with failure.

Getting Comfortable with Failure

Almost every time I give a talk, I like to ask people what they fear most about failure. There are three consistent answers: loss of reputation, loss of money, and loss of time. Reputation is a quality built through consistent performance and serial successes. One big failure can topple decades of effort. Money, a scare resource for most, comes more easily to those with a track record of success. And time is just plain irreplaceable. Blow your reputation on the front page of the newspaper, file for bankruptcy, or waste years chasing a bad idea, and you too are likely to become risk adverse.

Since the creation of abundance-related technologies requires taking risks, figuring out how to convert what Baba Shiv calls type 1 riskphobic individuals into type 2 riskphilic players is vital to this effort. There are a number of approaches now gaining favor.

Some companies are focusing on how to make their working environment more tolerant of failure. At the financial software company Intuit, for example, the team responsible for a particularly disastrous marketing campaign received an award from Chairman Scott Cook, who said, "It's only a failure if we fail to get the learning." Similarly, Ratan Tata, CEO of the Indian conglomerate the Tata Group, told the *Economist* "failure is a goldmine" when explaining why his company instituted a prize for the best failed idea that taught the company an important lesson.

Another way that companies have begun strengthening their fearlessness muscles is rapid prototyping: the process of brainstorming wild new ideas, then quickly developing a physical model or mock-up of the solution. "This process," says Shiv, "allows people to move quickly from the abstract to the concrete, and lets them visualize the outcome of their ideas. Because not all prototypes end up as the best or final solution, rapid prototyping also teaches that failure is actually a necessary part of the process."

Michael Schrage, a research fellow with MIT's Center for Digital Business and MIT's Entrepreneurship Center, has developed the 5x5x5 Rapid Innovation Method, a very concrete way of putting Shiv's notion into practice. "The idea is fairly simple and straightforward," he says. "A company looking to drive breakthroughs in a particular area sets up five teams of five people and gives each team five days to come up with a portfolio of five 'business experiments' that should take no longer than five weeks to run and cost no more than five thousand dollars each to conduct. These teams are fully aware that they are 'competing' with their colleagues to come up with the best possible portfolios to present to their bosses, perhaps winning the chance to implement the best performing concept."

Schrage's methodology makes use of two ideas discussed earlier: the power of constraints and the power of small groups. If conducted in a friendly, riskphilic environment—in which everyone understands that most ideas will fail—participants will not fear ramifications to their reputations. Under these circumstances, there's no downside to having a crazy idea, and a tremendous upside if that crazy idea turns out to be revolutionary, so people are much more willing to take risks. Because each idea takes only five days and $5,000 to implement, no one worries too much about a significant loss of time or capital.

Will this process always lead to breakthroughs? Doubtful. But it does create a safe environment where people can practice stretching their imaginations, taking bigger risks, and learning to see failure as a building block of innovation rather than its anathema.

CHAPTER NINETEEN

WHICH WAY NEXT?

The Adjacent Possible

At the very beginning of this book, we argued that the true promise of abundance was one of creating a world of possibility: a world where everyone's days are spent dreaming and doing, not scrapping and scraping. Never before has such promise really been in the offing. For most of human history, life was a constrained affair. Just finding ways to survive took most of our energy. The gap between one's day-to-day reality and one's true potential was vast indeed. But in these extraordinary days, that chasm is beginning to close.

On a certain level, change is being driven by a fundamental property of technology: the fact that it expands into what theoretical biologist Stuart Kauffman calls "the adjacent possible." Before the invention of the wheel, the cart, the carriage, the automobile, the wheelbarrow, the roller skate, and a million other offshoots of circularity were not imaginable. They existed in a realm that was off-limits until the wheel was discovered, but once discovered, these pathways became clear. This is the adjacent possible. It's the long list of first-order possibilities that open up whenever a new discovery is made.

"The strange and beautiful truth about the adjacent possible is that its boundaries grow as you explore them," wrote author Steven Johnson in the *Wall Street Journal*. "Each new combination opens up the possibility of other new combinations. Think of it as a house that magically expands with each door you open. You begin in a room with four doors, each leading to a new room that you haven't visited yet. Once you open one of those doors and stroll into that room, three new doors appear, each leading to a brand-new room that you couldn't have reached from your original starting point. Keep opening new doors, and eventually you'll have built a palace."

Our path of adjacent possibles has led us to a unique moment in time. We have wandered into a world where the expansive nature of technology has begun to connect with our inner desires. In *What Technology Wants,* Kevin Kelly explains it this way: "For most of history, the unique mix of talents, skills, insights, and experiences of each person had no outlet. If your dad was a baker, you were a baker. As technology expands the possibility space, it expands the chance that someone can find an outlet for their personal traits . . . When we enlarge the variety and reach of technology, we increase options, not just for ourselves and not for others living, but for all generations to come."

A half century ago, Abraham Maslow pointed out that people whose basic needs were not being met had little time to spend on self-fulfillment. If you're trying to feed yourself or find medications for your children or survive other, similar threats, then living a life of possibility is not much of a probability. But this is exactly, as economist Daniel Kahneman figured out, where the adjacent possible meets the road to abundance and produces some spectacular leverage.

The Pursuit of Happiness

A few years ago, Kahneman set aside the question of cognitive biases and turned his attention to the relationship between income level and well-being. By analyzing the results of the Gallup-Healthways Well-Being Index, which asked some 450,000 Americans what brings them joy, he discovered, as the *New York Times* aptly put it, "Maybe money does buy you happiness after all."

With *maybe* being the operative word.

What the data show is that one's emotional satisfaction moves in lockstep with one's income—as income rises, well-being rises—but only to a point. Before the average American earns $75,000 a year, there is a direct correlation between money and happiness. Above that number, the correlation disappears. This tells us something interesting: that in the United

States, the freedom to flourish—to truly enjoy a life of possibility—costs roughly $75,000 a year in 2008 dollars. But what's really important is what that money buys.

The typical American spending breakdown shows that 75 percent to 80 percent of the money we earn goes to meet basic needs such as water, food, clothing, shelter, health care, and education. It's over 90 percent in most developing countries. But many of the technologies investigated in this book have dematerializing properties: they service fundamental needs without costing us much beyond an Internet connection. Take health care. In today's world, quality health care is about access. Access to transportation to the hospital, access to the right people—doctors, nurses, specialists—and the doctor's access to the latest lab tests and equipment. But in our envisioned future, all of this goes away. You don't need transportation, since the system is ubiquitous. Access to the best medical care available means access to Dr. Watson living in the cloud. And the best labs in the world are built into your phone. More important, this dematerialized system, coupled with tomorrow's array of demonetized sensors, can be focused completely on prevention, working to keep people healthy in the first place.

In our abundant future, the dollar goes further. As does the yen, the peso, the euro, and so forth. This happens because of dematerialization and demonetization; because of exponential price-performance curves; because each step up prosperity's ladder saves time; because those extra hours add up to additional gains; because the close ties between categories in our abundance pyramid produce positive feedback loops, bootstrapping potential, and the domino effect, and for a thousand other reasons. So you have to wonder: what does it take to make a real difference?

Not much, actually. Daniel Kahneman's calculation has lately been extended to the rest of the planet. On average, across the globe, the point on the chart where well-being and money diverge is roughly $10,000. That's how much the average global citizen needs to earn to fulfill his or her basic needs and gain a toehold toward much greater possibility.

There is no debate that life has gotten considerably better at the bottom over the last four decades. During that stretch, the developing world has seen longer life expectancies, lower infant mortality rates, better access to information, communication, education, potential avenues out of poverty, quality health care, political freedoms, economic freedoms, sexual freedoms, human rights, and saved time. But what that $10,000 figure tells us is that we've actually come much further.

Twenty years ago, most well-off US citizens owned a camera, a video camera, a CD player, a stereo, a video game console, a cell phone, a watch, an alarm clock, a set of encyclopedias, a world atlas, a *Thomas* Guide, and a whole bunch of other assets that easily add up to more than $10,000. All of which come standard on today's smart phones, or are available for purchase at the app store for less than a cup of coffee. In this, our exponentially enabled world, that's how quickly $10,000 worth of expenses can vanish. More importantly, these things can vanish without too much outside intervention. No one set out to zero the costs of two dozen products, inventors set out to make better cell phones, and the path of the adjacent possible did the rest.

But this time around, we can squeeze a bit of randomness out of the equation. We don't have to wait for history to help our cause, we can help it ourselves. We have our hard targets for abundance, we know which technologies need further development, and—if we can improve our appetite for risk and utilize the leverage of incentive prizes—we know how to go from A to B much faster than ever before. Unlike earlier eras, we don't have to wait for corporations to get interested in solutions, or for governments to get around to our problems. We can take matters into our own hands. Today's technophilanthropist crowd seems determined to provide the necessary seed capital (and often much more than that), and today's DIY innovators have proven themselves more than capable of getting the job done. Meanwhile, the one-quarter of humanity that has forever been on the sidelines—the rising billion—has finally gotten into the game.

Most importantly, the game itself is no longer zero-sum. For the first time in forever, we don't need to figure out how to divide our pie into more slices, because we now know how to bake more pies. Everyone can win.

Proverbs 29:18 tells us: "Where there is no vision, the people will perish." Perhaps that's true, but it's also myopic. Abundance is both a plan and a perspective. This second bit is key. One of the more important points made throughout this book is that our perspective shapes our reality. The best way to predict the future is to create it yourself. So while the Bible offers a warning, it's helpful to remember that the inverse is also true: where there is vision, the people flourish. The impossible becomes the possible. And abundance for all becomes imagine what's next.

AFTERWORD

NEXT STEP—JOIN THE ABUNDANCE HUB

One of the most difficult tasks in finishing this book was deciding when to stop incorporating the latest and greatest breakthroughs into our story. In the weeks and months following the completion of this manuscript, a barrage of new technologies supporting our case for abundance continued to appear at an ever-increasing rate. We think it's critically important for you to have access to this ongoing evidence for abundance. Therefore, we've created five different ways for you to stay plugged in, interact with the authors, and join an ongoing conservation about radical advances in energy, food, water, health, education, technophilanthropy, DIY innovation, and all the rest.

- Visit our website **http://www.AbundanceHub.com**, where you can sign up for a free newsletter and participate in any future initiatives. In partnership with Singularity University, we will continue to provide news on critical developments driving us toward a much better future.
- We invite you to visit and contribute to **http://videos.AbundanceHub.com**, where you can view and submit video content inspired by this book.
- Become a fan at our Facebook page **http://www.AbundanceHub.com**.
- Follow our Twitter feed **@AbundanceHub** to receive the latest breaking news.

- Become part of the global team identifying abundance-related breakthroughs by simply sending a tweet to the community with the hashtag ***#Abundance***.

If you enjoyed learning about Singularity University (SU) and would like to participate in one of our programs, you are welcome to get involved. Graduate and post graduate students can apply for the ten-week Graduate Studies Program (GSP). Others, including executives, investors and entrepreneurs, can apply for the four-day or seven-day executive programs held on a regular basis at the SU Campus in Mountain View, California. Details on both programs are available at **www.SingularityU.org**. Or for more information simply email us at **Abundance@SingularityU.org**.

Philanthropist and corporate executives interested in the design or funding of an X PRIZE or X CHALLENGE can learn more at **www.xprize.org**. Or for more information simply email us at **abundance@xprize.org**.

To learn more about the authors, or to engage either of them to speak on the subject of Abundance, please visit **www.Diamandis.com** and **www.StevenKotler.com**.

Thank you for taking the time to read *Abundance*. We hope our contrarian view of the future has provided an antidote to some of today's dark pessimisms. Providing abundance is humanity's grandest challenge—one that together, with intention and action, we can make happen within our lifetime.

Reference Section Raw Data

CONTENTS

1 Abundance Pyramid

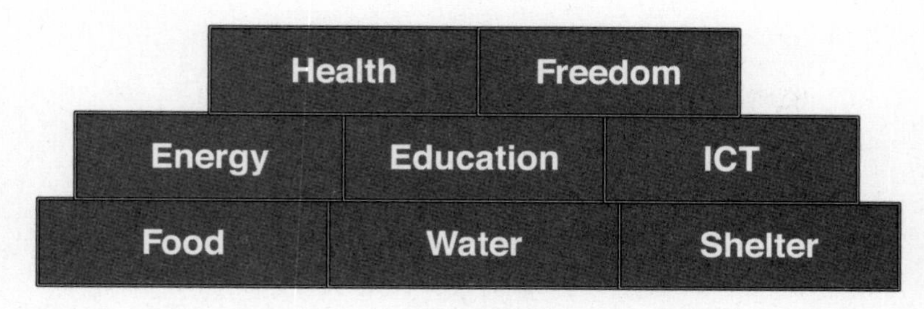

The Abundance Pyramid outlines the increasing levels of needs enabled by technology. This is loosely based on Maslow's (pyramid) hierarchy of needs.

2 Growth of World Population and the History of Technology

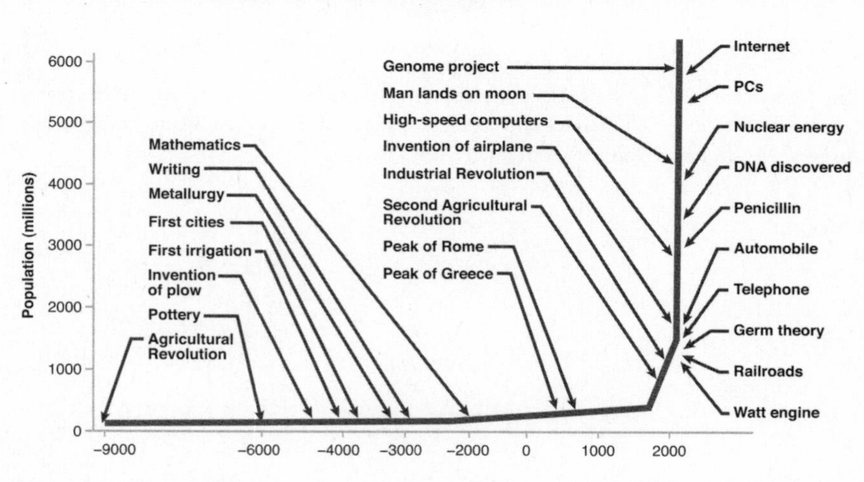

This graph shows how the rate of technological innovation has dramatically increased at the same time that the human population has increased. (Note: Selected technological milestones are subjective.)
Source: Robert Fogel, University of Chicago.

Water and Sanitation

3 Distribution of Water on Earth

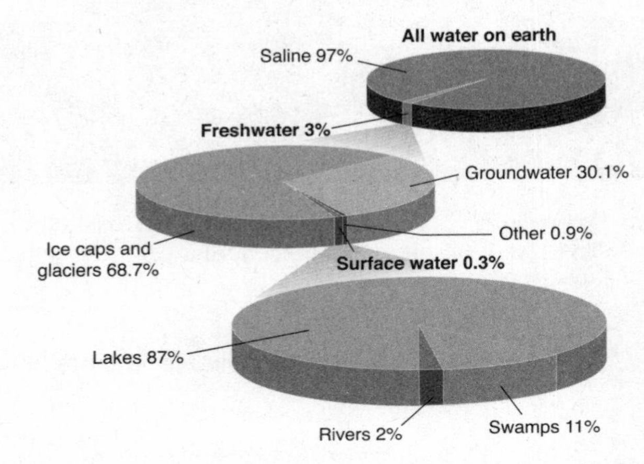

The freshwater that humanity depends upon makes up less than 1 percent of the water on Earth. 97 percent is saltwater and 2 percent is locked up in the ice caps and glaciers. **Source:** World Fresh Water Resources via USGS.

4 Daily Time Spent Fetching Water from Sources Outside the Home

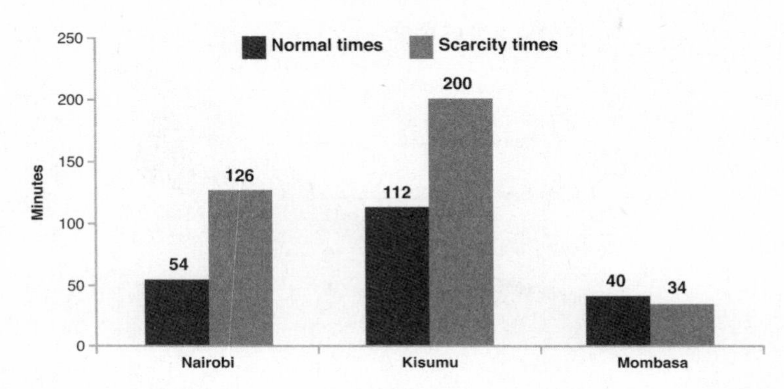

In urban areas, a larger fraction of households have access to piped water, yet many are forced to rely on water kiosks (15 percent in Nairobi; 45 percent in Kisumu and in Mombasa) (CRC 2009). This situation presents a huge burden to households, as fetching water is time consuming. A typical household makes four to six trips daily to fetch water. In Nairobi, a typical household spends 54 minutes going to the kiosk in normal times, and more than twice that (126 minutes) in times of water scarcity.
Source: Citizen Report Card, 2007; www.twaweza.org/uploads/files/Its%20our%20water%20too_English.pdf.

5 Average Price for Water Service in Fifteen Largest Cities, by Type of Provider

	House connection	Small-piped network	Standpost	Household reseller	Water tanker	Water vendor
Average price (US$ per cubic meter)	0.49	1.04	1.93	1.63	4.67	4.00
Markup over house connection (%)	100	214	336	402	1,103	811

Non-piped water services can cost 200 to 1100 percent of what a house connection would cost (study of fifteen large cities in Africa).
Source: Keener, Luengo, and Banerjee 2009; www.infrastructureafrica.org/system/files/Africa%27s%20Water%20and%20Sanitation%20Infrastructure.pdf.

6 Estimated Annual World Water Use

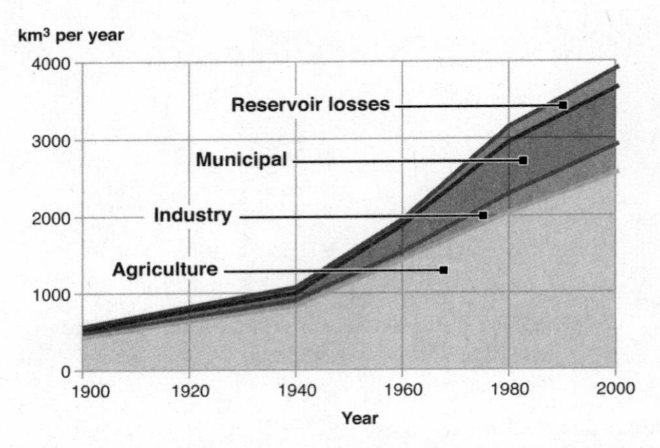

Source: http://blogs.princeton.edu/chm333/f2006/water/2006/11/how_does_water_use_in_developing_countries_differ.html.

7 Virtual Water Footprint for Various Products

Product	Virtual water content (liters)
1 sheet of A4 paper (80 g/m²)	10
1 tomato (70 g)	13
1 potato (100 g)	25
1 microchip (2 g)	32
1 cup of tea (250 ml)	35
1 slice of bread (30 g)	40
1 orange (100 g)	50
1 apple (100 g)	70
1 glass of beer (250 ml)	75
1 slice of bread (30 g) with cheese (10 g)	90
1 glass of wine (125 ml)	120
1 egg (40 g)	135
1 cup of coffee (125 ml)	140
1 glass of orange juice (200 ml)	170
1 bag of potato crisps (200 g)	185
1 glass of apple juice (200 ml)	190
1 glass of milk (200 ml)	200
1 cotton T-shirt (250 g)	2000
1 hamburger (150 g)	2400
1 pair of shoes (bovine leather)	8000

Global average virtual-water content (in liters) for some selected products, per unit of product (in 2007).

Source: http://www.waterfootprint.org/Reports/Hoekstra_and_Chapagain_2007.pdf.

8 Losses Due to Water Scarcity and Poor Sanitation

Problem	Description
Child deaths	1.8 million children die each year as a result of diarrhea—4,900 die each day, equivalent to the under-five population of London and New York combined. Together, unclean water and poor sanitation are the world's second biggest killer of children. Deaths from diarrhea in 2004 were some six times greater than the average annual deaths in armed conflict for the 1990s.
School days	The loss of 443 million school days each year from water-related illness.
Overall health	Close to half of all people in developing countries suffer at any given time from a health problem caused by water and sanitation deficits.
Lost time	Millions of women spending several hours a day collecting water.
Lost opportunities	Life cycles of disadvantage affecting millions of people, with illness and lost educational opportunities in childhood leading to poverty in adulthood.
Economic impact	Losses are greatest in some of the poorest countries. Sub-Saharan Africa loses about 5 percent of GDP, or some $28.4 billion annually, a figure that exceeds total aid flows and debt relief to the region in 2003. In one crucial respect these aggregate economic costs obscure the real impact of the water and sanitation deficit. Most of the losses are sustained by households below the poverty line, retarding the efforts of poor people to produce their way out of poverty.

Source: http://hdr.undp.org/en/media/HDR06-complete.pdf.

9 Use of Improved Sanitation Facilities in Africa and Asia, 2008

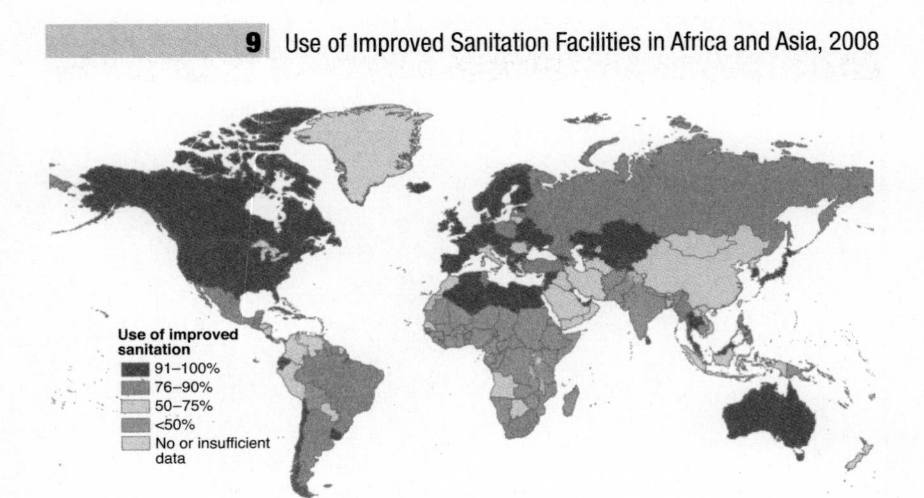

Improved sanitation facilities are used by less than two thirds of the world population; 1.2 billion people still practice open defecation.
Sources: http://www.unicef.org/wash/files/JMP_report_2010.pdf and http://is662ict4sd14.blogspot.com.

10 Sanitation Coverage in Africa: 1990–2008

	1990	2000	2008
Population ('000)	517681	674693	822436
Percentage urban population	28	33	37
Urban			
Improved sanitation	43	43	44
Shared sanitation	29	30	31
Unimproved facilities	17	17	17
Rural			
Improved sanitation	21	23	24
Shared sanitation	10	11	13
Unimproved facilities	22	23	25
Open defecation	47	43	38
Total			
Improved sanitation	28	29	31
Shared sanitation	16	18	20
Unimproved facilities	20	21	22
Open defecation	36	32	27

Source: http://www.unhabitat.org/pmss/getElectronicVersion.aspx?nr=3074&alt=1; compilation from WHO/UNICEF (2010) Progress on Water and Sanitation: 2010.

Food and Agriculture

11 Global Area of Biotech Crops, 1996 to 2010 (Million Hectares)

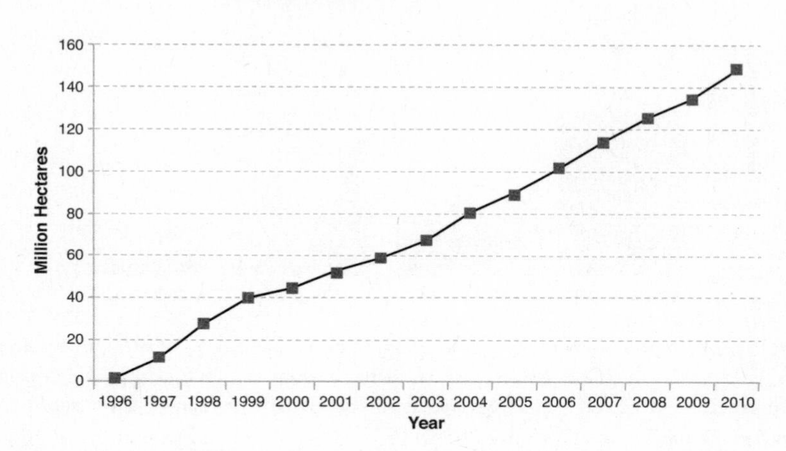

In 2010, the global market value of biotech crops was US$11.2 billion, representing 22 percent of the US$51.8 billion global crop protection market in 2010 and 33 percent of the approximately US$34 billion 2010 global commercial seed market. Of the US$11.2 billion biotech crop market, US$8.9 billion (80 percent) was in the industrial countries and US$2.3 billion (20 percent) was in the developing countries. This graph shows the consistency of global adoption and growth.

Source: Clive James, 2010; http://www.isaaa.org/resources/publications/pocketk/16/default.asp.

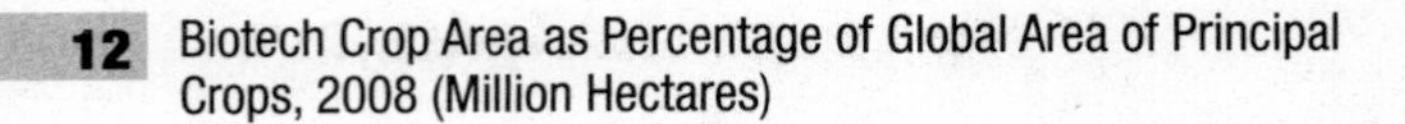

12 Biotech Crop Area as Percentage of Global Area of Principal Crops, 2008 (Million Hectares)

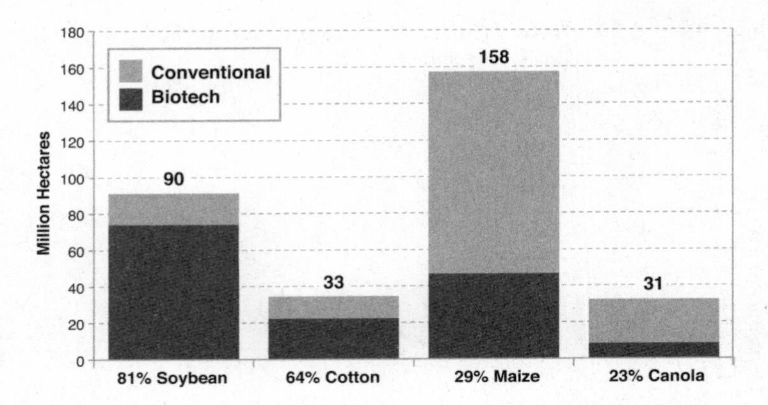

The future of biotech crops looks encouraging. Commercialization of drought-tolerant maize is expected in 2012; golden rice in 2013; and Bt rice before the Millennium Development Goal (MDG) of 2015, which will potentially benefit 1 billion poor people in Asia alone.

Source: http://www.isaaa.org/resources/publications/pocketk/16/default.asp.

13 Past and Projected Trends in Consumption of Meat and Milk in Developed and Developing Countries

	Developing countries				
	1980	1990	2002	2015	2030
Food demand					
Annual per capita meat consumption (kg)	14	18	28	32	37
Annual per capita milk consumption (kg)	34	38	46	55	66
Total meat consumption (million tonnes)	47	73	137	184	252
Total milk consumption (million tonnes)	114	152	222	323	452

	Developed countries				
	1980	1990	2002	2015	2030
Food demand					
Annual per capita meat consumption (kg)	73	80	78	83	89
Annual per capita milk consumption (kg)	195	200	202	203	209
Total meat consumption (million tonnes)	86	100	102	112	121
Total milk consumption (million tonnes)	228	251	265	273	284

There is a growing demand for both meat and milk in both the developed and the developing world.

Source: FAO 2006, "Livestock's Long Shadows: Environmental Issues and Options"; ftp://ftp.fao.org/docrep/fao/010/a0701e/a0701e.pdf.

14 Past and Projected Food Consumption of Livestock Products (1960–2050)

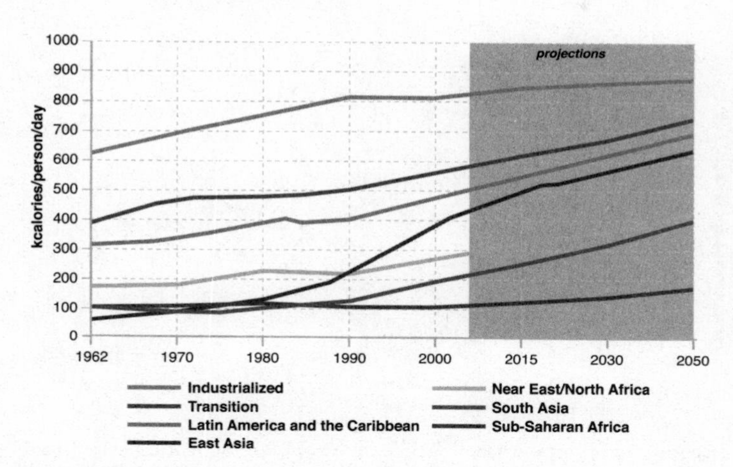

There is a growing demand for livestock products worldwide.
Source: FAO 2006, "Livestock's Long Shadows: Environmental Issues and Options"; ftp://ftp.fao.org/docrep/fao/010/a0701e/a0701e.pdf.

15 Household Food Spending Worldwide

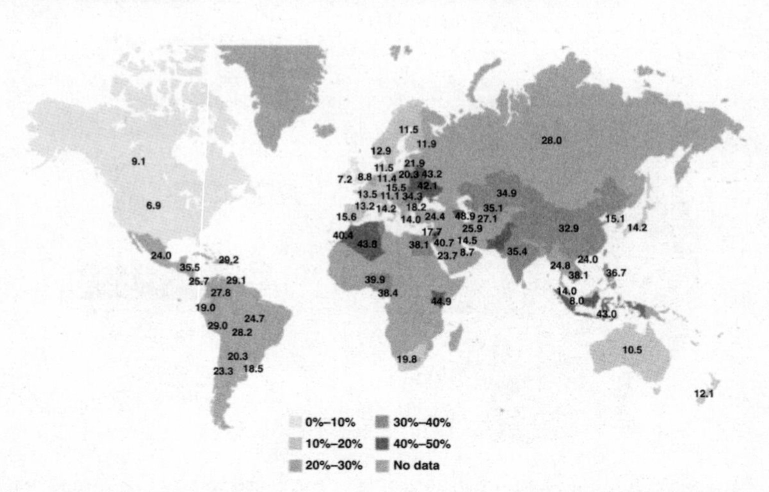

Less than 7 percent of the money Americans spend goes to buy food, the lowest of any country that keeps such data. Each number on the map represents a country and the percentage of people's total expenditures spent on food in that country.
Sources: http://civileats.com/2011/03/29/mapping-global-food-spending-infographic/data, http://www.ers.usda.gov/briefing/cpifoodandexpenditures/Data/Table_97/2009table97.htm.

16 Proportion of Undernourished People in the Developing World, 1969–2010

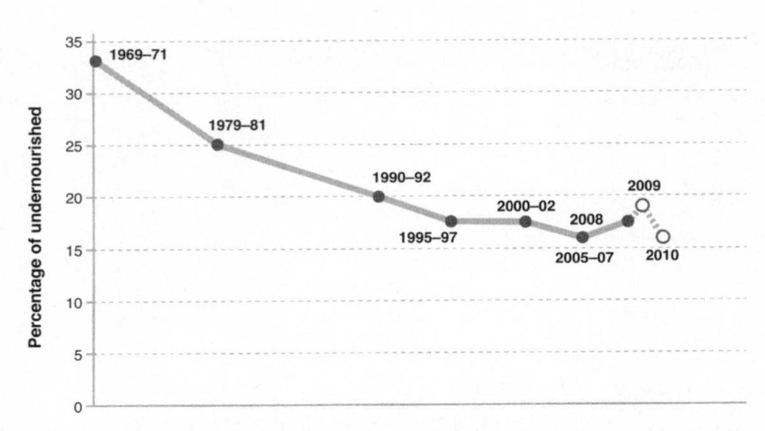

The percentage of undernourished people in the developing world has dropped by over 50 percent since 1969.
Source: http://www.fao.org/docrep/013/i1683e/i1683e00.htm.

17 Undernourished People in the World Today by Region

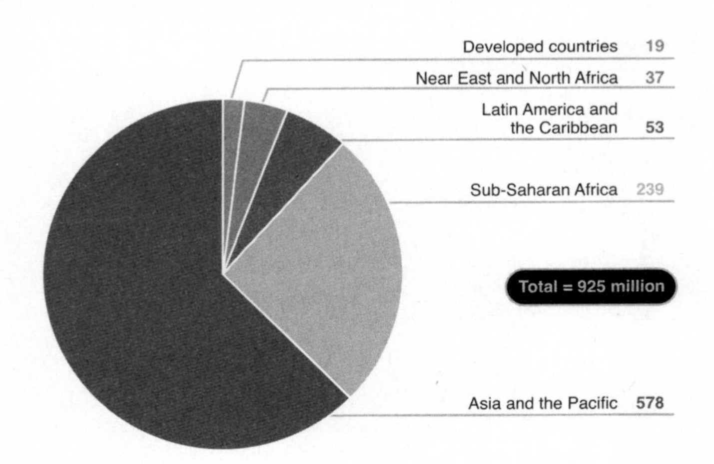

There are 925 million undernourished people in the world today. This means that one out of every seven of us does not get enough food for a healthy, active life.
Source: http://www.fao.org/docrep/012/al390e/al390e00.pdf http://www.wfp.org/hunger.

18 Regional and Global Burden of Nutrition-Related Disease Risk Factors

Population/Risk Factor	Africa	Americas	East Mediterranean	Europe	Southeast Asia	Western Pacific	World
				(thousand)			
Total population	**639,593**	**827,345**	**481,635**	**873,625**	**1,535,625**	**1,687,287**	**6,045,017**
Childhood and maternal-undernutrition-related diseases							
				(DALYs as percent of regional and world population)			
Underweight	9.82	0.24	3.58	0.09	3.06	0.48	2.28
Iron deficiency	1.59	0.21	0.77	0.12	0.91	0.26	0.58
Vitamin A deficiency	2.57	0.04	0.61	0	0.42	0.03	0.44
Zinc deficiency	2.15	0.06	0.67	0.01	0.35	0.03	0.46
Malaria							0.74
HIV/AIDS							1.49
Respiratory infections							1.67
Iodine deficiency							0.04
Measles							0.4
Diarrhea							1.19
Other nutrition-related risks							
High blood pressure	0.69	0.78	1.02	2.22	0.98	0.83	1.06
High cholesterol	0.31	0.55	0.67	1.51	0.8	0.31	0.67
High BMI	0.23	0.89	0.6	1.35	0.27	0.35	0.55
Low fruit and vegetable intake	0.24	0.36	0.34	0.76	0.57	0.3	0.44
Diabetes							0.25

The table shows the estimated disease burden for each risk factor. These risks act on their own and jointly with others. Consequently, the burden due to groups of risk factors will usually be less than the sum of individual risks. The disability-adjusted life year is a measure of the burden of disease. It reflects the total amount of healthy life lost to all causes.
Source: http://www.millenniumassessment.org/documents/document.277.aspx.pdf; adapted from Ezzati et al. 2002; Ollila n.d.; and WHO 2002a.

19 Energy Loss in Food (Field to Fork)

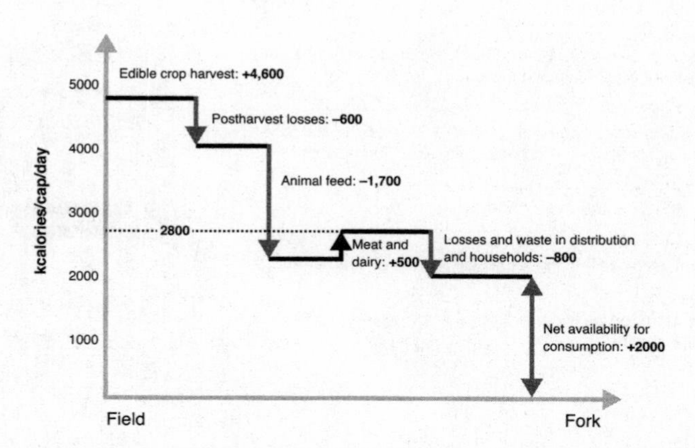

A schematic summary of the amount of food produced globally at field level and estimates of the losses, conversions, and wastage in the food chain. This is part of the argument for the creation of Vertical Farms.
Source: "From Field to Fork: Curbing Losses and Wastage in the Food Chain," Stockholm International Water Institute; http://www.siwi.org/documents/Resources/Papers/Paper_13_Field_to_Fork.pdf.

20 Energy Loss in Food (Harvest to Home)

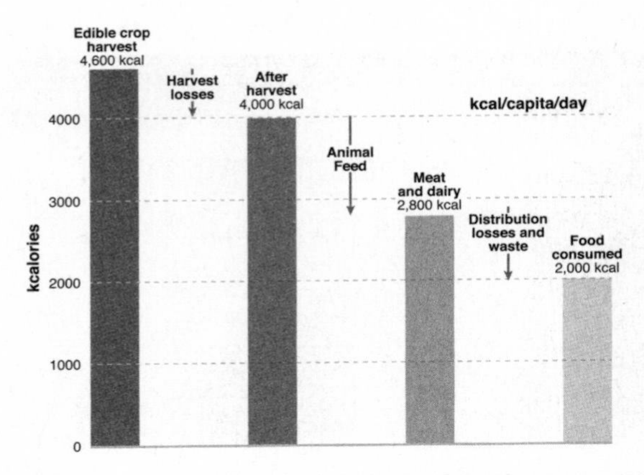

Losses in the food chain from field to household consumption. Over 50 percent of the kcal (energy) of the food harvested in the field is lost by the time it gets to your table. This is part of the argument for the creation of Vertical Farms.
Source: http://maps.grida.no/go/graphic/losses-in-the-food-chain-from-field-to-house hold-consumption.

21 Vertical Farming

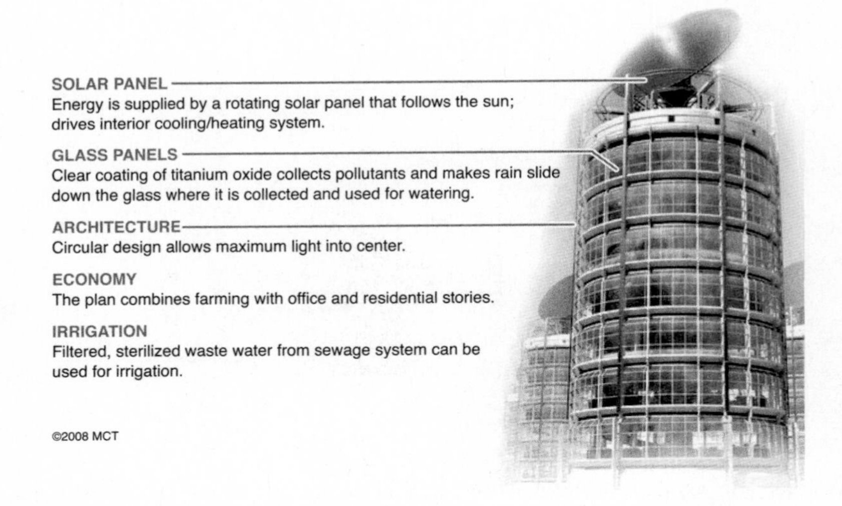

While this image depicts only a few potential vertical farming technologies, it does explore the integration of the system into urban environments.
Source: Vertical Farm Project; http://www.the-edison-lightbulb.com/2011/03/09/ vertical-farms-the-21st-century-agricultural-revolution.

22 Evidence of Overfishing (1950–2003)

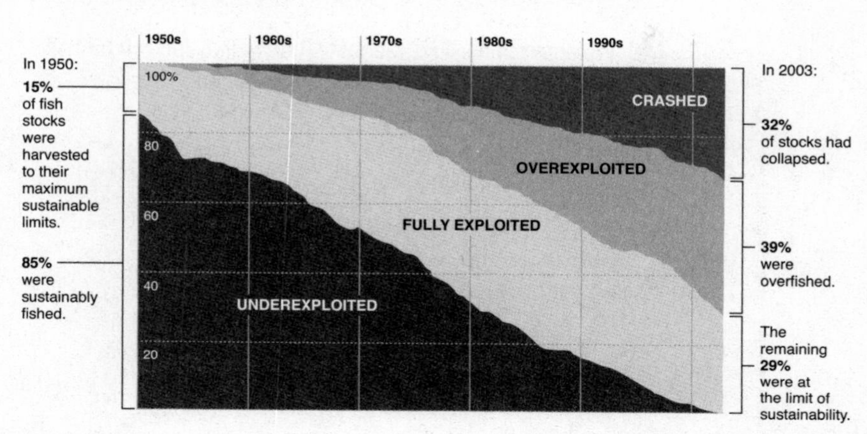

The condition of the world's fisheries has declined drastically because of overfishing. Today's fisheries are at the breaking point.
Source: http://simondonner.blogspot.com/2008/11/farming-oceans.html.

23 Growth of Seafood Aquaculture vs. Wild Caught, 1950–2008

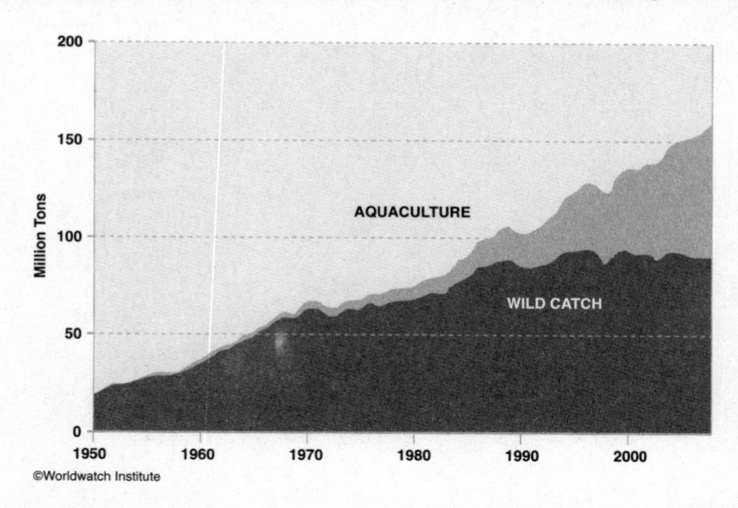

Growth of Seafood Aquaculture between 1950 and 2008 has helped to make up for the decimation of our natural fisheries.
Source: FAO; http://peakwatch.typepad.com/.a/6a00d83452403c69e201538f2305b2970b-pi.

Health and Health Care

24 Projected Under-Five Mortality Rate (per 1,000 live births) Globally and by WHO Region, 1980–2010

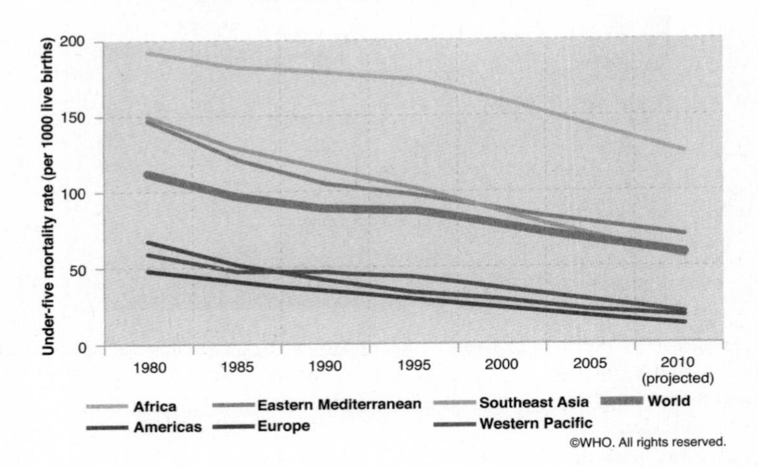

Note that health care improvements have decreased early childhood (under five years) mortality rates by nearly 50 percent in many regions of the world.

Source: http://www.who.int/gho/child_health/mortality/mortality_under_five/en/index.html.

25 Cause of Deaths Among Children Under Five Years of Age

Global distribution of deaths among children under five years of age		
	% distribution of under-five deaths (2000–03)	Total number of under-five deaths (2006)
Neonatal causes	37	3,600,000
Pneumonia	19	1,800,000
Diarrheal diseases	17	1,600,000
Other	10	970,000
Malaria	8	780,000
AIDS	3	290,000
Measles	4	390,000
Injuries	3	290,000
Total	**100**	**9,700,000**

a. Note that the totals may not sum due to rounding.

b. Neonatal causes refer to deaths in the first 28 days of life, which include preterm birth, severe infections, birth asphyxia, congenital anomalies, neonatal tetanus, diarrheal diseases, and other neonatal causes.

Neonatal causes refer to deaths in the first 28 days of life. These include: preterm birth, severe infections, birth asphyxia, congenital anomalies, neonatal tetanus, and diarrheal diseases. Many of these diseases and conditions are preventable with modern health care technologies.

Source: http://www.unicef.org/media/files/Under_five_deaths_by_cause_2006_estimates 3.doc. 53% from World Health Organization, *The World Health Report 2005: Make Every Mother and Child Count,* WHO, Geneva, 2005.

26 Percentage of Disability-Adjusted Life Years (DALYs) by Income Level (2004)

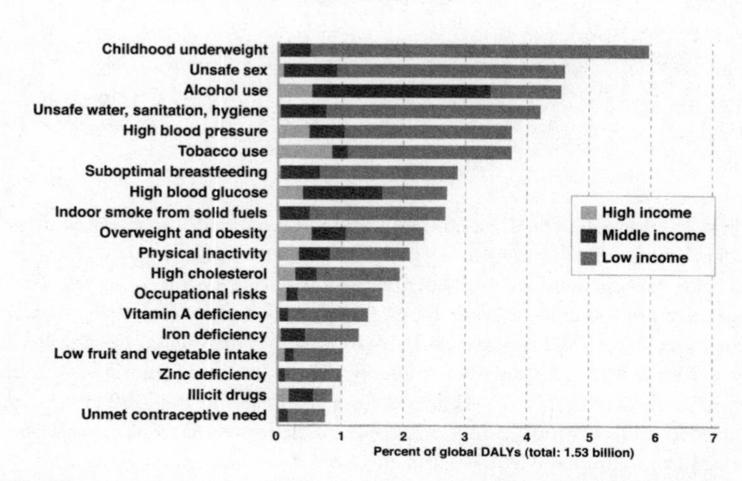

According to this chart the following categories—childhood underweight, unsafe sex, unsafe water, suboptimal breastfeeding, indoor smoke, vitamin A deficiency, iron deficiency, zinc deficiency, and unmet contraceptive needs—are all conditions of poverty. These are primary areas for near-term improvement.

Source: WHO, 2009. Global health risks.

27 Health and Indoor Pollution

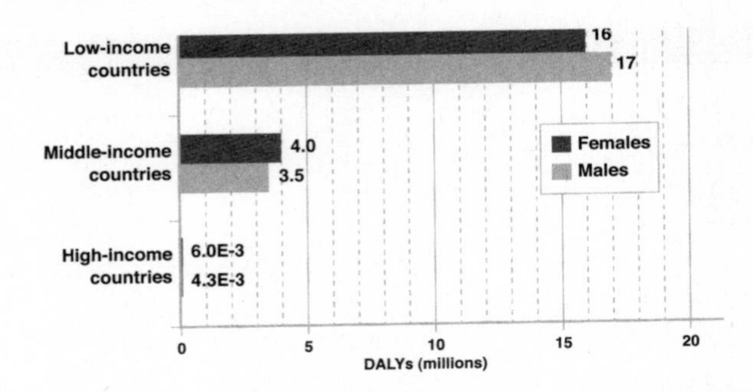

This chart shows the disease burden due to indoor air pollution by level of development. In 2004, indoor air pollution from solid fuel use was responsible for almost 2 million deaths and 2.7 percent of the global burden of disease (in disability-adjusted life years, or DALYs). This makes it the second largest contributor to ill health. Acute lower respiratory infections, in particular pneumonia, continue to be the biggest killer of young children, causing more than 2 million annual deaths.
Source: http://www.who.int/indoorair/health_impacts/burden_global/en.

28 Health and Water-Related Disease (1999)

Disease	Deaths[a]	DALYs[a]
Schistosomiasis	14	1,932
Trachoma	0	1,239
Ascariasis	3	505
Trichuriasis	2	481
Hookworm disease	7	1,699
Total	26	5,856

[a]x 1,000.

For the year 1999, worldwide disease burden caused by selected water-related diseases other than infectious diarrhea (figures × 1000). Safe water supplies, hygienic sanitation, and good water management are fundamental to global health. Almost one tenth of the global disease burden could be prevented by: (i) increasing access to safe drinking water; (ii) improving sanitation and hygiene; and, (iii) improving water management to reduce risks of waterborne infectious diseases. Annually, safer water could prevent 1.4 million child deaths from diarrhea, 500,000 deaths from malaria, and 860,000 child deaths from malnutrition. In addition, 5 million people can be protected from being seriously incapacitated from lymphatic filariasis and another 5 million from trachoma.
Sources: http://ehp.niehs.nih.gov/realfiles/members/2002/110p537-542pruss/pruss-full .html; http://www.who.int/features/qa/70/en/index.html.

29 Exponential Decrease in DNA Sequencing Costs

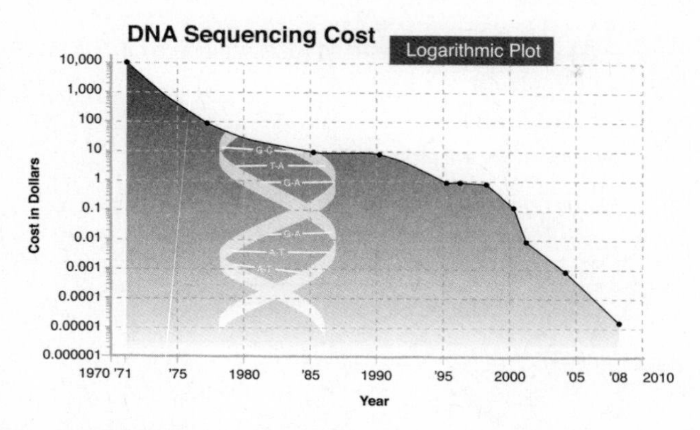

Source: Kurzweil, *The Singularity Is Near.*

30 Worldwide Life Expectancy Growth

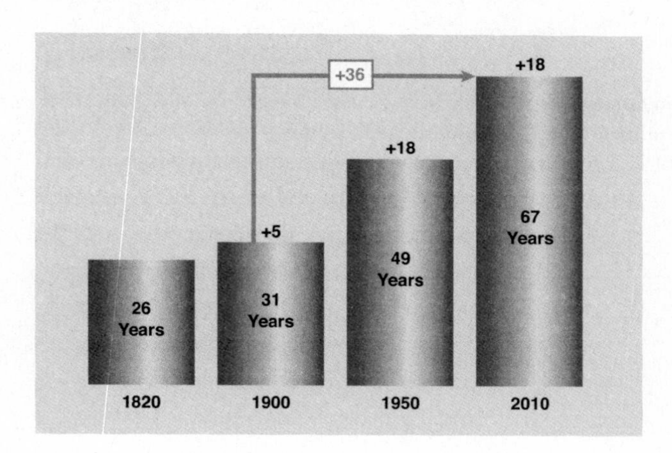

Global lengthening of average human life span over the past 190 years.
Source: United Nations Development Program.

Energy

31 Sources and Demand (Uses) of Energy in the United States (2009)

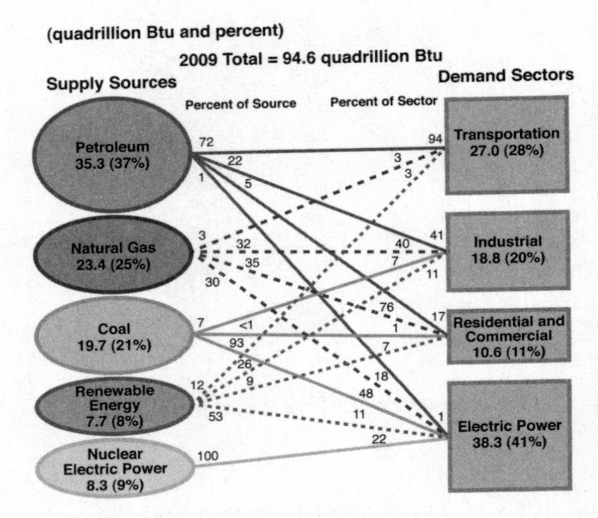

This graphic shows the complex web of energy sources and uses in the United States (in 2009). Numbers along lines indicate percentages.

Source: http://www.eia.gov/totalenergy/data/annual/pecss_diagram2.cfm.

32 U.S. 2009 Sources of Energy

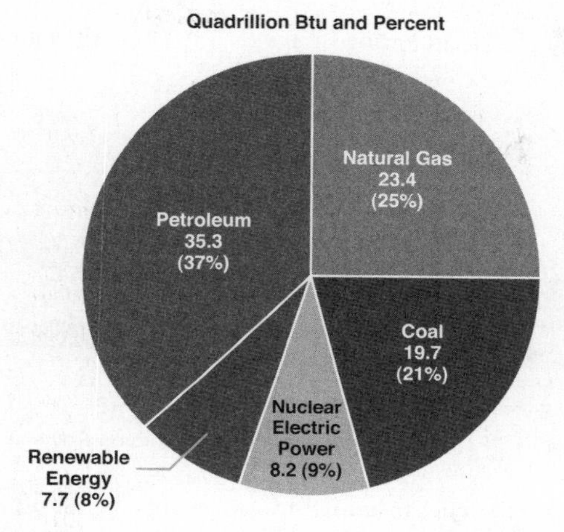

This pie chart indicates the sources of energy powering the United States in 2009. The first number is quadrillion Btu and the second number is percentage.
Source: http://www.eia.gov/energy_in_brief/major_energy_sources_and_users.cfm.

33 GDP per Capita vs. Power Consumption (each dot represents a country)

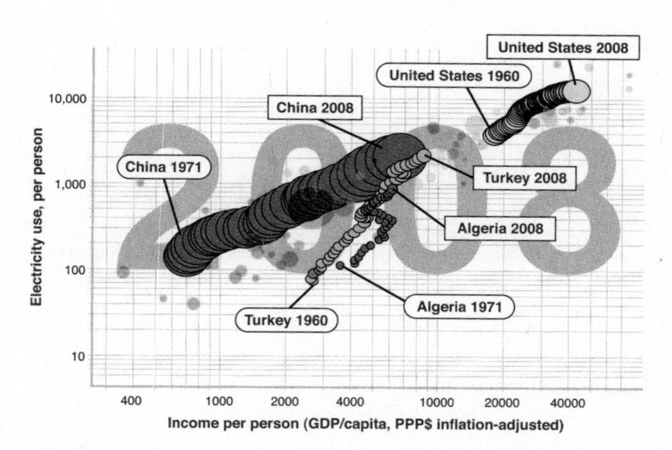

The trend is pretty clear: the wealthier a nation becomes (GDP per capita), the more energy it consumes (KWh per capita). This Gapminder chart shows the progress of a nation between 1960 and 2008 (data for China and Algeria is available only starting in 1971). The size of the circle represents the population size. The four countries chosen are for representation purposes only.
Source: http://www.inference.phy.cam.ac.uk/withouthotair/c30/page_231.shtml.

34 Primary Sources of Energy in Africa (2008)

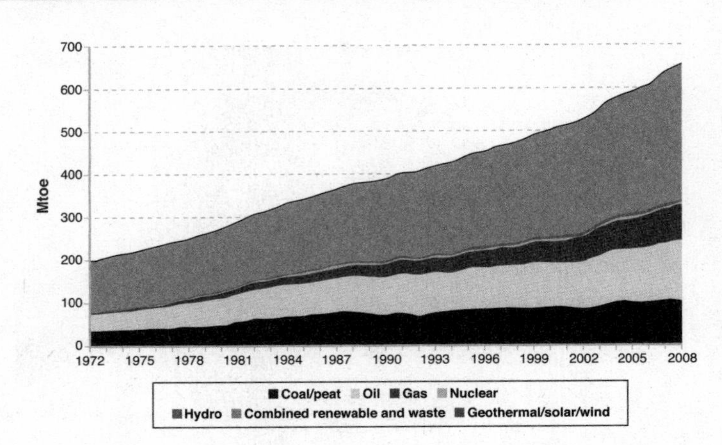

Africa's primary power supply measured in million of tons of oil equivalent (Mtoe), broken down by source.
Source: http://www.iea.org/stats/pdf_graphs/11TPES.pdf.

35 Average Price of US Electricity over Time ($ per kWh at 1990 prices)

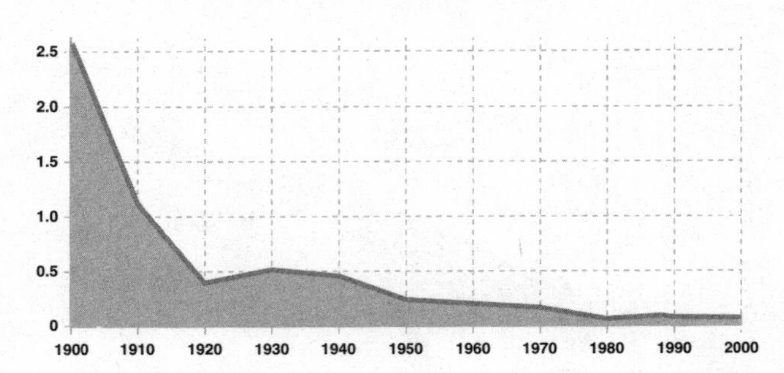

During the past hundred years there has been a constant decrease in the cost of electricity ($ per KWh).
Source: Bill Gates TED Talk, 2010.

36 Deaths per TWh by Energy Generation Source

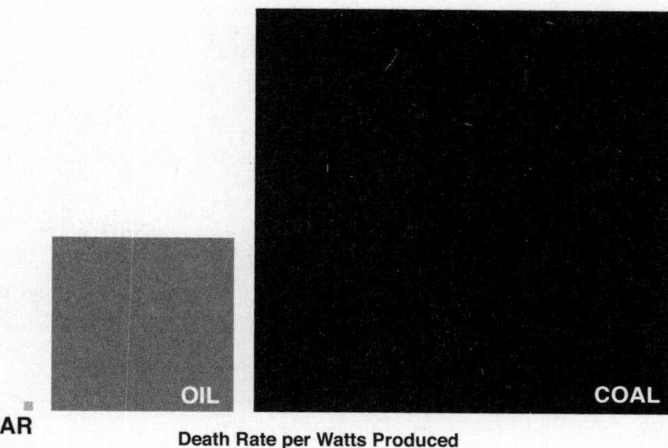

Note the tiny dot labeled "Nuclear" to the far left. For every one person killed by nuclear power generation, 4,000 die because of coal.

Sources: Seth Godin at http://sethgodin.typepad.com/seths_blog/2011/03/the-triumph-of-coal-marketing.html. Using Brian Wang's data: http://nextbigfuture.com/2011/03/deaths-per-twh-by-energy-source.html.

37 Energy Storage: Specific Power vs. Specific Energy

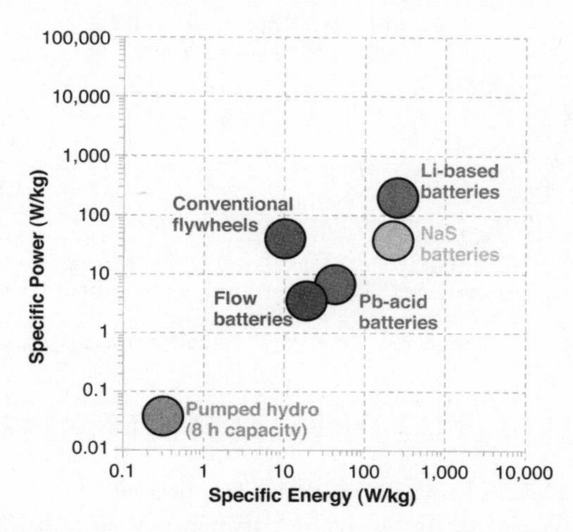

The graph shows the relative specific power (amount of current the battery can deliver) versus specific energy (energy per unit mass).

Source: Professor Don Sadoway, MIT, LMBC.

38 Installed Capacity vs. Capital Costs

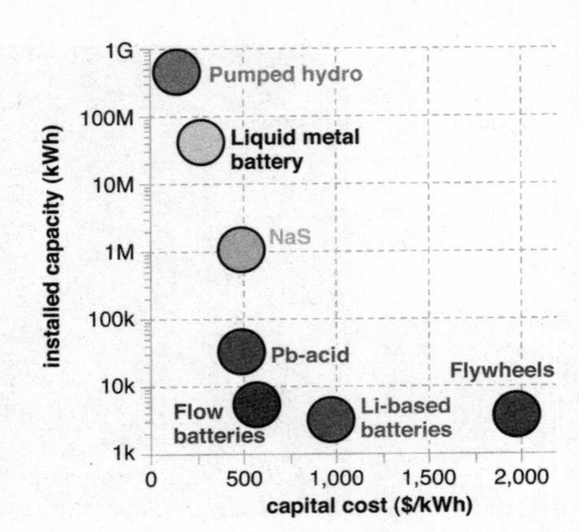

This figure shows the various energy storage methods plotted on an Installed Capacity versus Capital Cost graph. According to Professor Sadoway, the key metrics required for grid-scale storage are (i) cost (<$150/kWh); life span (>10 years); and energy efficiency (>80 %)—all of which are achievable by liquid metal batteries.
Source: Professor Don Sadoway, MIT, LMBC.

39 Solar PV Cost per Watt (1980–2009)

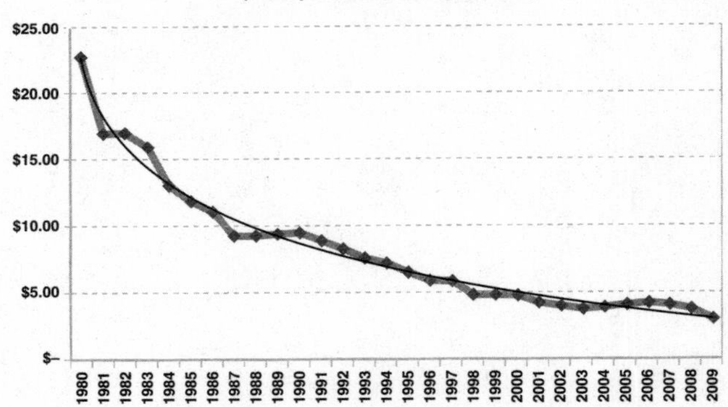

The cost of solar PV cells have been decreasing exponentially.
Source: DOE NREL Solar Technologies Market Report, Jan. 2010. Ramez Naam, "The Exponential Gains in Solar Power per Dollar," http://unbridledspeculation.com/2011/03/17/the-exponential-gains-in-solar-power-per-dollar.

40 Watts Produced per Constant $100 (1980–2010)

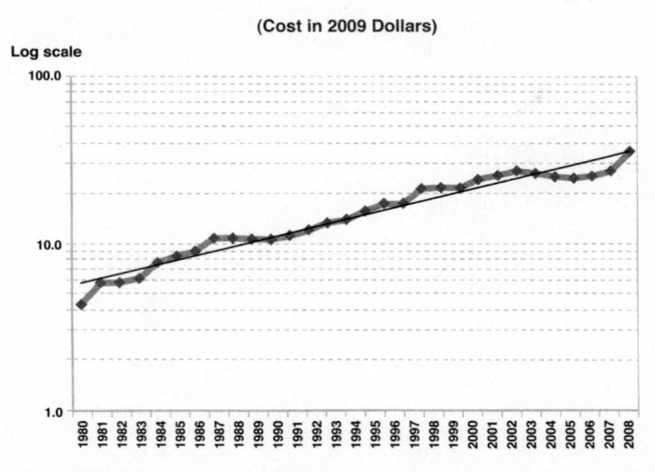

The efficiency of PV (Watts produced per constant $100) has been increasing exponentially. Note that the Y axis is on a log scale.
Sources: DOE NREL Solar Technologies Market Report, Jan. 2010. Ramez Naam, "The Exponential Gains in Solar Power per Dollar," http://unbridledspeculation.com/2011/03/17/the-exponential-gains-in-solar-power-per-dollar.

41 PV Cost Reduction Road Map (2007–2014)

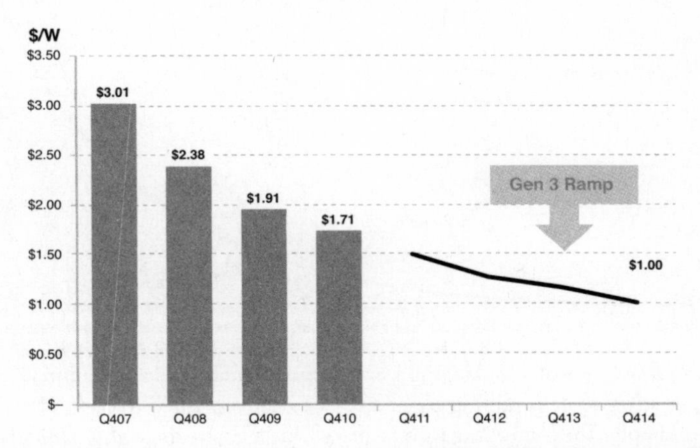

The continued and projected decrease in the cost per watt ($/W) for photovoltaic panels, according to the SunPower Corporation, one of the leading manufacturers of PVs.
Source: © 2010 SunPower Corporation.

42 Learning Curve for Solar Power

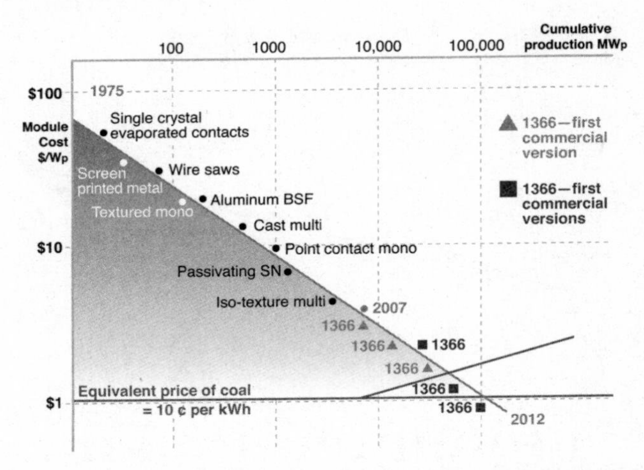

The dropping cost of solar and corresponding increase in cumulative production is essentially a graphic representation of the industry's learning curve.
Source: Presentation by Frank van Mierlo, CEO, and Ely Sachs, CTO, of 1366 Technologies. Data is from Greg Nemet at UC Berkeley.

43 Global Annual Wind Power Capacity vs. Time

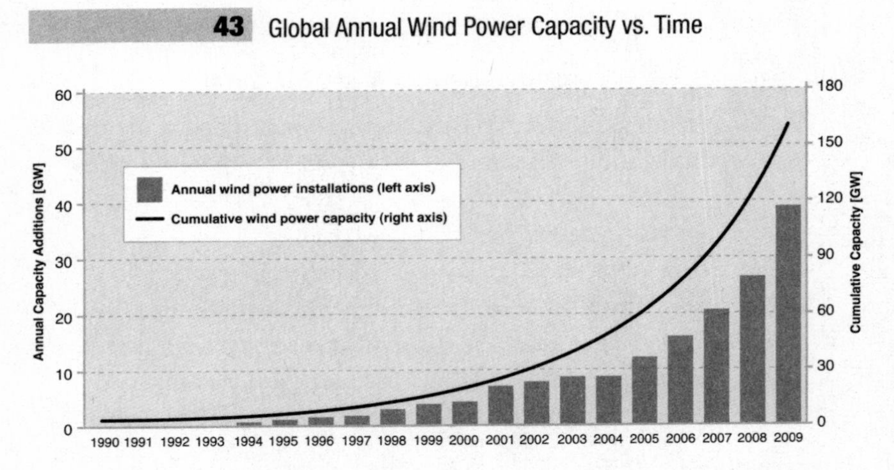

This graph shows global annual wind power capacity additions and cumulative capacity. Despite these trends, wind energy remains a relatively small fraction of worldwide electricity supply. The total wind power capacity installed by the end of 2009 would, in an average year, meet roughly 1.8 percent of worldwide electricity demand.
Source: Special Report on Renewable Energy Sources and Climate Change Mitigation (SRREN). http://srren.ipcc-wg3.de/report/IPCC_SRREN_Ch07.

44 Maximum Power Contained in Renewable Sources

Energy source	Maximum power	Percent of Solar
Solar	85,000 TW	100.000
Ocean thermal	100 TW	0.120
Wind	72 TW	0.080
Geothermal	32 TW	0.380
River hydroelectric	7 TW	0.008
Biomass	6 TW	0.008
Tidal	3 TW	0.003
Coastal wave	3 TW	0.003

No other renewable scales like solar. It has nearly 850 times the potential of ocean thermal, its nearest competitor.
Source: Derek Abbott, Fellow, IEEE, "Keeping the Energy Debate Clean: How Do We Supply the World's Energy Needs?" *Proceedings of the IEEE* 98, no. 1 (January 2010).

45 2007 World Energy Consumption

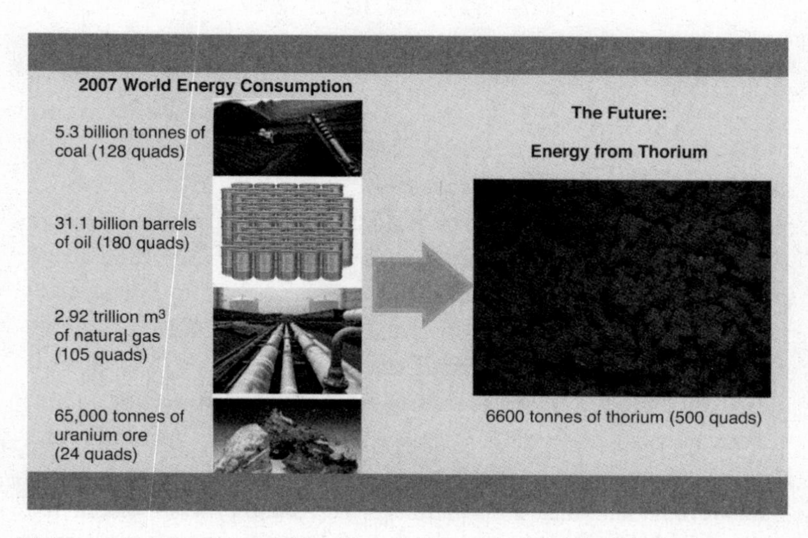

In 2007, just 6,600 tonnes of thorium could have supplied all of the world's energy.
Source: Bill Gates, TED Talk, 2010.

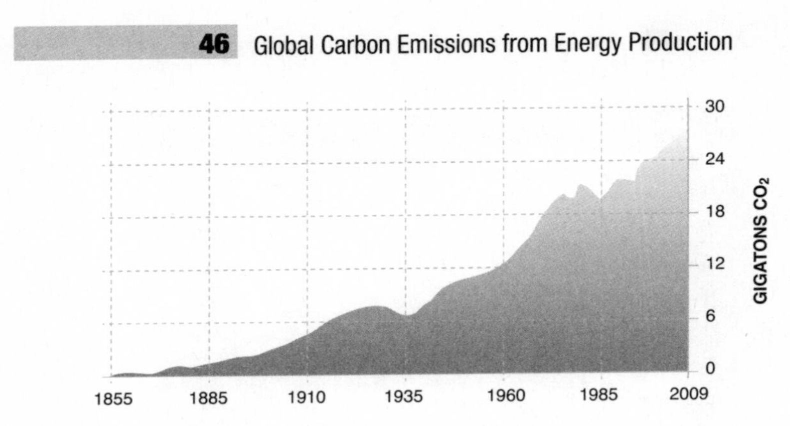

The growth of global CO_2 emissions (gigatons) over the past 150 years.
Source: Bill Gates, TED Talk, 2010.

Education

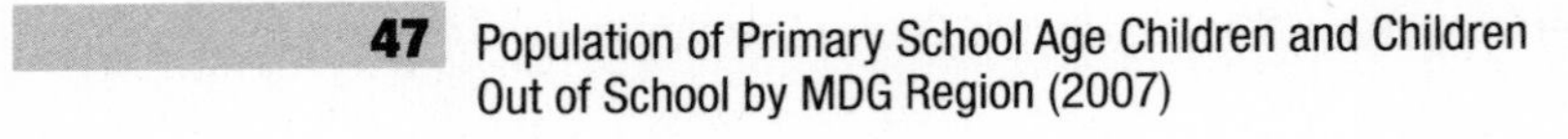

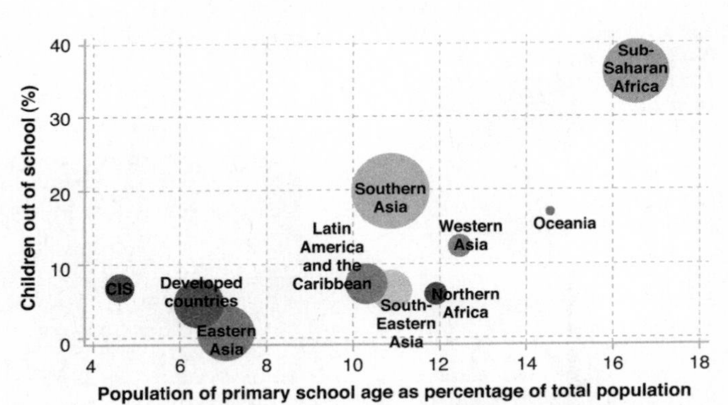

The link between the population structure and the number of children out of school is shown above. Sub-Saharan Africa is the region with the highest percentage of children out of school. At the same time, the population of most countries in sub-Saharan Africa is increasing and children of primary school age constitute a large and growing share of the population. The share of children of primary school age in a region's population is plotted along the horizontal axis and the share of children out of school along the vertical axis.
Source: Population structure and children out of school. http://huebler.blogspot.com/2009/02/coos.html.

48 Primary School Net Enrollment Ratio and GDP per capita (2002)

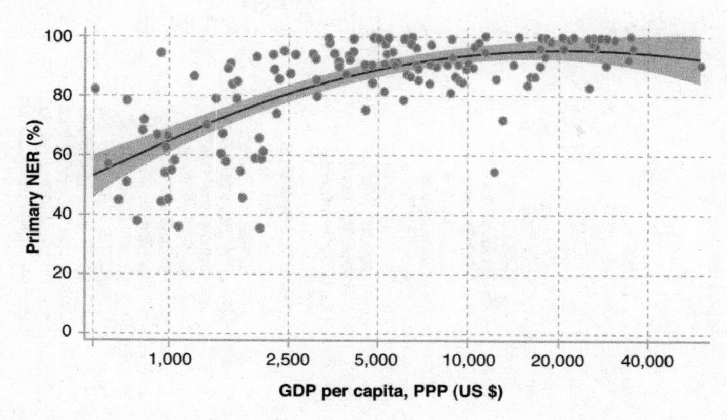

The relationship between poverty and education is stark. Most countries with a GDP per capita of $2,500 or less have net enrollment ratios below 80 percent. Almost all countries above this level of GDP have NER values of more than 80 percent.
Source: http://huebler.blogspot.com/2005/09/national-wealth-and-school-enrollment .html.

49 Technology Engagement in Children 10–12

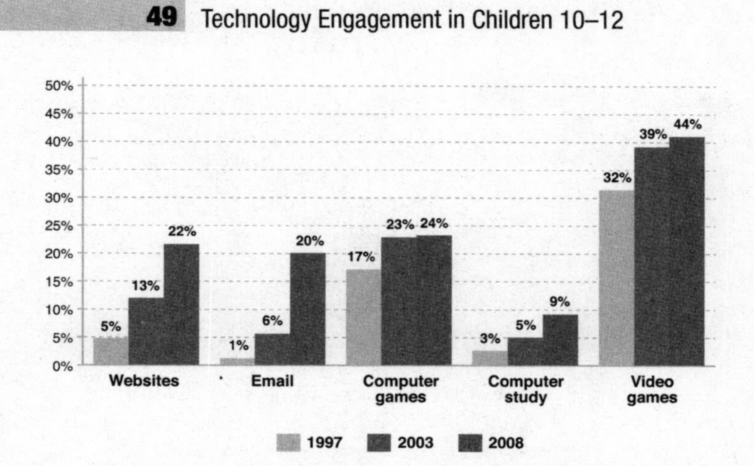

Source: http://newsdesk.umd.edu/bigissues/release.cfm?ArticleID=2229; www.popcenter .umd.edu.

Democracy

50 Freedom in the World—Population Trends

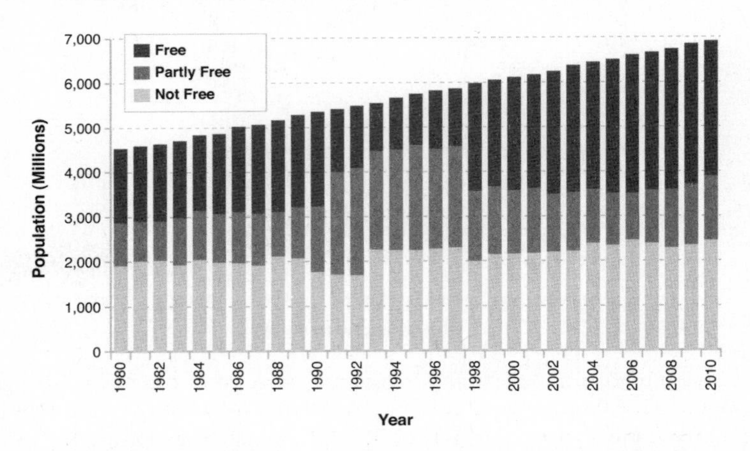

Source: http://www.freedomhouse.org/images/File/fiw/historical/PopulationTrendsFIW 1980–2011.pdf.

51 The Democracy Index (2010)

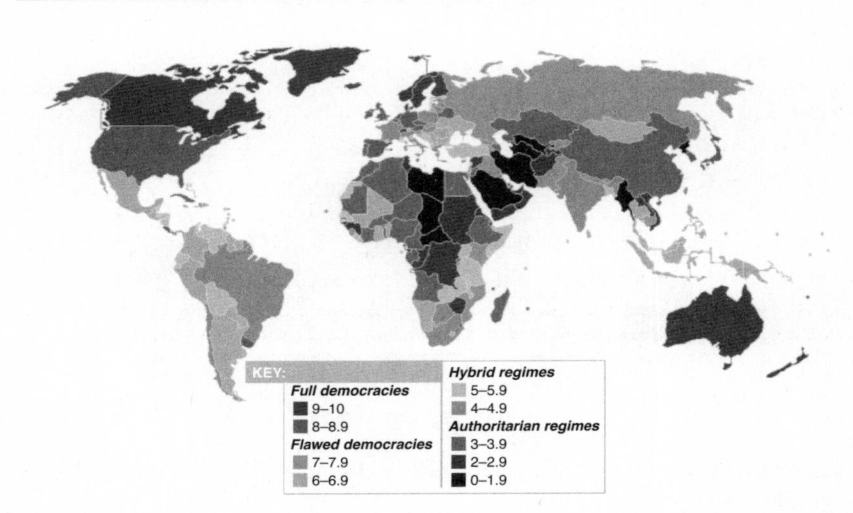

Source: *Economist* via http://en.wikipedia.org/wiki/File:Democracy_Index_2010_green_and_red.svg.

Population and Urbanization

52 The True Size of Africa

COUNTRY	AREA x 1000 km²	COUNTRY	AREA x 1000 km²
China	9,597	Germany	357
USA	9,629	Norway	324
India	3,287	Italy	301
Mexico	1,964	New Zealand	270
Peru	1,285	United Kingdom	243
France	633	Nepal	147
Spain	506	Bangladesh	144
Papua New Guinea	462	Greece	132
Sweden	441	TOTAL	30,102
Japan	378	AFRICA	30,221

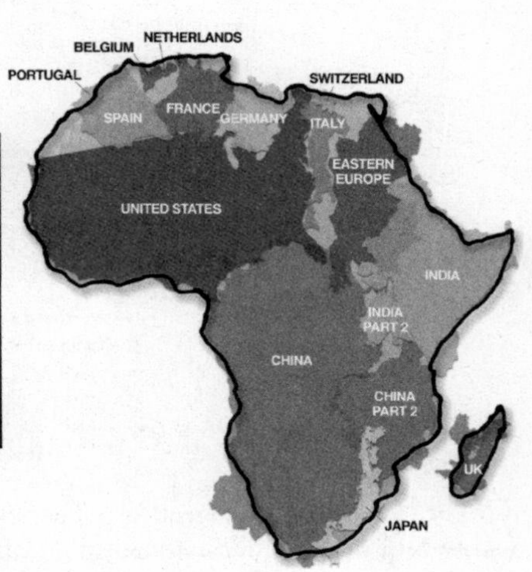

Graphic layout for visualization only (some countries are cut and rotated), but the conclusions are very accurate: refer to table at left for exact data.

In addition to the well-known social issues of *illiteracy and innumeracy*, there also should be such a concept as "*immappancy*," meaning *insufficient geographical knowledge*.

A survey with random American schoolkids let them guess the population and land area of their country. Not entirely unexpected, but still rather unsettling, the majority chose "*1–2 billion*" and "*largest in the world*," respectively.

Even with Asian and European college students, geographical estimates were often off by factors of *2–3*. This is partly due to the highest distorted nature of the predominantly used mapping projections (such as *Mercator*).

A particularly extreme example is the worldwide misjudgement of the true size of Africa. This single image tries to embody the massive scale, which is larger than the *USA, China, India, Japan and all of Europe… combined!*
Source: Kai Krause, Creative Commons.

53 World Urbanization Prospects (2009)

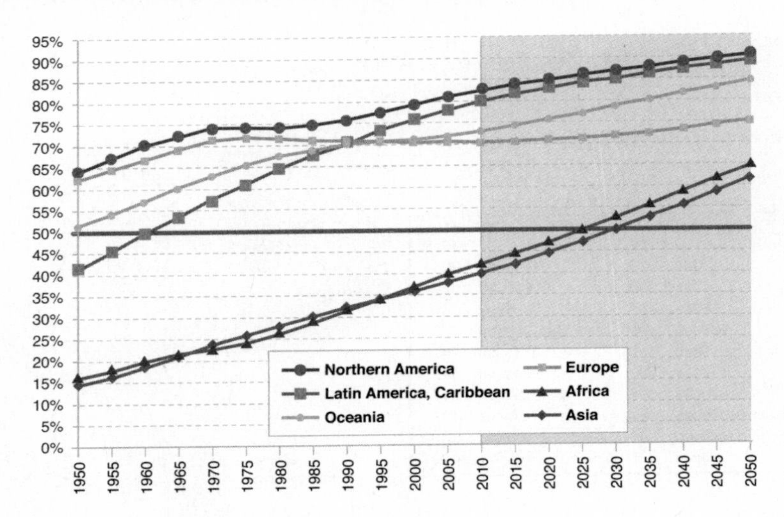

By 2050, 70 percent of the world's population will live in cities.
Source: http://esa.un.org/unpd/wup/Fig_1.htm.

54 Comparison of Urban-Rural Statistics (2003–2007) for India, Vietnam, and Tanzania

	INDIA		VIETNAM		TANZANIA	
	Urban	Rural	Urban	Rural	Urban	Rural
Under-five mortality (per 1,000 live births)	52	82	108	138	16	36
Access to adequate sanitation (percent of households)	77	23	53	43	92	50
Median years of schooling (men)	8	4	6	3	9	6
Access to electricity (percent of households)	93	56	38	1	99	87

Urbanization vs. Income

Percent of Population in Cities

100%
80%
60%
40%
20%
0%

$0 $9,000 $18,000 $27,000 $36,000 $45,000

GDP per Capita (current international $. PPP)

In most countries, city dwellers fare better than their rural counterparts. More urbanized countries in the developed world enjoy a higher per capita income. Within many developing countries, urban residents have more access to basic health and educational services. **Source:** http://earthtrends.wri.org/updates/node/287; UN (population) and World Bank (GDP).

55 World Population 1800–2009

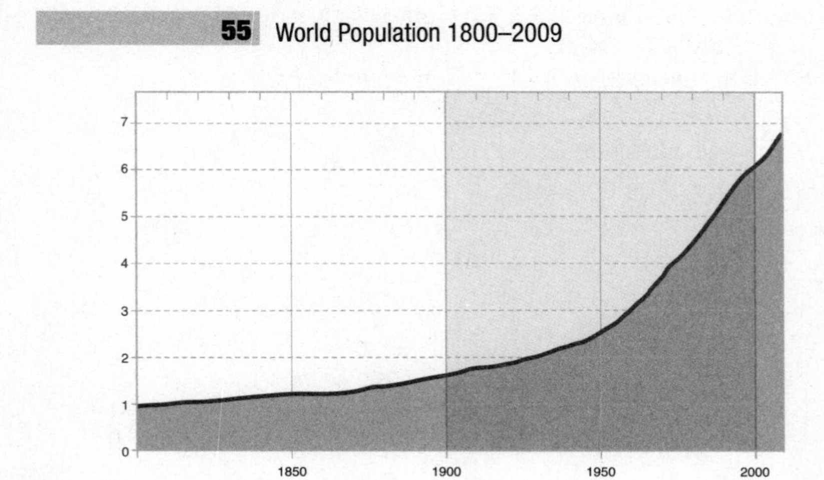

World population growth over the past 209 years in billions of people.
Source: Generated on Wolfram Alpha.

56 Estimated and Projected World Population Variants (1950–2100)

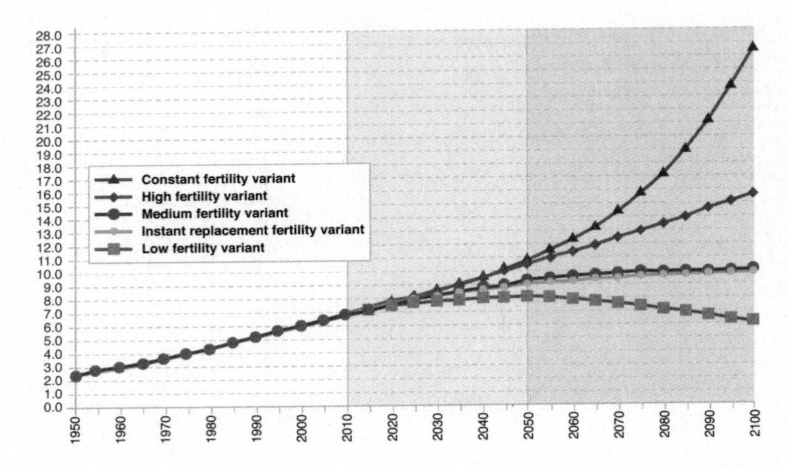

According to the medium variant of the *2010 Revision of World Population Prospects,* the world population is expected to increase from 6.9 billion in mid-2011 to 9.3 billion in 2050 and to reach 10.1 billion by 2100. Realization of this projection is contingent on the continued decline of fertility in countries that still have fertility above replacement level (that is, countries where women have, on average, more than one daughter) and an increase of fertility in the countries that have below-replacement fertility. In addition, mortality would have to decline in all countries. If fertility were to remain constant in each country at the level it had in 2005–2010, the world population could reach nearly 27 billion by 2100.

Source: http://esa.un.org/wpp/Analytical-Figures/htm/fig_1.htm.

57 Number of Countries by Total Fertility

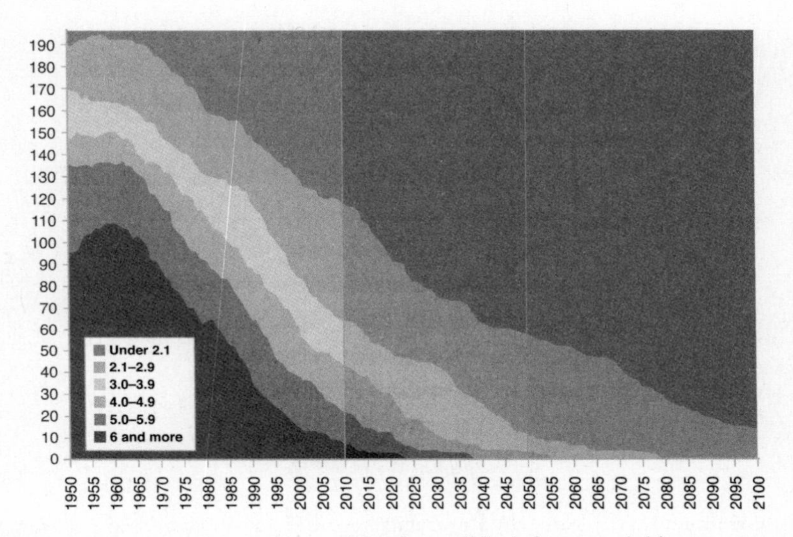

In the great majority of countries total fertility will be below 2.1 children per woman in 2100. This figure displays the number of countries by level of total fertility from 1950 to 2100.

Source: http://esa.un.org/unpd/wpp/Analytical-Figures/htm/fig_9.htm.

58 Children per Women vs. Child Mortality over Time

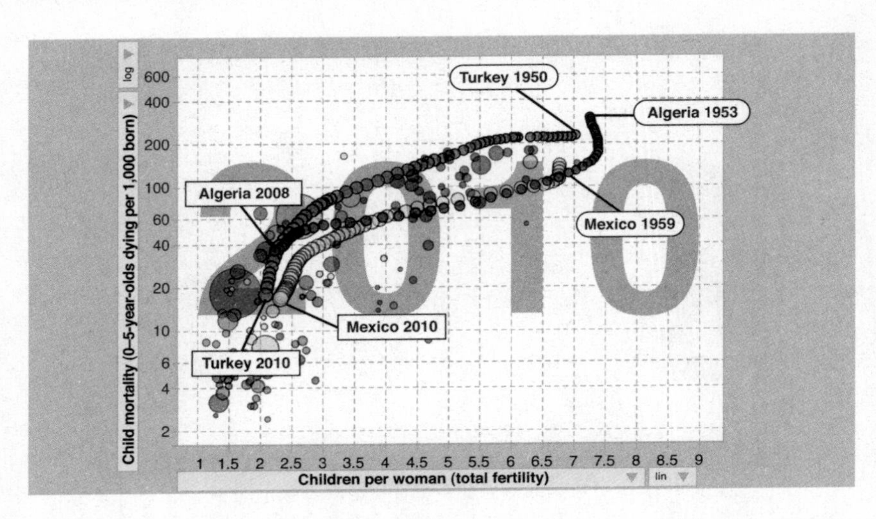

This Gapminder chart plots child mortality (age 0–5) against children per woman, demonstrating a direct correlation between the two. Specifically, as the childhood mortality rate decreases, so too does the number of children born to each woman. The chart shows the progress of a nation between 1950's and 2008. The size of the circle represents the nation's population size. The three countries chosen are for representation purposes only.

Source: Gapminder, Hans Rosling.

59 Children per Women vs. Child Mortality (2009)

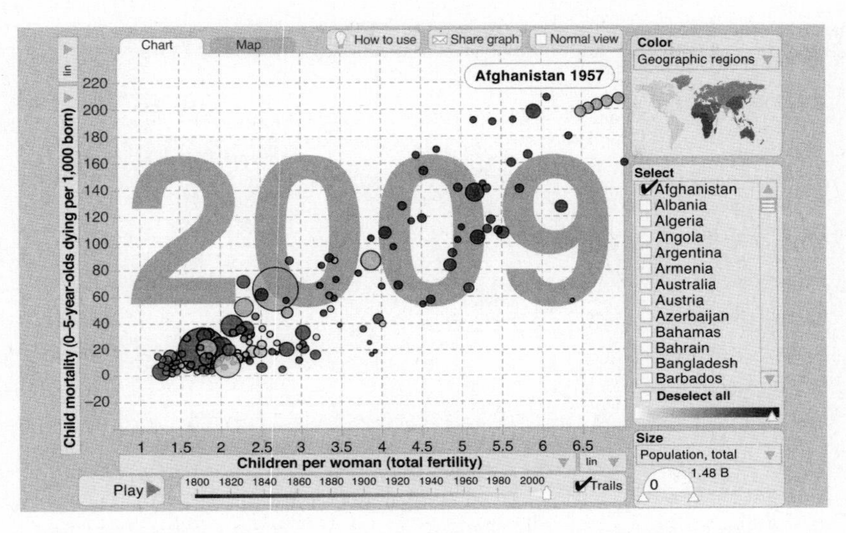

This Gapminder chart plots child mortality (age 0–5) against children per woman, demonstrating a direct correlation between the two. Specifically, as the childhood mortality rate decreases, so too does the number of children born to each woman.
Source: Gapminder, Hans Rosling.

60 Population Change Between 2010 and 2100 by Major Region (millions)

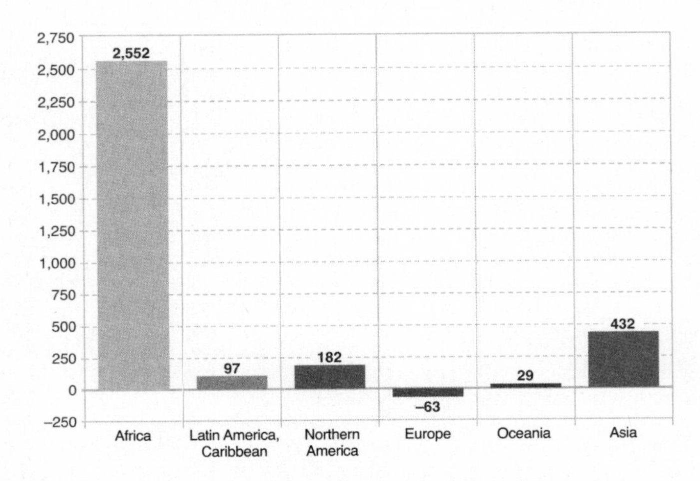

Source: http://esa.un.org/unpd/wpp/Analytical-Figures/htm/fig_13.htm.

Information and Communications Technologies

61 Exponential Growth of Computing for 110 Years

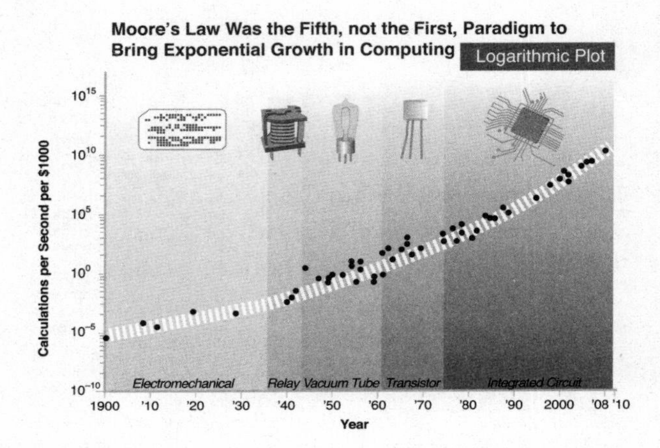

Moore's Law in action. Note how smooth this exponential curve has been over the past hundred years regardless of world wars, depressions, and recessions. Also the curve is actually trending upward (toward vertical), demonstrating the rate of exponential growth itself is increasing over time.
Source: Kurzweil, *The Singularity Is Near.*

62 The Exponential Growth of Computing on a *Logarithmic Plot*

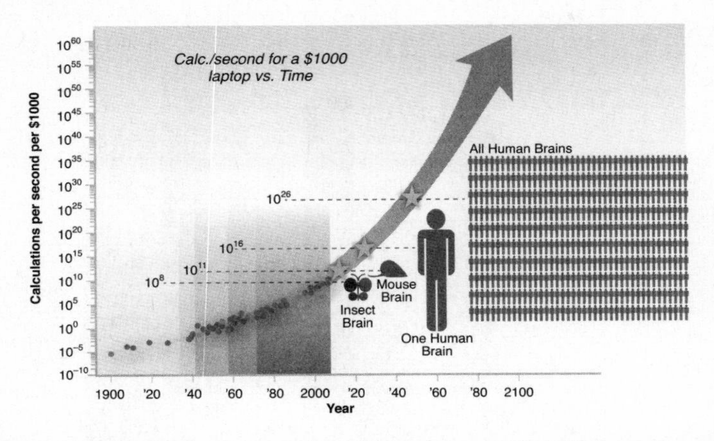

This curve from *Singularity Is Near* projects the continuation of Moore's Law over the next century. It indicates that by roughly 2023 the average $1,000 laptop will be able to communicate at the rate of the human brain, and another ~25 years later, at the rate of the entire human race.

Source: Kurzweil, *The Singularity Is Near.*

63 Exponentially Falling Cost of Memory (1950–2008) Dollars per Megabyte

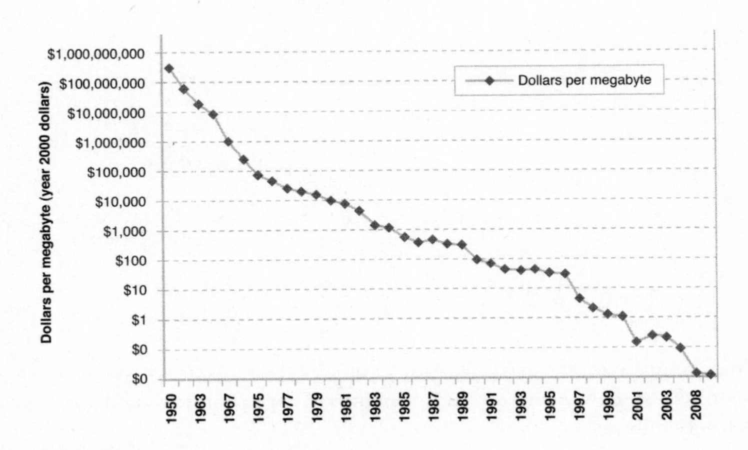

Source: Kurzweil, *The Singularity Is Near.*

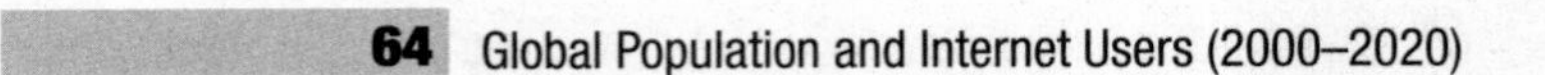

64 Global Population and Internet Users (2000–2020)

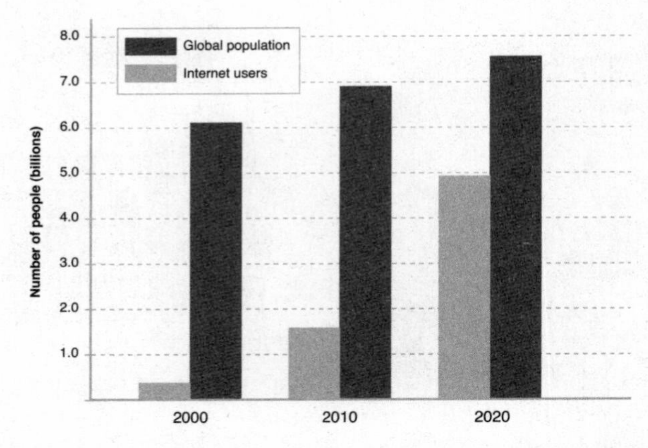

Source: http://www.futuretimeline.net/21stcentury/2020-2029.htm#ref3.

65 Hours of YouTube Video Uploaded per Minute

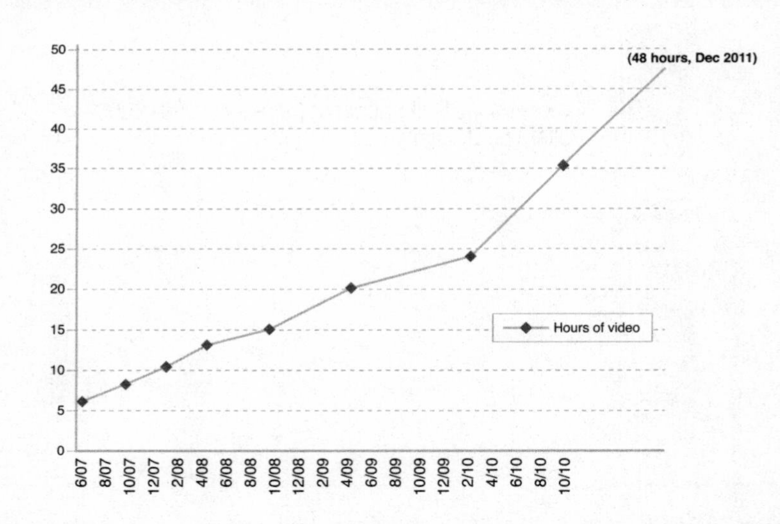

Nothing demonstrates the explosive growth of digital data better than the rise in content on YouTube. By the end of 2011, 48 hours of video content will be uploaded to the site every minute.

Sources: http://www.youtube.com/t/press_statistics; http://youtube-global.blogspot.com/2010/11/great-scott-over-35-hours-of-video.html; http://youtube-global.blogspot.com/2011/05/thanks-youtube-community-for-two-big.html.

66 Mobile Cellular Subscription Growth 2000–2010

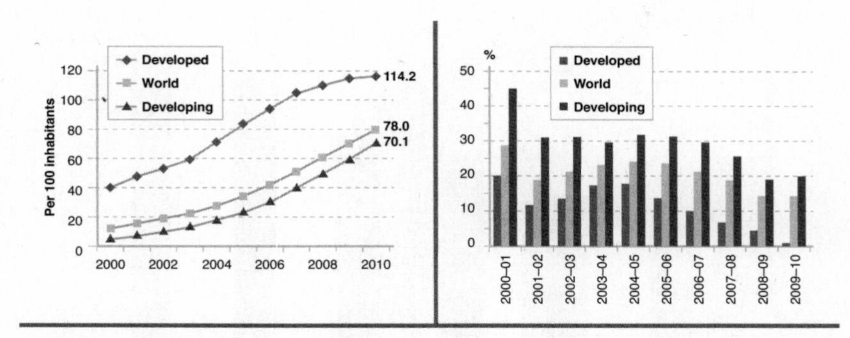

The graph on the left shows the rapid growth of mobile/cellular subscriptions in the developed and developing world. In the developed world, a number greater than 100 indicates individuals have more than one handset. The graph on the right shows annual growth rate over time.

Sources: http://www.itu.int/ITU-D/ict/publications/idi/2011/Material/MIS_2011_without_annex_5.pdf; http://www.itu.int/ITU-D/ict/publications/idi/2010/Material/MIS_2010_without_annex_4-e.pdf.

67 Mobile Broadband Subscription Penetration and Growth (2007–2010) by Level of Development

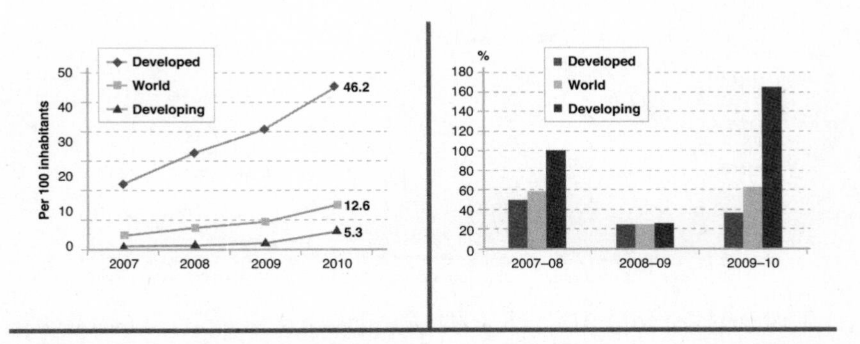

These charts specifically show the growth in wireless-broadband Internet access, rather than cellular phones. The single, most dynamic ICT development over the past year has been the surge in mobile broadband subscriptions.

Source: http://www.itu.int/ITU-D/ict/publications/idi/2011/Material/MIS_2011_without_annex_5.pdf.

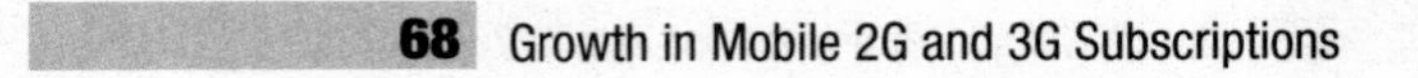

68 Growth in Mobile 2G and 3G Subscriptions

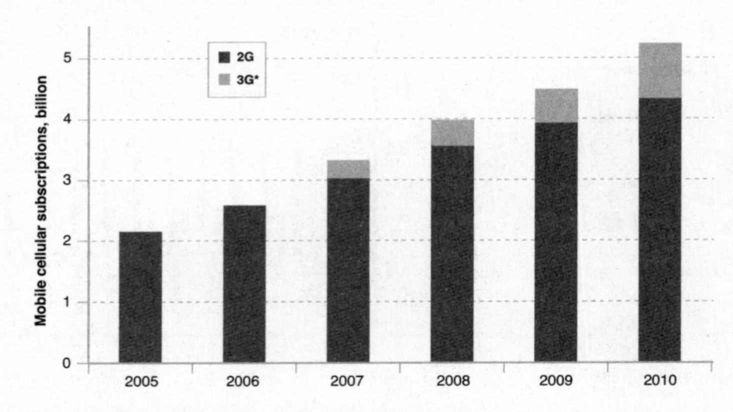

Source: http://www.itu.int/ITU-D/ict/publications/idi/2011/Material/MIS_2011_without_annex_5.pdf.

69 Total International Internet Bandwidth (Gbits/sec): 2000–2010

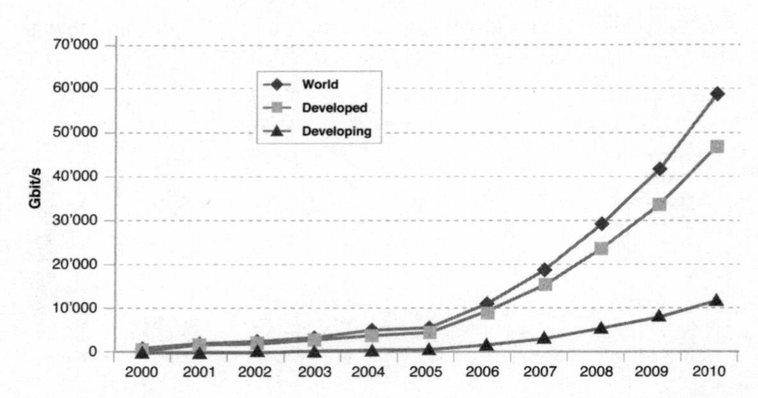

Between 2008 and 2010, Africa has made great progress in international Internet connectivity. Many countries have doubled or tripled their international bandwidth capacity; some have witnessed a tenfold increase. If accompanied by effective policy measures that ensure competitive access to the newly available bandwidth, this increase may have a positive impact on broadband affordability—one of the major issues in the region.

Sources: http://www.itu.int/ITU-D/ict/publications/idi/2011/Material/MIS_2011_without_annex_5.pdf; http://www.itu.int/ITU-D/ict/publications/idi/2010/Material/MIS_2010_without_annex_4-e.pdf.

70 2G and 3G Mobile Phone penetration in Africa 2011–2015

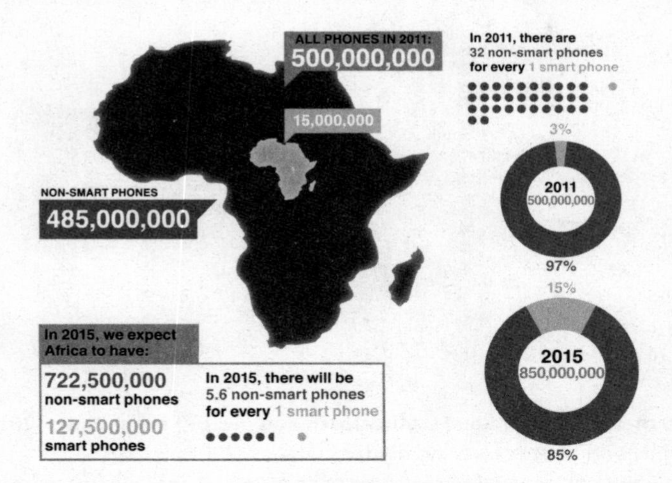

Projected five-year growth of 2G and 3G mobile devices. Note the disproportionately high rate of growth for smart phones.

Source: http://afrographique.tumblr.com/post/7087562485/infographic-depicting-smart-and-dumb-mobile.

71 Internet Users (2005–2010) and per 100 Inhabitants (2010)

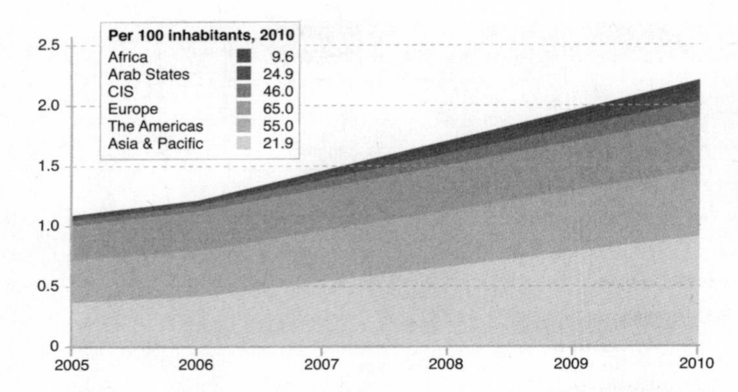

ITU: Internet users in billions (2005–2010) and per 100 Inhabitants (2010)

- The number of Internet users has doubled between 2005 and 2010.
- In 2010, the number of Internet users will surpass the 2 billion mark, of which 1.2 billion will be in developing countries.
- A number of countries (Estonia, Finland, and Spain) have declared access to the Internet as a legal right for citizens.
- With more than 420 million Internet users, China is the largest Internet market in the world.
- While 71 percent of the population in developed countries are online, only 21 percent of the population in developing countries are online. By the end of 2010, Internet user penetration in Africa will reach 9.6 percent, far behind both the world average (30 percent) and the developing country average (21 percent).

Source: http://www.itu.int/ITU-D/ict/material/FactsFigures2010.pdf.

72 Number of Mobile-Only Internet Users

	2010	2011	2012	2013	2014	2015
Global	13,976,859	31,860,295	78,855,662	188,375,368	487,426,725	788,324,804
Asia Pacific	2,448,932	6,768,196	20,543,294	67,012,433	240,350,642	420,277,951
Latin America	1,329,853	4,040,217	12,720,259	26,665,349	49,199,321	71,548,055
North America	2,615,787	4,218,310	6,550,322	14,257,565	38,783,886	55,646,710
Western Europe	5,237,113	10,348,319	21,163,143	33,524,429	58,670,609	83,364,841
Japan	441,060	1,021,441	3,322,664	10,780,236	21,462,108	31,876,998
Central and Eastern Europe	1,156,893	3,140,746	8,252,679	20,303,462	38,480,441	58,717,045
Middle East and Africa	747,221	2,323,065	6,303,302	15,831,895	40,479,719	66,893,204

These tables provide details on the projected growth of mobile-only Internet users, meaning those accessing the net over a smart phone.
Source: Cisco VNI Mobile, 2011.

Philanthropy

73 Concentration of High-Net-Worth Individuals per 1,000 people, 2010

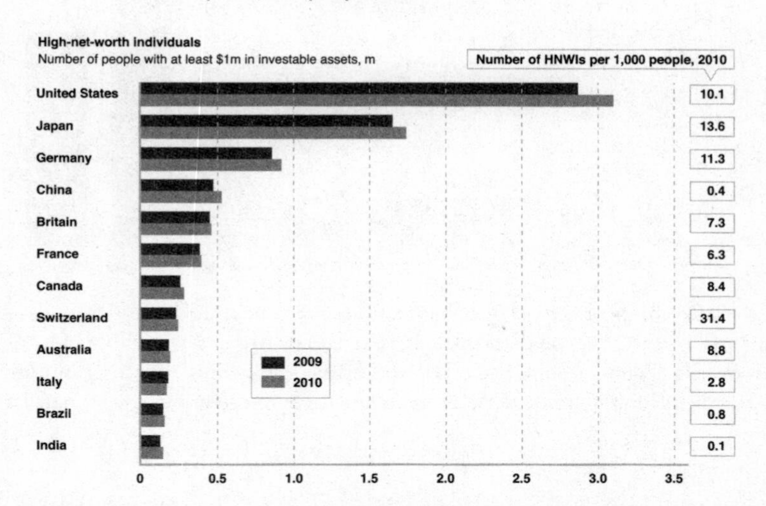

Sources: http://www.economist.com/blogs/dailychart/2011/06/rich from http://www.capgemini.com/services-and-solutions/by-industry/financial-services/solutions/wealth/worldwealthreport.

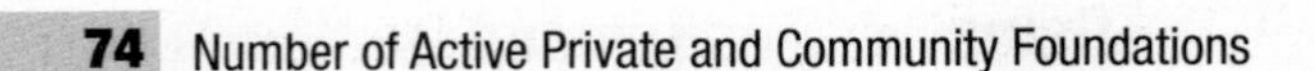

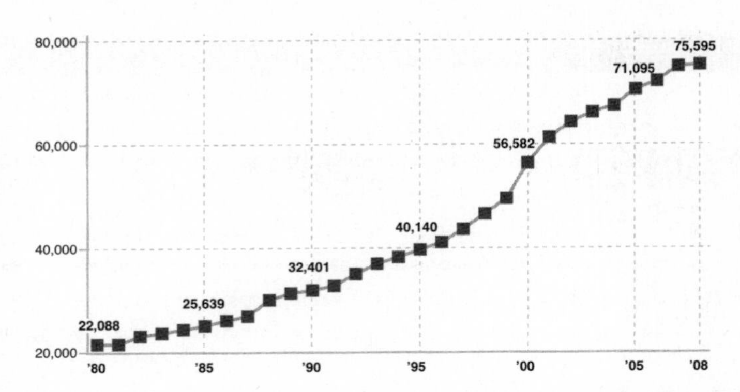

The number of active foundations has almost quadrupled in the past two decades. **Sources:** US Foundation Center (2010), http://foundationcenter.org/findfunders/statistics; http://foundationcenter.org/gainknowledge/research/pdf/fgge10.pdf.

75 Number of Active Private and Community Foundations

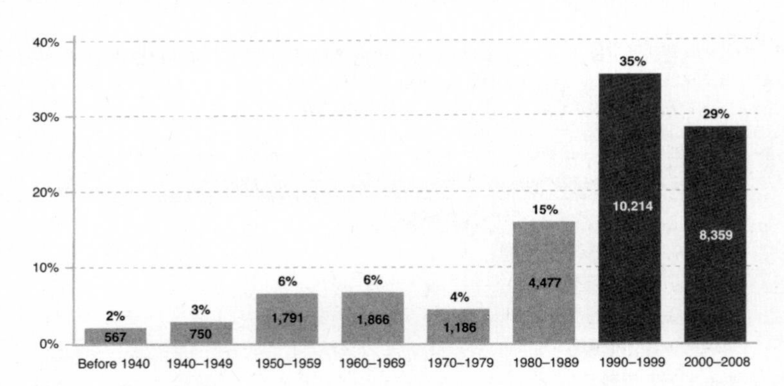

Close to two-thirds of active larger foundations were established after 1989. Based on Foundation Center data on grant-making foundations with assets of at least $1 million. **Sources:** US Foundation Center (2010), http://foundationcenter.org/findfunders/statistics; http://foundationcenter.org/gainknowledge/research/pdf/fgge10.pdf.

Dematerialization and Demonetization

76 Dematerialization

>$900,000 worth of applications in a smart phone today

	Application	$ (2011)	Original Device Name	Year*	MSRP	2011's $
1	Video conferencing	*free*	Compression Labs VC	1982	$250,000	$586,904
2	GPS	*free*	TI NAVSTAR	1982	$119,900	$279,366
3	Digital voice recorder	*free*	SONY PCM	1978	$2,500	$8,687
4	Digital watch	*free*	Seiko 35SQ Astron	1969	$1,250	$7,716
5	5 Mpixel camera	*free*	Canon RC-701	1986	$3,000	$6,201
6	Medical library	*free*	e.g. CONSULTANT	1987	Up to $2,000	$3,988
7	Video player	*free*	Toshiba V-8000	1981	$1,245	$3,103
8	Video camera	*free*	RCA CC010	1981	$1,050	$2,617
9	Music player	*free*	Sony CDP-101 CD player	1982	$900	$2,113
10	Encyclopedia	*free*	Compton's CD Encyclopedia	1989	$750	$1,370
11	Videogame console	*free*	Atari 2600	1977	$199	$744
	Total	**free**				**$902,065**

*Year of Launch

People with a smart phone today can access tools that would have cost thousands a few decades ago.

Sources: (1) http://www.nefsis.com/Best-Video-Conferencing-Software/video-conferencing-history.html
(2) http://www.americanhistory.si.edu/collections/surveying/object.cfm?recordnumber=998407
(3) http://www.videointerchange.com/audio_history.htm
(4) http://www.shvoong.com/humanities/1714780-history-digital-watch
(5) http://www.digicamhistory.com/1986.html
(6) http://www.tnyurl.com/63ljueq
(7) http://www.mrbetamax.com/OtherGuys.htm
(8) http://www.cedmagic.com/museum/press/release-1981-02-12-1.html
(9) http://www.digicamhistory.com/1980_1983.html
(10) http://www.mba.tuck.dartmouth.edu/pdf/2000-2-0007.pdf
(11) http://www.thegameconsole.com/atari-2600/

77 iPad 2 as Fast as a Supercomputer in 1985

	Cray 2 (1985)	iPad2 (2011)	Difference (iPad vs. Cray 2)
Weight	5,500 pounds	610-613 g / 1.33-1.35 pounds (Wi-Fi+3G)	1/4000th the weight
Size	45 inches high, 53 inches diameter 99318 inches^3	9.5x7.31x.34 inches 23.6 inches^3	1/4000th the volume
Cost	$17.5M (1985) $36.2M (2011)	$699 (64 GB, 2011) $338 (1985)	1/51,775 the cost
Processing power (CPU)	244 MHz	1 GHz	4 times the processing speed
Memory	2 GB RAM	512 MB DDR2	1/4 the memory
Power (watts)	150-200 kW	10 Watts	1/15,000 the power

Sources: http://bits.blogs.nytimes.com/2011/05/09/the-ipad-in-your-hand-as-fast-as-a-supercomputer-of-yore; http://archive.computerhistory.org/resources/text/Cray/Cray.Cray2.1985.102646185.pdf; http://en.wikipedia.org/wiki/Cray-3; 2 GB; RAM; http://www.cs.umass.edu/~weems/CmpSci635A/Lecture16/L16.16.html15,000; http://books.google.com/books?id=LkrTkAa10McC&pg=PA61-IA8; Cray 2 Brochure; http://www.craysupercomputers.com/downloads/Cray2/Cray2_Brochure001.pdf.

78 iPhone (2007) vs. Osborne Executive (1982)

	Osborne Executive (1982)	iPhone	Difference
Weight	12.9 kg / 28.34pounds	135 g / 4.8 oz / .3 lbs	95.5 times less weight
Size	9x20.5x13 inches 2430 inches^3 23x52x33 cm 39470 cm^3	4.5x2.4x.46 inches 5 inches^3 11.5x6.1x1.16 cm 81 cm^3	Nearly 1/500th the volume
Cost	$2,495 (1982) $5,759 (2011)	$599/$399 (8 GB, 2007) $279/$186 (1982)	10–14 times less (constant dollars)
Processing power (CPU)	4.0Mhz	620 Mhz	155 times more
Storage	Up to 720 KB	Up to 8 GB Flash (2007)	11,650 times more
Memory	Up to 384 kibibytes	128 MB eDRAM	341 times more
Display	80 char x 24 lines, monochrome	320 x 480 18-bit LCD	Not applicable (OE is not pixel-based.)
Camera and video	NA	2.0 megapixel camera	Not applicable
Software	Multiple types, on floppy disks	Multiple to start	Not applicable
Communications	300 baud modem (0.3kbits/s)	Wi-Fi (802.11b/g, 11Mb/s), Bluetooth, GPS	26,666 times more

Sources: http://www.computermuseum.li/Testpage/OsborneExecSpecs.htm; http://en.wikipedia.org/wiki/Osborne_Executive; http://en.wikipedia.org/wiki/IPhone.

Exponential Curves

79 Exponential vs. Linear Curves

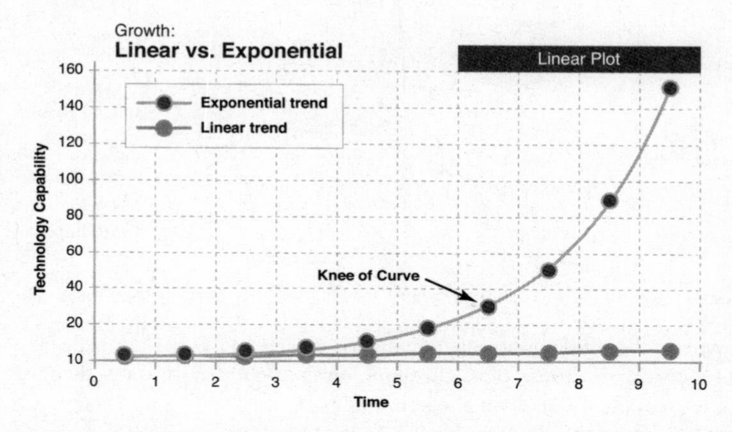

This graph shows the fundamental difference between exponential growth and linear growth. In the early period of exponential doublings, before the front edge of the curve is reached, exponential and linear growth are difficult to distinguish.

Source: Ray Kurzweil, *The Singularity Is Near*.

80 Exponential Curves

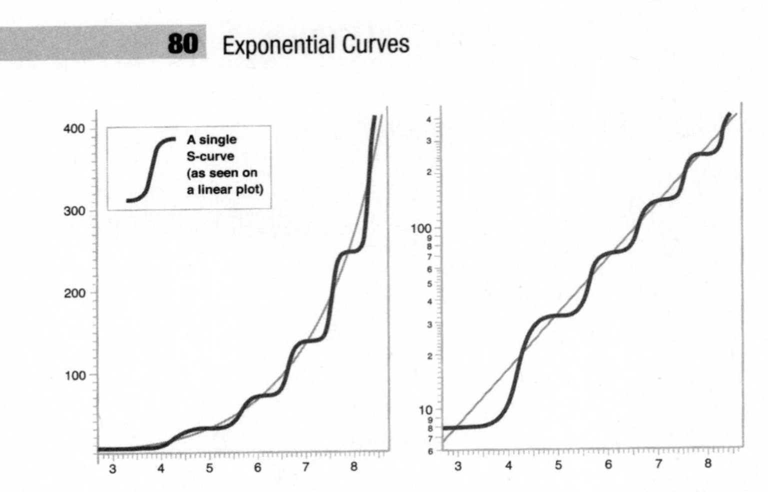

The graph on the left demonstrates an ongoing exponential sequence made up of a cascade of S-curves on a linear plot. The graph on the right demonstrates the same exponential sequence of S-curves on a logarithmic plot.
Source: Ray Kurzweil, *The Singularity Is Near.*

81 The Gartner Hype Curve

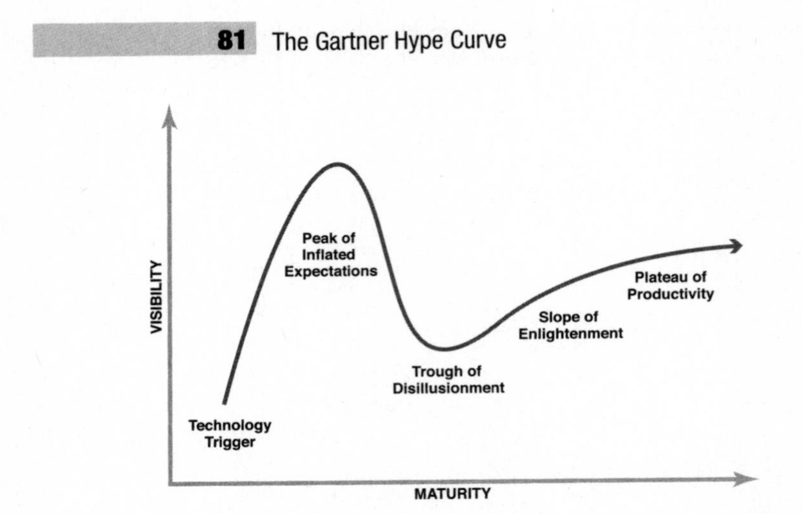

Each hype cycle drills down into the five key phases of a technology's life cycle. Early in the life cycle of a new technology there is an overestimation of the technology's potential, which leads to the peak of inflated expectations, followed by a dismissal of its abilities and the trough of disillusionment, and ultimately by the technology's true fulfillment and its plateau of productivity.
Source: http://www.gartner.com/it/page.jsp?id=1124212.

APPENDIX

DANGERS OF THE EXPONENTIALS

Why the Future Doesn't Need Us

One of the first well-constructed examinations of the dangers of exponential technology appeared in April 2000 in *Wired*, when Bill Joy (then chief scientist at Sun Microsystems) wrote his now famous article "Why the Future Doesn't Need Us." Joy's argument is that the most powerful twenty-first-century technologies—robotics, nanotech, and genetic engineering—all threaten the human species, leaving us only one clear course of action:

> The experiences of the atomic scientists clearly show the need to take personal responsibility, the danger that things will move too fast, and the way in which a process can take on a life of its own. We can, as they did, create insurmountable problems in almost no time flat. We must do more thinking up front if we are not to be similarly surprised and shocked by the consequences of our inventions . . . We are being propelled into this new century with no plan, no control, no brakes.
>
> . . . The only realistic alternative I see is relinquishment: to limit development of the technologies that are too dangerous, by limiting our pursuit of certain kinds of knowledge.

While I disagree with Joy's prescription (for reasons we'll get to), he's not wrong in his appraisal. Exponential technologies can pose grave dangers. Although those dangers are not the focus of this work, it would be a significant oversight to pass them by without discussion. This, then, is the portion of the text devoted to examining those issues. I will warn you in advance that the discussion of these threats and potential mitigating factors put forward here are woefully inadequate, given the importance of the subject. My goal is simply to make you aware of the major concerns and challenges, and provide a macroscopic overview to stimulate your further reading.

Imagining these dangers isn't hard, as Hollywood has already done much of the heavy lifting. Films like *I, Robot, The Terminator*, and *The Matrix* are classic stories of evil, intelligent robots dominating humanity, while *Blade Runner, Gattaca*, and *Jurassic Park* focus on the downside of genetic manipulation. Nanotech, it seems, is slightly less cinematic, and shows up only in the 2008 remake of *The Day the Earth Stood Still*. But the film gives us a fairly accurate version of Eric Drexler's "grey goo" scenario, wherein self-replicating nanobots get free and consume everything in their path. While it is true that Hollywood has played fast and loose with the facts, it does a pretty fair job in assessing the dangers. Simply put: the wrong technology in the wrong hands leads nowhere good.

Every year at SU, I lead a series of workshops discussing this topic. In these sessions, we try to list and prioritize near-term and medium-term doomsday scenarios. Three near-term concerns consistently rise to the top and are therefore going to be our focus here: the fear of biotechnology in the hands of terrorists; the continued rise of cyber crime; and the loss of jobs resulting from advances in robotics and AI. We'll take them one at a time.

Bioterrorism

Earlier in this book, I described how high school and college students participating in today's International Genetically Engineered Machine (iGEM) competition are using genetic engineering to manipulate simple life forms to do useful or interesting things. For example, previous competition winners have built life forms that blink fluorescent green, consume oil spills, or manufacture ulcer-preventing vaccines. But that's where we are today. Tomorrow is quite a different story.

"There is a new generation of biohackers coming online who will use genetic engineering to start amazing companies," says Andrew Hessel, cochair of the Biotechnology at Singularity University and an eloquent advocate for today's DIY-bio movement. "At the same time, however, as the technology becomes easier to use and cheaper to access, biological attacks and hacks are inevitable."

And the technology is already cheap enough. DNA sequencing and synthesizing machines are available to anyone who can afford a used car. This might be fine, save for the fact that some pretty nasty nucleotide sequences such as the Ebola virus and the 1918 influenza (which killed over fifty million worldwide) are accessible on line. British cosmologist and astronomer royal Lord Martin Rees thinks the danger so grave that in 2002 he placed a $1,000 bet with *Wired* magazine that "by the year 2020, an instance of bio-error or bio-terror will have killed a million people."

Rees and Hessel have every right to sound the alarm. Dr. Larry Brilliant—who helped lead the WHO team that successfully eradicated smallpox and now runs Jeff Skoll's Urgent Threats Fund (which focuses, among other things, on pandemics and bioterrorism)—summed up everyone's fears in a recent article for the *Wall Street Journal*: "Genetic engineering of viruses is much less complex and far less expensive than sequencing human DNA. Bioterror weapons are cheap and do not need huge labs or government support. They are the poor man's WMD."

And terrorists won't even have to actually create the virus to cause the damage. "The widespread media frenzy around the H1N1 [flu virus] in 2009 panicked the public and saw pharmaceutical companies waste billions to make vaccines that were ultimately ineffective," explains Hessel. "Fear and ignorance of biological agents can lead to reactive and disruptive societal responses with real-world consequences, even if the agent itself isn't that harmful." In effect, just the threat of a biological attack can be severely damaging, producing negative economic, societal, and psychological impact.

One instinctive reaction to this threat has been a call for more regulation on the distribution of technology and reagents, but there's little proof that such measures will have the desired effect. The first problem is that banning anything tends to create a black market and a criminal workforce dedicated to exploiting that market. In 1919, when America made the manufacture, sale, and transportation of intoxicating liquors illegal, organized crime was the main result. Prison populations soared by 366 percent; total expenditures on penal institutions jumped 1,000 percent; even drunk driving went up by 88 percent. All told, as John D. Rockefeller Jr. (once a vocal proponent of the idea) pointed out: "[D]rinking has generally increased; the speakeasy has replaced the saloon; a vast army of lawbreakers has appeared; many of our best citizens have openly ignored Prohibition; respect for the law has been greatly lessened; and crime has increased to a level never seen before."

Currently beyond those drugs that increase athletic performance, there isn't much of a black market for biologicals. Stricter regulation would change that in a hurry. It would also create a brain drain, as researchers interested in these areas would move to places were the work wasn't illegal—something we already saw with stem cells. Moreover, there are serious economic considerations. Regulation hurts small businesses most, and it is small businesses that make most economies run. Industrial biotech is a rapidly growing market sector, but that will taper off if we start hamstringing these operations with too many rules—and this decline would hurt more than just our wallets.

"Our greatest resource to combat emerging natural and artificial biological threats is an open and broadly distributed technological capability," writes synthetic biology pioneer Rob Carlson in a recent overview of the field: "Synthetic Biology 101." "Regulation that is demonstrably ineffective in improving security could easily end up stifling the technological innovation required to improve security. And make no mistake: we desperately require new technologies to provide for an adequate bio-defense capability."

Beyond this dark prognosis, a few bright spots are beginning to emerge. For starters, viruses spread only at the speed of human travel—going from infected host to the soon-to-be-infected target. Simulations show that a pandemic, even in a local region, can take months to peak. Meanwhile, warnings and news can spread at the speed of Twitter, Facebook, and CNN. Already systems like Google Flu Trends monitor search data for terms like "flu," "coughing," "influenza," and so on, and can identify early outbreaks. In the near future, Lab-on-a-Chip technologies, which can be used to detect, sequence, and effectively serve as a pandemic early-warning system, will feed data to organizations such as the Centers for Disease Control.

"If regional facilities are put in place to rapidly manufacture and distribute vaccines and antiviral drugs in towns and cities worldwide," continues Hessel, "then we can imagine providing an effective treatment along the same lines that Norton Antivirus broadcasts an update to protect our computers at home."

Work on exactly these sorts of facilities is already starting. In May 2011 the UCLA School of Public Health launched a state-of-the-art, $32 million, high-speed, high-volume automated laboratory designed to be the next weapon against bioterrorism and infectious diseases. This global biolab is designed to test high volumes of deadly agents very quickly. "For example," says UCLA School of Public Health dean Linda Rosenstock, "to find out where an agent came from. Did it originate in Mexico? Did it start in Asia? How's it changing over time? How might we develop a vaccine to protect against it? Really, the possibilities are endless."

This is only a piece in what will have to be a much larger puzzle. Larry Brilliant imagines a scenario wherein air filters in major public facilities such as airports and concert halls will be attached to biological monitoring systems. Sneeze in a restroom at Yankee Stadium, and the system will automatically analyze your germs for known and unknown pathogens. Making Brilliant's idea that much more feasible, in August 2011, researchers at MIT's Lincoln Laboratory invented a new kind of biosensor that can detect airborne pathogens like anthrax, plague, and smallpox in less than three minutes—a vast improvement over previous efforts.

Despite such progress, a thoroughly robust pathogen-monitoring system will take a few years, maybe even a few decades. In the meantime, another important defense against biological attack may be the telltale electronic droppings that a would-be terrorist generates in his efforts to acquire equipment, supplies, and information. For this reason, the loss of privacy arising from social media and web searches may turn out, ironically, to be a major protector of our freedom and health.

The fact remains that any new technology carries a novel risk. Mostly we live with these trade-offs. The automobile kills about forty thousand Americans a year, while dumping one and a half billion tons of CO_2 into the atmosphere, but we have little inclination to ban these machines. The most potent painkillers we've developed have both saved lives and ended lives. Even something as straightforward as processed sugar is a double-edged sword, giving us a brilliant array of new foods yet contributing to a bevy of killer diseases. As comic book artist Stan Lee pointed out so many years ago in the first issue of *Spider-Man*: "With great power there must also come great responsibility." One thing of which we are certain: biotechnology is a very great power.

Cyber Crime

Marc Goodman is a cyber crime specialist with an impressive résumé. He has worked with the Los Angeles Police Department, Interpol, NATO, and the State Department. He is the chief cyber criminologist at the Cybercrime Research Institute, founder of the Future Crime Institute, and now head of the policy, law, and ethics track at SU. When breaking down this threat, Goodman sees four main categories of concern.

The first issue is personal. "In many nations," he says, "humanity is fully dependent on the Internet. Attacks against banks could destroy all records. Someone's life savings could vanish in an instant. Hacking into hospitals could cost hundreds of lives if blood types were changed. And there are already 60,000 implantable medical devices connected to the Internet. As the integration of biology and information technology proceeds, pacemakers, cochlear implants, diabetic pumps, and so on, will all become the target of cyber attacks."

Equally alarming are threats against physical infrastructures that are now hooked up to the net and vulnerable to hackers (as was recently demonstrated with Iran's Stuxnet incident), among them bridges, tunnels, air traffic control, and energy pipelines. We are heavily dependent on these systems, but Goodman feels that the technology being employed to manage them is no longer up to date, and the entire network is riddled with security threats.

Robots are the next issue. In the not-too-distant future, these machines will be both commonplace and connected to the Internet. They will have superior strength and speed and may even be armed (as is the case with today's military robots). But their Internet connection makes them vulnerable to attack, and very few security procedures have been implemented to prevent such incidents.

Goodman's last area of concern is that technology is constantly coming between us and reality. "We believe what the computer tells us," says Goodman. "We read our email through computer screens; we speak to friends and family on Facebook; doctors administer medicines based upon what a computer tells them the medical lab results are; traffic tickets are issued based upon what cameras tell us a license plate says; we pay for items at stores based upon a total provided by a computer; we elect governments as a result of electronic voting systems. But the problem with all this intermediated life is that it can be spoofed. It's really easy to falsify what is seen on our computer screens. The more we disconnect from the physical and drive toward the digital, the more we lose the ability to tell the real from the fake. Ultimately, bad actors (whether criminals, terrorists, or rogue governments) will have the ability to exploit this trust."

While we have not yet discovered any silver bullet solutions, Goodman does believe there are a few steps that would greatly reduce our peril. The first is better technology and more responsibility. "It's insane that we allow developers to release crappy software," he says. "We're making life hard on consumers and easy on criminals. We have to accept the fact that in today's world, our lives depend on software, and to allow companies to release products riddled with security flaws in today's climate doesn't make any sense."

The next issue is how we handle the security flaws that still make it through. Right now the responsibility for patching old code is left up to the consumer, but people don't get around to it as often as they should. Goodman explains: "Ninety-five percent of all hacks exploit old security flaws—flaws for which patches already exist. We need software that automatically updates itself, plugs holes, and thwarts hackers. You have to automate this stuff, put the responsibility on the developer and not the consumer."

Goodman also feels that it's time to start considering some type of global liability law that covers software security. To this end, on September 9, 2011, Connecticut Democrat Richard Blumenthal introduced the Personal Data Protection and Breach Accountability Act in the Senate. This would enable the US Justice Department to fine companies with more than ten thousand customers $5,000 per day (for a maximum of $20 million in violations) for lax security. If the bill passes, standards would be set, and businesses would be required to test their security systems on a regular basis—although who performs the testing and how, and who owns the resulting data, remain thorny concerns.

An international net-based police force able to operate across borders in the same way that the Internet enables criminals to operate across borders is Goodman's last suggestion. "The Internet has made the world a borderless place," he says, "but all of our law-enforcement agencies are trapped in the old world—the one where borders still matter a great deal. This makes it almost impossible for law enforcement to deal with cyber criminals. I don't think we'll ever completely defeat cyber crime, but if the playing field remains this uneven, we don't even have a fighting chance."

Goodman is aware that this proposal make many uneasy. "Everybody's main concern is a cop from El Salvador being able to arrest people in Switzerland, but if you made it a net-based policing mechanism (and left arrests up to home country officers), you could sidestep this issue. Certainly there's still a lot of international law to consider—spewing Nazi propaganda, for example, is free speech in America and illegal in Germany—but we live in a globally connected world. These problems are going to keep on coming up. Isn't it time we get ahead of the curve?"

Robotics, AI, and the Unemployment Line

There are some curves we might not be able to get ahead of. It won't be long now before robots make up the majority of the blue-collar workforce. Whether it's shelf-stocking robots maintaining inventory at Costco or burger-slinging robots serving lunch at McDonald's, we're less than a decade away from their arrival. Afterward, humans are going to have a hard time competing. These robots work 24/7, and they don't get sick, make mistakes, or go on strike. They never get too drunk on Friday night to come to work Saturday morning, and—bad news for the drug-testing industry—have no interest in mind-altering substances. Certainly there will be companies that continue to employee humans out of principle or charity, but it's hard to envision a scenario where they remain cost competitive for long. So what becomes of these millions of blue-collar workers?

No one is entirely certain, although it's helpful to remember that this isn't the first time automation changed the employment landscape. In 1862, 90 percent of our workforce were farmers. By the 1930s, the number was 21 percent. Today it's less than 2 percent. So what happened to the farm jobs that were displaced by automation? Nothing fancy. The old low-skill jobs were replaced by new higher-skilled jobs, and the workforce was trained to fill them. This is the way of progress. In a world of ever-increasing specialization, we are constantly creating anew. "At a high level," says Second Life creator Philip Rosedale, "humans have consistently demonstrated an ability to find new things to do that are of greater value when jobs have been outsourced or automated. The industrial revolution, outsourced IT work, China's low-cost labor force all ultimately created more interesting new jobs than they displaced."

Vivek Wadhwa, director of research at the Center for Entrepreneurship at Duke University, agrees. "Jobs that can be automated are always at risk. Society's challenge is to keep moving up the ladder, into higher realms. We need to create new jobs that use human creativity rather than human labor. I admit that it's difficult to conceive of the jobs of the future because we have no idea what technology will emerge and change the world. I doubt anyone could have predicted two decades ago that countries like India would go from being seen as lands of beggars and snake charmers to an employment threat for the developed world. Americans no longer tell their children to think about starving Indians before wasting the food on their plates, they tell them to study math and science or the Indians will take their white-collar jobs away."

In addition to training up, others might simply retire. SU AI expert Neil Jacobstein explains, "Exponential technologies may eventually permit people to not need jobs to have a high standard of living. People will have many choices with how they utilize their time and develop a sense of self-esteem—ranging from leisure normally associated with retirement, to art, music, or even restoring the environment. The emphasis will be less on making money and more on making contributions, or at least creating an interesting life."

This may seem a fairly future-forward opinion, but in a 2011 special report for CNN, media specialist Douglas Rushkoff argued that this transition is already under way:

> I understand we all want paychecks—or at least money. We want food, shelter, clothing, and all the things that money buys us. But do we all really want jobs?
>
> We're living in an economy where productivity is no longer the goal, employment is. That's because, on a very fundamental level, we have pretty much everything we need. America is productive enough that it could probably shelter, feed, educate, and even provide health care for its entire population with just a fraction of us actually working.
>
> According to the UN Food and Agriculture Organization, there is enough food produced to provide everyone in the world with 2,720 kilocalories per person per day. And that's even after America disposes of thou-

> sands of tons of crop and dairy just to keep market prices high. Meanwhile, American banks overloaded with foreclosed properties are demolishing vacant dwellings to get the empty houses off their books.
>
> Our problem is not that we don't have enough stuff—it's that we don't have enough ways for people to work and prove that they deserve this stuff.

Part of the problem is that most contemporary thinking about money and markets and such has its roots in the scarcity model. In fact, one of the most commonly used definitions of economics is "the study of how people make choices under conditions of scarcity, and the results of those choices for society." As traditional economics (which believes that markets are equilibrium systems) gets replaced by complexity economics (which both fits the data significantly better and believes that markets are complex, adaptive systems), we may begin to uncover a postscarcity framework for assessment, but there's no guarantee that such thinking will result in either more jobs or a different resource allocation system.

And this is merely where we are today. The bigger question is what happens once strong AI, ubiquitous robotics, and the Internet of everything—a combination that many feel will be able to handle every job in every market—comes online? Strong AI brings the possibility of computers with intelligence superior to that of humans, meaning that even the creative jobs that remains for us humans may soon be in jeopardy. "When you look at the possibility of us creating beings more intelligent than we are," says Philip Rosedale, "there is a fear that if we are enslaved by our descendant machines, we will be forced to do things we like less than what we are doing now, but it seems hard to imagine exactly what those things would be. In an age of abundance, where we are increasingly exploiting cheaper and cheaper ways of creating and modeling the world around us (virtual reality or nanotech, for example), is there really anything we could do to help the machines, even if we are left behind as their ancestors? I would suggest that the most likely outcome is that even if we are faced with smarter machines than us being a part of our lives, we may exist on two sides of a sort of digital IQ divide, with our lives being relatively unaffected."

So what's left for the humans? I see two clear possibilities. In one future, society takes a turn for the Luddite. We take Bill Joy's advice, follow the designs of the slow food movement, and begin to backtrack with the Amish. But this option will work only for those willing to forgo the vast benefits afforded by all this technology. This desire for the "good old days" will be tempered by the realities of disease, ignorance, and missed opportunities.

In the second future, the majority of humanity will end up merging with technology, enhancing themselves both physically and cognitively. A lot of people recoil at the sound of this, but this transformation has been going on for eons. The act of writing, for example, is simply the act of using technology to outsource memory. Eyeglasses, contact lenses, artificial body parts (stretching from the wooden peg leg to Scott Summit's 3-D printed prosthetics), cosmetic implants, cochlear implants, the US Army's "super soldier" program, and a thousand more examples have only continued this trend. As AI and robotics guru Marvin Minsky writes in *Scientific American,* "In the past, we have tended to see ourselves as a final product of evolution, but our evolution has not ceased. Indeed, we are now evolving more rapidly, though not in the familiar, slow Darwinian way. It is time that we started to think about our new emerging."

Soon the vast majority of us will be augmented in one way or another, and this will thoroughly change the economic landscape. This newly enhanced self, plugged into the net, working in both virtual and physical worlds, will generate value for society in ways we cannot even imagine today. Right now four thousand people are making a living designing clothing for Second Life avatars, but the day is not far off when a great many of us will be using digital doppelgängers. So while four thousand people doesn't sound like much of a market, what happens when avatars are representing us at international conferences and major business meetings? How much money are we spending on virtual clothing and accessories then?

Unstoppable

Considering the issues explored in the past few sections, Bill Joy's suggestion "to limit development of the technologies that are too dangerous" doesn't sound so bad. But the tools of yesterday are not designed to meet the problems of tomorrow. Considering the gravity of these concerns and the continued march of technology, reigning in our imaginations seems the worst possible plan for survival. We're going to need those future tools to solve future problems if we're serious about future survival.

Moreover, putting the brakes on technology just won't work. As the Bush administration's ban on human embryonic stem cells bore out, attempting to silence technology in one place only drives it elsewhere. In an interview about the impact of that ban, Susan Fisher, a professor at the University of California, San Francisco, said recently, "Science is like a stream of water, because it finds its way. And now it has found its way outside the United States." All the Bush pronouncement did was outsource what was originally a domestic product to countries like Sweden, Israel, Finland, South Korea, and the United Kingdom. What did the White House ban achieve? Only a reduction in US scientific preeminence.

There are also psychological reasons why it's nearly impossible to stop the spread of technology—specifically, how do you squelch hope? Ever since we figured out how to make fire, technology has been how humans dream into the future. If 150,000 years of evolution is anything to go by, it's how we dream *up* the future. People have a fundamental desire to have a better life for themselves and their families; technology is often how we make that happen. Innovation is woven into the fabric of who we are. We can no more stomp it out than we could shut off our instinct to survive. As Matt Ridley concludes in the final pages of *The Rational Optimist,* "It will be hard to snuff out the flame of innovation, because it is such an evolutionary, bottom-up phenomenon in such a networked world. So long as human exchange and specialization are allowed to thrive somewhere, then culture evolves where leaders help it or hinder it, and the result is that prosperity spreads, technology progresses, poverty declines, disease retreats, fecundity falls, happiness increases, violence atrophies, freedom grows, knowledge flourishes, the environment improves, and wilderness expands."

Sure, there are always going to be a few holdouts (again, the Amish), but the vast majority of us are here for the ride. And, as should be clear by now, it's going to be quite a ride.

Notes

PART ONE PERSPECTIVE

CHAPTER ONE: OUR GRANDEST CHALLENGE

The Lesson of Aluminum

3 *Gaius Plinius Cecilius Secundus, known as Pliny the Elder*: There's a ton of information out there on Pliny. A good starting point is John Healy, *Pliny the Elder: Natural History, A Selection* (Penguin Classics, 1991).

3 *This shiny new metal was aluminum*: If all you're looking for is a brief history, try the International Aluminum Institute's website: www.world-aluminum.org/Home. If you want a very thorough look: Joseph William Richards, *Aluminum: Its History, Occurrence, Properties, Metallurgy, and Application, Including Its Alloys* (Nabu Press, 2010).

4 *To expand on this a bit, let's take a look at the planned city of Masdar*: Nicolai Ouroussoff, "In Arabian Desert, A Sustainable City Rises," *New York Times*, September 25, 2010.

5 *a conceptual foundation known as One Planet Living (OPL)*: www.oneplanetliving.org/index.html.

6 *The amount of solar energy that hits our atmosphere*: NASA is responsible for first calculating what is now known as the "Earth's Energy Budget." Its website covers the topic pretty thoroughly: http://earthobservatory.nasa.gov/Features/EnergyBalance.

The Limits to Growth

6 *British scholar Thomas Robert Malthus realized*: Thomas Malthus, Geoffrey Gilbert, *An Essay on the Principle of Population* (Oxford University Press, 2004).

6 *"The power of population is indefinitely greater"*: Ibid., chapter 7, p. 61.

7 *"Unlike the plagues of the dark ages"*: Dr. Martin Luther King Jr., May 5, 1966. The Planned Parenthood Federation of America gave Dr. King the Margaret Sanger Award. This quote comes from his acceptance speech.

7 Paul Ehrlich, *The Population Bomb* (Sierra Club-Ballantine, 1970).

7 *The Club of Rome*: For all things Club of Rome, try its website: www.clubofrome.org.

7 The Limits to Growth *became an instant classic*: Donella H. Meadows, *Limits to Growth* (Signet, 1972).

7 *One in four mammals now faces extinction*: Julie Eilperin, *Washington Post*, October 7, 2008.

7 *90 percent of the large fish are already gone*: Ransom A. Myers, Boris Worm, *Nature* 423 (May 15, 2001), pp. 280–83.

7 *Our aquifers are starting to dry up*: Mathew Power, "Peak Water," *Wired*, April 21, 2008.

7 *We're running out of oil*: Marion King Hubbert, "Nuclear Energy and the Fossil Fuels," Spring Meeting of the Southern District, American Petroleum Institute (June 1956). With a pdf available here: www.hubbertpeak.com/hubbert/1956/1956.pdf.

7 *running low on uranium*: For a good review paper on peak uranium, try www.theoil drum.com/node/5060.

7 *Even phosphorus . . . is in short supply*: Patrick Dery, Bart Anderson, "Peak Phosphorus," *Energy Bulletin*, August 13, 2007. Available at: www.energybulletin.net/node/33164.

7 *one child will die of hunger*: The United Nations's World Food Programme has a great overview at: www.wfp.org/hunger.

7 *another will be dead from thirst*: Water.org has a great overview here: water.org/learn-about-the-water-crisis/facts.

7 *Scientists who study the carrying capacity of the Earth*: Over sixty estimates of the Earth's carrying capacity have been conducted. For a great look at all the research: Joel E. Cohen, *How Many People Can the Earth Support?* (W. W. Norton & Company, 1996).

7 *Dr. Nina Fedoroff*: One Planet, BBC World Service, March 31, 2009.

8 *Nazis' eugenics program*: Susan Bachrach, "In the Name of Public Health—Nazi Racial Hygiene," *New England Journal of Medicine,* vol. 351 (July, 2004), pp. 417–20.

8 *India performed tubal ligations and vasectomies*: "The Indira Enigma," *Frontline,* May 11, 2001.

8 *China, meanwhile, has spent thirty years under a one-child-per-family policy*: Laura Fitzpatrick, "A Brief History of China's One-Child Policy," *Time,* July 27, 2009.

8 *According to Amnesty International*: "Women in China," Amnesty International, June 1995.

8 *the X PRIZE Foundation*: www.xprize.org; complete list of board of trustees available at: www.xprize.org/about/board-of-trustees; a complete list of our major Vision Circle Benefactors is available at: www.xprize.org/about/vision-circle.

The Possibility of Abundance

9 *Mobile phone penetration is growing exponentially*: Bob Tortora and Magali Rheault, "Mobile Phone Access Varies Widely in Sub-Saharan Africa," Gallup, September 16, 2011. Also: "Mobile Phone Penetration in Indonesia Triples in Five Years," *Nielsen Wire*, February 23, 2011; Jagdish Rebello, "India Cell Phone Penetration Reaches 97 Percent in 2014," *iSuppli*, September 22, 2010; "The World in 2010: The Rise of 3G," International Telecommunications Union, available: www.itu.int/ITU-D/ict/material/FactsFigures2010.pdf; Jenny C. Aker and Isaac M. Mbiti, "Mobile Phones and Economic Development in Africa," *Journal of Economic Perspectives* 24, no. 3 (summer 2010), pp. 207–32.

10 *Burt Rutan flew into space*: A lot's been written about Burt, and we go into much greater detail later in the book, but also see: Dan Linehan, *Burt Rutan's Race to Space: The Magician of the Mojave and His Flying Innovations* (Zenith Press, 2011).

10 *Craig Venter tied the mighty US government*: Jamie Shreeve, "The Blueprint of Life," *US News & World Report*, October 31, 2005.

CHAPTER TWO: BUILDING THE PYRAMID

The Trouble with Definitions

12 *The US Government defines poverty*: www.census.gov/hhs/www/poverty/about/overview/measure.html.

12 *in 2008 the World Bank revised its international poverty line*: Martin Ravallion, Shaohua Chen, Prem Sangraula, "Dollar a Day Revisited," *World Bank Policy Research Institute Working Paper No. 4620,* June 22, 2008.

12 *the US government claimed the 39.1 million individuals*: 2008 Health and Human Service Poverty Guide: http://aspe.hhs.gov/poverty/08poverty.shtml.

13 *Today 99 percent of Americans living below the poverty line have*: Matt Ridley, *The Rational Optimist* (HarperCollins Books, 2010), pp. 16–17.

A Practical Definition

13 *Feeding the hungry, providing access to clean water, ending indoor air pollution, and wiping out malaria*: World Health Organization, *The World Health Report 2004-Changing History*. WHO, 2004.

14 *Abraham Maslow's now-famous pyramid*: Edward Hoffman, *The Right to Be Human: A Biography of Abraham Maslow* (St. Martin's Press, 1988).

14 *Maslow created his "Hierarchy of Human Needs"*: A. H. Maslow, "A Theory of Human Motivation," *Psychological Review* 50, no. 4 (1943), pp. 370–96.

The Base of the Pyramid

15 *Having three to five liters of clean drinking water per person per day*: 2005 Dietary Guidelines Advisory Committee, *Nutrition and Your Health: Dietary Guidelines for Americans,* USDA & HHS, January 2005, available: www.heath.gov/dietary guidelines/dga2005/report/default.htm.

15 *2,000 calories or more of balanced and nutritious food*: The figure is based on FDA research into daily caloric intake. For a look at where the FDA got it: Marion Nestle, "Where Did the 2,000-Calorie Diet Idea Come From," *Foodpolitics.com,* August 3, 2011.

15 *Vitamin A removes the leading cause of preventable blindness*: World Health Organization, United Nations Children's Fund, VAVG Task Force, *Vitamin A Supplements* (World Health Organization, 1997).

15 *an additional twenty-five liters of water is necessary*: This is something of a complicated calculation, as there is very poor data about water use in developing countries. All of our data are based on work done by Peter Gleick at the Pacific Institute for Studies in Development, Environment, and Security, with a caveat. Gleick's calculations vary depending on what type of bathroom facilities are available (pit latrine, toilet, and so on). Ours ignore old technologies and are based on what happens once the waterless toilets discussed in chapter 8 come into use. See: Peter Gleick, "Basic water requirements for human activities: Meeting Basic Needs." *Water International* 21, no. 2 (1996), pp. 88–92.

15 *837 million people now live in slums*: Harvey Herr, Guenter Karl, "Estimating Global Slum Dwellers: Monitoring the Millennium Development Goal 7, Target 11," UN-HABITAT working paper, Nairobi, 2003, p. 19.

15 *Thomas Friedman's Flat World*: Thomas Friedman, *The World Is Flat* (Farrar, Straus & Giroux, 2005).

The Upside of Water

15 *a billion people lack*: water.org/learn-about-the-water-crisis/facts.

16 *According to the World Health Organization*: "Burden of Disease and Cost-Effectiveness Estimates," World Health Organization, available here: www.who.int/water_sanitation_health/disease/burden/en.

16 *access to a cell phone than a toilet*: water.org/learn-about-the-water-crisis/facts.

16 *Peter Gleick at the Pacific Institute*: Peter Gleick, *Dirty Water: Deaths from Water-Related Disease 2000–2020,* Pacific Institute Report, August 15, 2002.

16 *sub-Saharan Africa no longer loses the 5 percent*: Kevin Watkins, *Beyond Scarcity: Power, Poverty and the Global Water Crisis,* Human Development Report 2006, United Nations Development Programme, 2006, p. 6.

16 *until 1900, there was only one country with an infant mortality*: M. Abouharb and A. Kimball, "A New Dataset on Infant Mortality, 1816–2002," *Journal of Peace Research* 44, no. 6 (2007), pp. 745–56.

17 *As Microsoft cofounder Bill Gates*: This line comes in the Q&A following Gates's TED talk "Bill Gates on Mosquitoes, Malaria, and Education," February 2009. See: www.ted.com/index.php/talks/bill_gates_unplugged.html.

17 *Morocco, for example, is now a young nation*: Anthony Ham, *Lonely Planet Country Guide: Morocco* (Lonely Planet Publications, 2007), p. 47.

17 *John Oldfield*: personal interview with John Oldfield, 2010.

The Pursuit of Catallaxy

18 *Friedrich Hayek called catallaxy*: Friedrich A. Hayek, *Law, Legislation, and Liberty,* vol. 2 (University of Chicago Press, 1978), pp 108–9.

18 *"'If I sew you a hide tunic today'"*: Ridley, *The Rational Optimist*, p. 57.

18 *. . . the average household has five people living in a single room . . .* : "Social Statistics in Nigeria," National Bureau of Statistics, Federal Republic of Nigeria, 2009, pp. 23-26.

19 *average US household*: www.physics.uci.edu/~silverma/actions/HouseholdEnergy.html; US average: Total average household power usage is 6,000 KWh of electricity and an additional 12,000 KWh of gas equivalent. Looking at electricity alone, the average household consumes 16.4 KWh per day of electricity.

19 *3.5 billion people who . . . burning biomass*: World Health Organization, *Health and Environment in Sustainable Development: Five Years After the Earth Summit* (WHO, Geneva, 1997), Table 4.4, p. 87.

19 *According to a 2002 WHO report*: N. G. Bruce, R. Perez-Padilla, R. Albalak. *The Health Effects of Indoor Air Pollution Exposure in Developing Countries,* World Health Organization, 2002.

19 *2007 UN report found that 90 percent of all wood removals*: *UN FAO 2007 Forest Report*, fttp://ftp.fao.org/docrep/fao/009/a0773e/a0773e09.pdf. p. 27.

19 *Ecosystem services*: The term has been used by scientists for years, but it became widespread after the publication of: Millennium Ecosystem Assessment, *Ecosystems and Human Well-Being: Synthesis* (Island Press, 2005).

19 *the value of the ecosystems services: . . . a figure roughly equal to the entire annual global economy*: Robert Constanza, et al., "The Value of the World's Ecosystem Services and Natural Capital," *Nature* 387, May 15, 1997, pp. 253–60.

19 *$200 million experiment that was Biosphere 2*: Paul Hawken, "Natural Capitalism," *Mother Jones*, April 1997.

Reading, Writing, and Ready

20 *literacy, mathematics, life skills, and critical thinking*: Bernie Trilling, Charles Fadel, *21st Century Skills: Learning for Life in Our Times* (Jossey-Bass, 2009).

20 *Sir Ken Robinson*: "Ken Robinson Says Schools Kill Creativity," TED Talk, June 2006, See: www.ted.com/talks/ken_robinson_says_schools_kill_creativity.html.

20 *Most of today's educational systems are built*: Ibid.

21 *As Nicholas Negroponte*: personal interview, but also see: http://laptop.org/en/laptop/software/index.shtml.

Turning on the Data Tap

21 *a job placement service known as KAZI 560*: See: http://oneworldgroup.org/mobile4good-kazi560-kenya.

21 *In Zambia, farmers without bank accounts*: The Worldwatch Institute, *State of the World 2011: Innovations That Nourish the Planet*, Worldwatch Institute, 2011.

22 *Isis Nyong'o*: Paul Mason, "Kenya in Crisis," BBC, January 8, 2007.

22 *In 2001, 134 million Nigerians were sharing 500,000 landlines*: Jack Ewing, "Upwardly Mobile in Africa," *Bloomberg Businessweek*, September 13, 2007.

22 *When Nokia's profits hit $1 billion in 2009*: Nokia press release: http://press.nokia.fr/2005/09/21/nokia-introduces-nokia2652-fold-deseign-for-new-growth-markets-major-milestone-reached-one-billionth-nokia-mobile-phone-sold-this-summer.

The Peak of the Pyramid

22 *Acute respiratory infections are one of the leading causes*: World Health Organization, *The World Health Report 2005: Make every mother and child count,* WHO, Geneva, 2005.

23 *2010 report by PricewaterhouseCoopers*: PricewaterhouseCoopers LLP, "The Science of Personalized Medicine: Translating the Promise into Practice," PricewaterhouseCoopers, 2010.

Freedom

24 *Amartya Sen pointed out that political liberty moves in lockstep with sustainable development*: Amartya Sen, *Development as Freedom* (Anchor, 2000), pp. 14–16.

24 *Jurgen Habermas*: Jurgen Habermas, *The Structural Transformation of the Public Sphere: An Inquiry into a Category of Bourgeois Society* (MIT Press, 1991).

24 *The Webby Awards . . . put the so-called Twitter Revolution*: See: www.cnn.com/2009/TECH/11/18/top.internet.moments/index.html.

25 *"By using new media to extend horizontal linkages"*: Patrick Quirk, "Iran's Twitter Revolution," *Foreign Policy in Focus,* June 17, 2009.

25 *"Access to and the strategic use of ICTs"*: Association for Progressive Communications, "ICTs for Democracy: Information and Communication Technologies for the Enhancement of Democracy," Swedish International Development Cooperation Agency (SIDA), 2009. The quote itself comes from a summary article published by SIDA at: www.digitalopportunity.org/feature/democracy-and-icts-in-africa.

CHAPTER THREE: SEEING THE FOREST THROUGH THE TREES

Daniel Kahneman

27 *Kahneman was born Jewish . . . "people were endlessly complicated"*: Kahneman won a Nobel Prize in economics in 2002. This is from his autobiography, available here: www.nobelprize.org/nobel_prize/economics/laureates/2002/kahnman-autobiography.

28 *The Israelis had developed . . . the illusion of validity*: Ibid.

Cognitive Biases

29 *Heuristics are cognitive shortcuts*: Daniel Kahneman, Paul Slovic, Amos Tversky, *Judgment Under Uncertainty: Heuristics and Biases* (Cambridge University Press, 1982), pp. 4–5.

29 *Kahneman discovered . . . leads to . . . "severe and systematic errors"*: Amos Tversky, Daniel Kahneman, "Judgment Under Uncertainty," *Science* 185, no. 4157 (1974), pp. 1124–31.

30 *"President Obama's health care proposals would create government-sponsored 'death panels'"*: Jim Rutenberg and Jackie Calmes, *New York Times,* August 13, 2009.

30 *"When people believe the world's falling apart"*: personal interview with Daniel Kahneman, 2010.

31 *"psychological immune system"*: D. T. Gilbert, S. J. Blumberg, E. C. Pinel, T. D. Wilson, T. P. Wheatley, "Immune Neglect: A Source of Durability Bias in Affective Forecasting," *Journal of Personality and Social Psychology* 75, no. 3 (1998), pp. 617–38.

31 *Cornell University psychologist Thomas Gilovich*: personal interview.

If It Bleeds, It Leads

32 *the first filter most of this incoming information encounters is the amygdala*: Obviously, a tremendous amount of work has been done on the amygdala, but the folks at HowStuffWorks provide a pretty fantastic overview: http://science.howstuffworks.com/environment/life/human-biology/fear.htm.

32 *"Imagine you're watching a short film with a single actor cooking an omelet"*: The quote is from a personal interview with David Eagleman, but he says the same thing in *Incognito: The Secret Lives of the Brain* (Pantheon Books, 2011), p. 26.

33 *Statistically, the industrialized world has never been safer*: Marc Siegel, *False Alarm: The Truth About the Epidemic of Fear* (Wiley, 2005), p. 15.

34 *once our primitive survival instincts take over, our newer, prosocial instincts stay sidelined*: John Naish, "Warning: Brain Overload," London *Sunday Times,* June 2, 2009.

"It's No Wonder We're Exhausted"

34 *Homo sapiens evolved in a world that was "local and linear"*: First proposed in: Ray Kurzweil, *The Age of Spiritual Machines* (Viking, 1999), has been reprinted as a separate article, see: Ray Kurzweil, "The Law of Accelerating Returns": www.kurzweil.net/the-law-of-accelerating-returns.

34 *A week's worth of the* New York Times *contains more information"*: Michiko Kakutani, "Data Smog: Created by Information Overload," *New York Times* online edition, June 8, 1997.

34 *"From the very beginning of time until the year 2003"*: Eric Schmidt, Abu Dhabi Media Summit Keynote, Abu Dhabi 2010 Media Summit, March 12, 2010.

35 *"Five hundred years ago, technologies"*: This first appeared on Kelly's website, www.kk.org/thetechnium/archives/2008/11/the_origins_of.php, and later in *What Technology Wants* (Viking, 2010), p. 88.

35 *billion-dollar companies like Kodak*: Suzy Jagger, "Kodak Faces Break-Up After Fall in Digital Product Sales," *London Times*, December 11, 2008.

35 *Blockbuster*: Michael J. de la Merced, "Blockbuster, Hoping to Reinvent Itself, Files for Bankruptcy," *New York Times*, September 23, 2010.

35 *Tower Records*: Yuki Noguchi, "A Broken Record Store," *Washington Post,* August 23, 2006.

35 *YouTube*: Andrew Ross Sorkin and Jeremy W. Peters, "Google to Acquire YouTube for $1.65 Billion," *New York Times*, October 9, 2000.

35 *Groupon*: Eric Savitz, "Groupon Says No to Google's $6 Billion Bid? Really?," *Forbes*, December 4, 2010.

35 *"hype cycle"*: Jackie Fenn, "Understanding Hype Cycles," *When to Leap on the Hype Cycle,* Gartner Group, 2008.

Dunbar's Number

36 *he found that people tend to self-organize in groups of 150*: Aleks Krotoski, "Robin Dunbar: We Can Only Have 150 Friends at Most," *Guardian*, March 14, 2010.

36 *traffic patterns from social media sites such as Facebook*: NPR Staff, "Don't Believe Facebook: You Only Have 150 Friends," National Public Radio, *All Things Considered,* June 4, 2011.

36 *humans evolved in groups of 150*: Robin Dunbar, *How Many Friends Does One Person Need?: Dunbar's Number and Other Evolutionary Quirks* (Harvard University Press, 2010), pp. 4–5.

36 *Gossip, in its earlier forms, contained information*: Robin Dubar, *Grooming, Gossip, and the Evolution of Language* (Harvard University Press, 1998).

CHAPTER FOUR: IT'S NOT AS BAD AS YOU THINK

This Moaning Pessimism

38 *"It's incredible," he says, "this moaning pessimism"*: personal interview with Matt Ridley, 2010.

39 *English scientist Robert Angus Smith in 1852*: www.epa.gov/region1/eco/acidrain/history.html.

39 *In 1982 Canada's minister of the environment, John Roberts*: Frederic Golden, Jay Branegan, John M. Scott, "Environment: Storm over a Deadly Downpour," *Time*, December 6, 1982.

39 *The results were a reduction*: Nina Shen Rastogi, "Whatever Happened to Acid Rain," *Slate*, August 18, 2009. See www.slate.com/id/2225509.

Saved Time and Saved Lives

40 A rural peasant woman in modern Malawi: C. M. Blackden and Q. Wooden, *Gender, Time Use, and Poverty in SubSaharan Africa*, World Bank, 2006.

40 *"Forget dollars, cowrie shells, or gold"*: Matt Ridley, "Cheer Up: Life Only Gets Better," *Sunday Times*, May 16, 2010.

41 *Light is a fabulous example*: Matt Ridley, *The Rational Optimist* (Harper, 2010), pp. 20–21.

41 *Six thousand years ago, we domesticated the horse*: Horse Genome Project: www.uky.edu/Ag/Horsemap/hgpfaq4.html.

41 *In the 1800s, going from Boston to Chicago via stagecoach*: Distance is approximately one thousand miles. Stagecoaches averaged between four to seven miles per hour, or fifty to seventy miles per day.

41 *Norwegian adventurer Thor Heyerdahl spent 101 days*: Thor Heyerdahl, *Kon Tiki* (Simon & Schuster, 1990).

42 *as Ridley explains, they turn up almost every place we look*: Ridley, ibid., p. 12.

42 *On August 1, 2010, India's National Council of Applied Economic Research*: National Council of Applied Economic Research, "How India Earns, Spends, and Saves," August 1, 2010.

42 *According to the World Bank, the number of people living on less than $1 a day*: "World Development Indicators," World Bank, 2004.

42 *at the current rate of decline, Ridley estimates*: Ridley, ibid., p. 15.

43 *Between 1980 and 2000, the consumption rate . . . than in the previous five hundred*: Ibid., p. 15.

43 *"Once the rise in the position of the lower classes gathers speed"*: F. A. Hayek, Ronald Hamowy, *The Constitution of Liberty: The Definitive Edition* (University of Chicago Press, 2010), p. 101.

43 *both political liberty and civil rights have also improved substantially*: Charles Kenny, *Getting Better: Why Global Development Is Succeeding—and How We Can Improve the World Even More* (Basic Books, 2011), pp. 85–86.

43 *Slavery, for example, has gone from a common global practice to one outlawed everywhere*: Steven Pinker, "A History of Violence," *New Republic*, March 19, 2007.

43 *global surveys find democracy the preferred form*: Kenny, ibid., p. 134.

43 *Harvard evolutionary psychologist Steven Pinker discovered*: Pinker, ibid.

Cumulative Progress

44 *Ridley likens this process to sex*: Matt Ridley, "When Ideas Have Sex," TED, July 2010. See: www.ted.com/talks/matt_ridley_when_ideas_have_sex.html.

44 *Isaac Newton meant when he said, "If I have seen further"*: Isaac Newton, Letter to Robert Hooke, February 15, 1676.

45 *This process, Ridley feels, creates a further feedback loop*: Ridley, ibid., p. 7.

45 *"A large proportion of our high standard of living today"*: J. Bradford DeLong, "Estimating World GDP, One Million B.C.-Present, Department of Economics, UC Berkeley, May 24, 1998.
46 *"This is the diagnostic feature of modern life"*: Ridley, ibid., p. 39.
46 *"trade is a zero-sum game"*: personal interview with Dean Kamen, 2010.

The Best Stats You've Ever Seen

46 *Hans Rosling is in his early sixties . . . "The Best Stats You've Ever Seen"*: Hans Rosling, "Hans Rosling Shows Best Stats You've Ever Seen," TED, June 2006.
48 *In a 2010 updated presentation*: Hans Rosling, "Han's Rosling's 200 Countries, 200 Years, 4 Minutes," BBC Four, November 26, 2010.

PART TWO: EXPONENTIAL TECHNOLOGIES

CHAPTER 5: RAY KURZWEIL AND THE GO-FAST BUTTON

Better Than Your Average Haruspex

51 *The Romans, for example, employed a haruspex*: Not for nothing, it was a haruspex who warned Julius Caesar about the Ides of March.
51 *Kurzweil was born in 1948 and didn't start out trying to be a technological prognosticator*: Most of the information in this section came from personal interviews with Ray Kurzweil, but Kurzweil Tech has a very good biography: www.hurzweiltech.com/raybio.html; personal interviews with Ray Kurzweil 2010.

A Curve on a Piece of Paper

52 *In the early 1950s, scientists began to suspect that there might be hidden patterns*: The original research was done by Damien Broderick in *The Spike: How Our Lives Are Being Transformed by Rapidly Advancing Technologies* (Tor Books, 2002), but was referenced by Kevin Kelly on his blog, The Technium: "Was Moore's Law Inevitable," Available: www.kk.org/thetechnium/archives/2009/07/was_moores_law.php.
53 *the most famed of all tech trends*: Gordon Moore, "Cramming More Components onto Integrated Circuits," *Electronics* magazine, April 19, 1965.
53 *In 1975 Moore altered his formulation*: "'Moore's Law' Predicts the Future of Integrated Circuits," Computer History Museum, see: www.computerhistory.org/semi conductor/timeline/1965_Moore.html.
53 *Osborne Executive Portable*: See www.computerhistorymuseum.li/Testpage?osborne ExecSpecs.htm.
54 *Now compare this to the first iPhone*: See http://support.apple.com/kb/SP2.
54 *where Kurzweil returns to our story*: personal interview with Ray Kurzweil, 2010.

Google on the Brain

54 *Kurzweil found dozens of technologies that followed a pattern of exponential growth*: Kurzweil, "The Law of Accelerating Returns," ibid.

55 In his first book, 1988's *The Age of Intelligent Machines*: Ray Kurzweil, *The Age of Intelligent Machines* (MIT Press, 1992).

55 *to make a handful of predictions about the future*: For a thorough list of Kurzweil's predictions: http://en.wikipedia.org/wiki/Predictions_made_by_Ray_Kurzweil.

55 *In his 1999 follow-up*: Ray Kurzweil, *The Age of Spiritual Machines: When Computers Exceed Human Intelligence* (Penguin, 2000).

55 *Today's average low-end computer calculates*: Most people have something like a Pentium computer running Windows, or a Macintosh. A computer like this can execute approximately one hundred million instructions per second (10^11th), http://computer.howstuffworks.com/question54.htm.

55 *Scientists approximate that the level of pattern recognition necessary*: Many estimates for the approximate processing speed are available. Hans Morvec, principal research scientist at the Robotics Institute of Carnegie Mellon University, estimates the human brain's probable processing power is around 100 teraflops, roughly 100 trillion calculations per second (www.wired.com/techbiz/it/news/2002/11/56459); Ralph C. Merkle, in his Foresight Institute paper (www.merkle.com/brainLimits.html), estimates the speed ranging between 1012 and 1016 calculations per second. For the purpose of this book, the most conservative figure is utilized.

55 *Google cofounder Larry Page describes the future of search in similar terms*: Steven Levy, *In the Plex: How Google Thinks, Works, and Shapes Our Lives* (Simon & Schuster, 2011), p. 67.

Singularity University

56 *Early universities were devoted to religious teachings*: Jeffrey E. Garten, "Really Old School," *New York Times*, December 9, 2006.

56 *when the Catholic Church was responsible for many of Europe's top universities*: Thomas E. Woods Jr., *How the Catholic Church Built Western Civilization* (Regnery Publishing, 2005), pp. 49–50.

57 *While I was at MIT studying molecular genetics*: My course major at MIT was Course 7, which went by the name Biology, but Molecular Genetics is more descriptive of my focus and research.

57 *Archimedes: "Give me a lever long enough"*: Archimedes, 230 BC.

57 *In 2008 I took this idea forward, partnering with Ray Kurzweil to found Singularity University (SU)*: The idea for Singularity University came while I was trekking through Patagonia, Chile, with my wife, Kristen. Upon our return, I discussed the concept with Robert D. Richards, cofounder of International Space University, and Michael Simpson, president of ISU. ISU was a university I cofounded with Richards and Todd B. Hawley in 1987. After receiving enthusiastic feedback from Richards and Simpson, I took the idea to Kurzweil, who embraced the concept during our first dinner conversation.

CHAPTER SIX: THE SINGULARITY IS NEARER

A Trip Through Tomorrowland

59 *J. Craig Venter Institute*: www.jvci.org/; personal interviews with Craig Venter, 2010 and 2011.

59 *the Human Genome Project . . . "the language with which God created life"*: Nicholas Wade, "Scientists Complete Rough Draft of Human Genome, *New York Times,* June 26, 2006.

60 *the creation of a synthetic life form*: For a solid overview of the project, see James Shreeve, "Craig Venter's Epic Voyage to Redefine the Origin of the Species," *Wired*, August 2004.

60 *the kind that can manufacture ultra-low-cost fuels*: For a brief overview of algae-based biofuels, see: Andrew Pollack, "Exploring Algae as Fuel," *New York Times,* July 26, 2010.

61 *sailing his research yacht*: Shreeve, *Wired*, ibid.

61 *Venter wants to use similar methods to create human vaccines*: personal interview with Craig Venter, 2010.

Networks and Sensors

61 *It's fall 2009, and Vint Cerf*: ICANN has a solid biography of Cerf here: www.icann .org/en/biog/cerf.htm.

61 *at Singularity University to talk about the future of networks and sensors*: If you'd like to see the talk: www.youtube.com/watch?v=KeALwlp9YmA.

62 *the future of networks and sensors is sometimes called the "Internet of things"*: The term was first used by Kevin Ashton, see Kevin Ashton, "That 'Internet of Things' Thing," *RFID Journal,* June 22, 1999.

62 *Mike Wing, IBM's vice president of strategic communications*: Mike Wing, "The Internet of Things," IBMSocialMedia, March 15, 2010.

62 *Now imagine its future: trillions of devices*: Bruce Sterling, "Spime Watch: The Internet of Things, a Window to Our Future," *Wired*, February 11, 2011.

63 *Cisco teamed up with NASA*: See www.planetaryskin.org.

63 *To take the Internet of things to the level predicted*: Bruce Sterling, "Spime Watch: Cisco and the IPV6 Internet-of-Things," *Wired*, May 7, 2011.

63 *"My only defense," says Cerf*: Singularity University speech, October 3, 2009.

Artificial Intelligence

63 *It's Saturday, July 2010, and Junior is driving me around Stanford University*: This test "ride" was actually taken by Steven Kotler. Peter Diamandis was afforded a very similar experience in the Google autonomous Prius.

63 *Junior is an artificial intelligence*: Stefanie Olsen, "Stanford Robot Passes Driving Test," CNET, June 14, 2007.

64 *In 2005 Stanley won DARPA's Grand Challenge*: W. Wayt Gibbs, "Innovations from a Robot Rally," *Scientific American*, December 26, 2005.

64 *In June 2011 Nevada's governor approved a bill*: Assembly Bill No. 511—Committee on Transportation; see: www.leg.state.nv.us/Session/76th2011/Bills?AB/AB511_EN.pdf.

64 *Sebastian Thrun, previously the director of the Stanford Artificial Intelligence Laboratory*: personal interview with Sebastian Thrun, 2010.

64 *Robocar evangelist Brad Templeton*: See: http://www.templetons.com/brad/robocars/.

65 *"Consider the man-versus-machine chess competition between Garry Kasparov and IBM's Deep Blue"*: personal interview with Ray Kurzweil, 2010.

65 *The first integrates electrical and optical devices on the same piece of silicon*: Stephen Shankland, "IBM Chips: Let There Be Light Signals," CNET, December 1, 2010.

65 *The second is SyNAPSE*: See: Ferris Jabr, "IBM Unveils Microchip Based On Human Brain," *New Scientist*, August 2011. And www.ibm.com/smarterplanet/us/en/business_analytics/article/cognitive_computing.html.

Robotics

66 *Hassan met Larry Page and Sergey Brin*: Levy, *In the Plex*, ibid., pp. 22–23.

66 *eGroups, which was then bought by Yahoo!*: David Kleinbard, "Yahoo! To Buy eGroup," CNN.com, June 28, 2000.

66 *Willow Garage's main project is a personal robot*: www.willowgarage.com.

67 *A quick tour of YouTube shows*: www.youtube.com/user/WillowGaragevideo.

67 *"Proprietary systems slow things down"*: personal interview with Scott Hassan, 2010.

67 *"In 1950 the global world product was roughly four trillion dollars"*: YouTube presentation by Scott Hassan (http://www.youtube.com/watch?v=OF7cr8kIRGI).

67 *Obama announced the National Robotics Initiative*: Jackie Calmes, "President Announces an Initiative in Technology," *New York Times*, June 24, 2011.

68 *Helen Greiner, president of the Robotics Technology Consortium, told*: "US Shifts Focus to Multipurpose Robotic Development," UPI, June 28, 2011.

Digital Manufacturing and Infinite Computing

68 *today's versions are quick and nimble*: Ashlee Vance, "3-D Printing Spurs a Manufacturing Revolution," *New York Times*, September 13, 2010.

68 *3-D printers to make everything from lamp shades and eyeglasses to custom-fitted prosthetic limbs*: For an amazing tour through this world, check out industrial designer Scott Summit's great talk at SU: www.youtube.com/watch?v=6IJ8vld4HF8.

69 *Biotechnology firms are experimenting with the 3-D printing*: "Making a Bit of Me: A Machine That Prints Organs Is Coming to Market," *Economist*, February 18, 2010.

69 *Behrokh Khoshnevis, an engineering professor at the University of Southern California*: http://craft.usc.edu/CC/modem.html.

69 *Made in Space*: www.madeinspace.us.

69 *"What gets me most excited"*: personal interview with Carl Bass, 2010.

69 *"Forget the traditional limitations imposed by conventional manufacturing"*: Hod Lipson, "3-D Printing: The Technology That Changes Everything," *New Scientist*, August 3, 2011.

69 *"For most of my life, computing has been treated as a scarce resource"*: personal interview with Carl Bass, 2011.

Medicine

70 *In 2008 the WHO announced that a lack of trained physicians*: Richard M. Scheffer, Jenny X Liu, Yohannes Kinfu, Mario R. Dal Poz, "Forecasting the Global Shortage of Physicians: An Economic-and-Needs-Based Approach," *Bulletin of the World Health Organization*, 2007.

70 *In 2006 the Association of American Medical Colleges reported*: Suzanne Sataline and Shirley S. Wang, "Medical Schools Can't Keep Up," *Wall Street Journal,* April 12, 2010.

71 *the Mayo Clinic used an "artificial neural network"*: "Artificial Intelligence Helps Diagnose Cardiac Infections," Mayo Clinic Newsletter, September 12, 2009. See: http://www.mayoclinic.org/news2009-rst/5411.html.

71 *reading computed tomography (CT) scans*: University of Chicago radiologist Kenji Suzuki is on the cutting edge of much of this work. His website is a great introduction: http://suzukilab.uchicago.edu. Or see Tom Simonite, "A Search Engine for the Human Body," *MIT Technology Review,* March 11, 2011.

71 *screening for heart murmurs in children*: Curt G. Degroff, et al., "Artificial Neural Network–Based Method of Screening Heart Murmurs in Children," *Circulation: Journal of the American Heart Association,* June 25, 2001.

Nanomaterials and Nanotechnology

71 *physicist Richard Feynman's*: Richard P. Feynman, "Plenty of Room at the Bottom," presented to the American Physical Society in Pasadena, California, December, 1959.

71 *K. Eric Drexler's 1986 book,* Engines of Creation: Eric Drexler, *Engines of Creation: The Coming Era of Nanotechnology* (Anchor, 1987).

72 *a "grey goo" scenario*: Ibid., pp. 172–73.

72 *Nanocomposites are now considerably stronger than steel*: George Elvin, "The Nano Revolution," *Architect*, May 2007.

72 *Single-walled carbon nanotubes exhibit very high electron*: Xiangnan Dang, Hyunjung YI, Moon-Ho Ham, Jifa Qi, Dong Soo Yun, Rebecca Ladewski, Michael Strano, Paula T. Hammond, Angela M. Belcher, *Nature Nanotechnology* 6, April 24, 2011, pp. 377–84.

72 *Buckminsterfullerenes (C60), or Buckyballs*: H. W. Kroto et al., "C(60): Buckminsterfullerene," *Nature* 318, November 14, 1985, pp. 162–63.

72 *National Science Foundation report on the subject*: Kelly Hearn, "The Next Big Thing (Is Practically Invisible)," *Christian Science Monitor*, March 24, 2003.

Are You Changing the World?

73 *I organized the founding conference for Singularity University*: the early team who helped to organize and support the founding conference included: S. Pete Worden, Chris Boshuizen, Will Marshall, Bob Richards, Michael Simpson, Susan Fonseca-Klein, and Bruce Klein (the last two were dubbed "founding architects" for their extraordinary assistance in curating the invitation list). Also supporting were Karen Bradford, Amara D. Angelica, Gary Martin, and Donald James.

73 *the founding conference for Singularity University*: Ray Kurzweil gave a TED talk on SU available here: www.ted.com/talks/ray_kurzweil_announces_singularity_university.html.

73 *There were representatives from*: Salim Ismail, who would become our first executive director; Barney Pell and Sonia Arrison, who would become both associate founders and trustees; and Moses Znaimer, Keith Kleiner, and Georges Harik, who were our first associate founders. Those participating who would return as faculty include: Neil Jacobstein (SU's past president), Ralph Merkle, Rob Freitas, Harry Kloor, George Smoot, Larry Smarr, Philip Rosedale, Dharmendra Modha, Aubrey de Grey, Stephanie Langhoff, Chris Anderson, and Jamie Canton.

73 *impromptu speech given by Google's cofounder Larry Page*: While there's no video on line of that first speech, in a brief talk given at the opening ceremony for the summer 2010 graduate program, Larry references his earlier speech: www.youtube.com/watch?v=eF1HAG3Ru91.

73 *"ten to the ninth-plus" (or 109+) companies*: http://singularityu.org.

PART THREE: BUILDING THE BASE OF THE PYRAMID

CHAPTER SEVEN: THE TOOLS OF COOPERATION

The Roots of Cooperation

77 *earliest single-cell life forms were called prokaryotes*: There are obviously millions of potential references here, but for a great look at the microbial world it's very hard to beat: Lynn Margulis and Dorian Sagan, *Microcosmos: Four Billion Years of Microbial Evolution* (University of California Press, 1997).

78 *individual entities that "decided" to work together toward a greater cause*: Lynn Margulis, *Symbiotic Planet: A New Look at Evolution* (Basic Books, 1999).

78 *whose ten trillion cells*: These numbers vary considerably, but ten trillion is a fairly conservative number. Bill Bryson, in his fabulous and very credible *A Short History of Nearly Everything*, puts the number in the quadrillions.

78 *"[H]ow ten trillion cells organize themselves"*: Paul Ingraham, "Ten Trillion Cells Walked Into a Bar," *Arts & Opinion* 6, no. 1 (2007).

78 *In the words of Robert Wright, author of* Nonzero: The Logic of Human Destiny: The quote came from Wright's TED talk (www.ted.com/talks/robert_wright_on_optimism.html), but the book the talk is based on is fantastic: Robert Wright, *Nonzero: The Logic of Human Destiny* (Vintage, 2001).

From Horses to Hercules

79 *In 1861 William Russell*: Christopher Corbett, *Orphans Preferred: The Twisted Truth and Lasting Legend of the Pony Express* (Broadway, 2004), p. 5.

79 *after he spoke them*: Ibid., p. 195.

80 *When famine struck the Sudan*: Michael Paterniti, "The American Hero in Four Acts," *Esquire*, 1998.

80 *economist Jeffery Sachs counts eight distinct contributions ICT*: Jeffery Sachs, *Common Wealth: Economics for a Crowded Planet* (Penguin Press, 2008), pp. 307–8.

Gold in Dem Hills

81 *Rob McEwen*: www.robmcewen.com and personal interview with Rob McEwen, 2010.

81 *when Linus Torvalds*: Gary Rivlin, "Leader of the Free World: How Linus Torvalds Became Benevolent Dictator of Planet Linux, the Biggest Collaborative Project in History," *Wired,* November 11, 2003.

82 *announced the Goldcorp Challenge*: Most of this section is based on a series of personal interviews with McEwen, but for a good history of the Goldcorp Challenge, see: Linda Tischler, "He Struck Gold on the Net (Really)," *Fast Company,* May 31, 2002.

83 *Sun Microsystems cofounder Bill Joy*: This principle, in keeping with a theme, is known in the management world as Joy's law.

83 *Clay Shirky uses the term "cognitive surplus"*: Clay Shirky, *Cognitive Surplus: Creativity and Generosity in a Connected Age* (Penguin Press, 2010).

83 *"Wikipedia took one hundred million hours of volunteer time"*: While Shirky writes about Wikipedia throughout *Cognitive Surplus,* this quote is from a talk he gave at SU in July 2011.

An Affordable Android

83 *the Chinese firm Huawei*: David Talbot, "Android Marches on East Africa," TechnologyReview.com, June 23, 2011.

83 *where 60 percent of the population*: *Human Development Report*, United Nations, 2007–2008, see hdr.undp.org.

84 *the Canada-based company Datawind*: "India Launches World's Cheapest Tablet," *International Business Times*, October 6, 2011.

84 *Hollywood produces five hundred films per year*: Shabnam Mahmood and Manjushri Mitra, "Bollywood Sets Sights on Wider Market," BBC Asian Network, June 24, 2011.

74 *YouTube users*: These numbers were announced by YouTube on May 25, 2011. See: http://youtube-global.blogspot.com/2011/05/thanks-youtube-community-for-two-big.html.

84 *In 2009 it received 129 million views a day*: And this was in 2009. In May 2011 YouTube announced that it had reached 3 *billion* views a day, a 50 percent increase over the previous year.

84 *Salim Ismail, SU's founding executive director and now its global ambassador*: personal interview with Salim Ismail, 2011.

CHAPTER EIGHT: WATER

Water for Water

85 *Peter Thum didn't intend to become a social entrepreneur*: This section is based mostly on personal interviews with Thum, but he tells this story in a piece for the website/video blog, *Big Think;* see: "The Prius of Bottled Water," www.bigthink.com/ideas/39293.

85 *In 2005 Howard Schultz*: Howard Schultz, *Onward: How Starbucks Fought for Its Life Without Losing Its Soul* (Rodale Books, 2011), p. 115.

86 *the global water crisis affects a billion people*: for a slightly dated but still really good breakdown of the issues, see Peter Gleick, "*Water in Crisis: A Guide to the World's Freshwater Resources* (Oxford University Press, 1993).

86 *Historically, because of the huge amount of infrastructure required*: Korinna Horta, "The World Bank's Decade for Africa: A New Dawn for Development Aid," *Yale Journal of International Affairs*, May 2005.

86 *as inventor Dean Kamen*: personal interview with Dean Kamen, 2011.

86 *water is thoroughly embedded in our lives*: for a pretty comprehensive breakdown, see American Water's list: www.amwater.com/learning-center/water-101/what-is-water-used-for.html.

86 *70 percent of the world's water is used for agriculture*: is according to the Food and Agriculture Organization of the United Nations, see: http://www.fao.org/newsroom/en/news/2007/1000520/index.html

86 *one egg requires 120 gallons to produce*: A. Y. Hoekstra, A. K. Chapagain, "Water footprints of nations: Water use by people as a function of consumption pattern," *Water Resource Management,* Vol. 21, pp. 35–48.

86 *Meat is among our thirstiest commodities*: John Robbins, *A Diet for a New America* (Stillpoint Publishing, 1987), p. 367.

86 *443 million school days a year are lost to water-related disease*: James Hughes, *Fact Sheet: Foodborne and Water-Related Diseases: A National and Global Update,* National Foundation for Infectious Diseases News Conference and Symposium on Infectious Diseases, July 11, 2007.

86 *35 gallons of water are used to make one microchip*: Hoekstra and Chapagain, Ibid.

87 *the power production chain makes the world a dryer place*: *Energy Demands on Water Resources: Report to Congress on the Interdependencies of Energy and Water*, Sandia National Laboratory. This is an incredibly comprehensive look at the links between energy and water: www.sandia.gov/energy-water/docs/121-RptToCongress-EWwEIA comments-FINAL.pdf.

87 *UC Berkeley professor of economics Edward Miguel*: Edward Miguel, Shanker Satyanath, Ernest Sergenti, "Economic Shocks and Civil Conflict: An Instrumental Variables Approach," *Journal of Political Economy* 112, no. 4 (2004), pp. 725–53.

87 *two hundred rivers and three hundred lakes share international boundaries*: The UN has a good report on the subject here: www.un.org/waterforlifedecade/transboundary_waters.shtml.

87 *3.5 million people dying annually from water-related illnesses*: Peter Gleick, "Dirty Water: Estimated Deaths from Water-Related Diseases, 2000–2020," the Pacific Institute, 2002.

87 *we humans consume almost 50 billion liters of bottled water*: Peter Gleick, *Bottled and Sold: The Story Behind Our Obsession with Bottled Water* (Island Press, 2011), p. 5.

87 "*fossil water*": Ibid., pp. 64–68.

Dean vs. Goliath

88 *Dean Kamen is a self-taught physicist*: Several lengthy interviews were conducted with Kamen during 2010 and 2011, but for more information, *Wired* ran a solid profile: Scott Kirsner, "Breakout Artist," *Wired*, September 2000.

88 *DEKA Research and Development . . . development of prosthetic limbs*: John Markoff, "Dean Kamen Lends a Hand, or Two," *New York Times*, August 8, 2007.

89 *That different machine was finished in 2003*: For a live demonstration of the Slingshot on *The Colbert Report*: www.colbertnation.com/the-colbert-report-videos/164485/march-20–2008/dean-kamen.

90 *"It better work that well"*: personal interview with Jonathan Greenblatt, 2011.

90 *Rob Kramer, chairman of the Global Water Trust*: personal interview with Rob Kramer, 2010.

90 *America's infrastructure is so old*: Michael Cooper, "Aging of Water Mains Is Becoming Hard to Ignore," *New York Times*, April 17, 2009.

91 *The average microfinance loan in the water space*: personal interview with April Rinne, 2011.

91 *And Coca-Cola has agreed to try*: personal interview with Dean Kamen, 2011.

Prophylaxis

92 *Malthusians often use the word* cornucopians: For a good look at the whole debate, see: John Tierney, "Betting on the Planet," *New York Times Magazine*, December 2, 1990.

92 *Population is linked directly to fertility*: *World Population Prospects: The 2002 Revision*, United Nations Population Division.

92 *Urbanization actually lowers fertility rates*: "Linking Population, Poverty, and Development," United Nations Population Fund: www.unfpa.org/pds/urbanization.htm. Also see: Stewart Brand, *The Whole Earth Discipline: An Ecopragmatist Manifesto* (Viking, 2009), pp. 59–61.

92 *Thus child mortality among the rural poor is one of the largest factors driving population*: "Return of the Population Growth Factor: Its Impact on Millennium Development Goals," All Party Parliamentary Group on Development and Reproductive Health, January 2007, p. 24.

92 *85 percent of them live in the countryside*: *Rural Poverty Report 2011*, IFAD, See: www.ifad.org/rpr2011.

Getting Roomier at the Bottom

93 *An English engineer named Michael Pritchard*: www.ted.com/talks/Michael_prichard_invents_a_water_filter.html.

94 *The nanotechnology industry is exploding*: John F. Sargent Jr., "The National Nanotechnology Initiative: Overview, Reauthorization, and Appropriation Issues," Congressional Research Service, January 19, 2011.

94 *the National Science Foundation predicts that it will hit $1 trillion*: www.nsf.gov/news/news_summ.jsp?cntn_id=112234.

94 *there are now nanomaterials with increased affinity, capacity, and selectivity*: Mamadao Diallo, Jeremiah Duncan, Nora Savage, Anita Street, Richard Sustich, *Nanotechnology Applications for Clean Water* (William Andrew, 2009).

94 *researchers at IBM and the Tokyo-based company Central Glass*: Katherine Boutzac, "Getting Arsenic out of Water," *Technology Review*, June 1, 2009.

94 *On the sanitation front*: See: Duncan Graham-Rowe, "Self-Healing Pipelines," *Technology Review*, December 21, 2006. And: Nicole Wilson, "Nano Technology May

Make Cleaning Toilets a Thing of the Past," Best Syndication News, February 7, 2006.

94 DIME Hydrophobic Materials, a company based in the United Arab Emirates: Lisa Zyga, "Hydrophobic Sand Could Combat Desert Water Shortages," Phyorg.com, February 16, 2009.

94 With 40 percent of the Earth's population living within one hundred kilometers (62 miles) of a coast: Liz Creel, "Ripple Effects: Population and Coastal Regions," Population Reference Bureau, September 2003.

94 the majority of the world's seven thousand desalination plants rely on thermal desalination: Jennifer Chu, "Desalination Made Simpler," *Technology Review*, July 30, 2008.

95 the Los Angeles-based company NanoH2O: For a nice breakdown of the technology, Jonathon Fahey, "Water Wizardry," *Forbes*, August 26, 2009. Also, the *Guardian* published a full list of winners here: *www.guardian.co.uk/globalcleantech100.*

The Smart Grid for Water

95 "Distinguished Scientist": personal interview with Peter Williams, 2011.

95 Mark Modzelewski: "House Committee Discusses Smart Water Grid Plans," *Water & Wastes Digest*, March 2009.

95 IBM believes that the smart grid for water: Martin LaMonica, "IBM Dives Into 'Smart Grid for Water,'" CNET, September 4, 2009.

95 it has partnered with the Nature Conservancy: For a look at all of IBM's current smart grid projects: www.ibm.com/smarterplanet/za/en/water_management/ideas/index.html.

96 Hewlett-Packard has implemented a smart metering system: http://h10134.www.1.hp.com/news/features/5831.

96 researchers at Chicago's Northwestern University have created a "Smart Pipe": Yu-Fen Lin and Chang Liu, "Smart Pipe: Nanosensors for Monitoring Water Quantity and Quality in Public Water Systems," Illinois State Water Survey, August 2009.

96 Spain just installed a nationwide: Ciaran Giles, "Water Management 2.0," Associated Press, November 12, 2007.

96 "With precision agriculture," says Doug Miell: personal interview with Doug Miell, 2011.

Solving Sanitation

97 Apocrypha holds that it was Thomas Crapper: "Thomas Crapper: Myth and Reality," *Plumbing & Mechanical*, 1993.

97 credit is now given to Sir John Harington: "The Men That Made the Water Closet," *Plumbing & Mechanical*, 1994.

97 Archaeologists recently unearthed a Han dynasty latrine: "The Chain Is Pulled on Britain's Crapper," Reuters, July 26, 2000.

97 Imagine toilets that require no infrastructure: personal interview with Lowell Wood, 2011.

97 recently announced Bill & Melinda Gates Foundation program: "Gates Foundation Launches Effort to Reinvent the Toilet," Bill & Melinda Gates Foundation, July 19, 2011.

98 *Toilets account for 31 percent of all water use*: P. W. Mayer, W. B. Oreo, et al., "Residential End Uses of Water," American Waterworks Research Foundation, 1999.

98 *The US Environmental Protective Agency (EPA) estimates 1.25 trillion gallons of water*: Bob Swanson, "Leaks, Wasteful Toilet Causes Cascading Water Loss," *USA Today*, April 5, 2009.

A Pale Blue Dot

99 *astronomer Carl Sagan decided it might be interesting*: If you've never heard the entire Pale Blue Dot speech, get thee to YouTube: www.youtube.com/watch?v=2pfwY2TNehw.

CHAPTER NINE: FEEDING NINE BILLION

The Failure of Brute Force

100 *925 million people currently don't have enough to eat*: "The State of Food Insecurity in the World: Addressing Food Insecurity in Protracted Crisis," Food and Agriculture Organization of the United Nations, Rome, 2010.

100 *Each year, 10.9 million children die*: "Progress for Children: A Report Card on Nutrition," UNICEF, 2006.

100 *agriculture has mainly been a brute force equation*: There are dozens of accounts of this available, but two of the best are Richard Manning's *Against the Grain* (North Point Press, 2005) and Michael Pollan's *The Omnivore's Dilemma* (Penguin, 2006).

101 *10 calories of oil to produce 1 calorie of food*: Martin C. Heller and Gregory A. Keoleian. "Life Cycle-Based Sustainability Indicators for Assessment of the US Food System," Center for Sustainable Systems, University of Michigan, 2000.

101 *Major aquifers in both China and India*: "Groundwater in Urban Development: Assessing Management Needs and Formulating Policy Strategies," World Bank Technical Paper 390, 1998.

101 *resulting in dust bowls*: Daniel Zwerdling, "India's Farming 'Revolution' Heading for Collapse," National Public Radio, April 13, 2009.

101 *Toxic herbicides and pesticides have destroyed*: "Managing Nonpoint Source Pollution from Agriculture," Environmental Protection Agency, Pointer No. 6., EPA841-F-96-004F. See: http://water.epa.gov/polwaste/nps/outreach/point6.cfm.

101 *turned our coastal waters into dead zones*: S. Joyce, "The Dead Zones: Oxygen-Starved Coastal Waters," *Environmental Health Perspectives* 108, no. 3 (March 2000), pp. 120–25.

101 *must now import 80 percent of its seafood*: "Chinese Seafood Imports," Don Kraemer, Deputy Director Office of Food Safety, FDA, statement before US and China Economic and Security Review Commission, April 25, 2008.

101 *Bottom trawling destroys about six million square miles*: The impact is so great that it's visible from space. See: "Trail of Destruction," TreeHugger.com, February 22, 2008.

101 *the world will run out of seafood by 2048*: Julie Eilperin, "World's Fish Supply Running Out, Researchers Warn," *Washington Post*, November 3, 2006.

101 *we seem to be exhausting the potential of many of the technologies*: Lester Brown, "The Great Food Crisis of 2011," *Foreign Policy*, January 10, 2011.

101 *celebrated environmentalist Vandana Shiva*: Vandana Shiva, "The Green Revolution in the Punjab," *Ecologist* 21, no. 2 (March-April 1991).

101 *we farm 38 percent of all the land in the world*: Matt Ridley, *The Rational Optimist,* pp. 143–44.

Cooking for Nine Billion

102 *there were 1.7 million hectares of biotech*: For a fantastic discussion of all the issues, read Stewart Brand's *The Whole Earth Discipline* (Viking, 2009), pp. 117–205, and Pamela Ronald and R. W. Adamchak's *Tomorrow's Table* (Oxford University Press, 2008).

102 *farming is just a 12,000-year-old way of optimizing lunch*: Ridley, ibid., p. 153.

103 *The Kansas-based Land Institute is attempting*: www.landinstitute.org or see Robert Kunzig, "The Big Idea: Perennial Grains," *National Geographic*, April 2011, pp. 31–33.

103 *a great many of our GE fears*: Brand, ibid., p. 141.

104 *agricultural portion of the biotech industry*: "Global Status of Commercialized Biotech/GM Crops: 2010," International Service for the Acquisition of Agri-Biotech Applications, ISAAA Brief 42–2010.

104 *The Gates Foundation–led effort BioCassava Plus*: www.danforthcenter.org/science/programs/international_programs.bcp.

104 *As the wife-and-husband team*: *Tomorrow's Table*, ibid., p. 57.

105 *According to the Institute for Food and Development Policy/Food First*: Melissa Moore, "Backgrounder: The Myth—Scarcity; The Reality—There *Is* Enough Food," *Institute for Food and Development Policy/Food First,* February 8, 2005. See: www.foodfirst.org/fr/node/239.

Vertical Farming

105 *During the tail end of the Second World War . . . Pan American Airways grew veggies*: Jeffrey Windterborne, *Hydroponics: Indoor Horticulture* (Pukka Press, 2005), p. 180, or see: http://hydroponicsdictionary.com/hydroponic-technology-used-in-wwii-to-feed-troops.

106 *NASA, which wanted to know how to feed*: www.nasa.gov/missions/science/biofarming.html.

106 *Hydroponics is 70 percent more efficient . . . Aeroponics*: Dickson Despommier, *Vertical Farming: Feeding Ourselves and the World in the 21st Century* (St. Martin's Press, 2009), pp. 164–69.

106 *"It's a PR problem"*: personal interviews with Dickson Despommier, 2010–11.

108 *the average American foodstuff now travels 1,500*: Vertical Farming, ibid., p. 7.

108 *pilot projects in the United States*: www.growingpower.org.

108 *Japan, while it hasn't switched yet from horizontal to vertical production*: David Derbyshire, "Is This the Future of Food? Japanese 'Plant Factory' Churns Out Immaculate Vegetables," *Mail Online,* June 3, 2009.

108 *Sweden's Plantagon*: personal interview with Plantagon CEO Hans Hassle, 2011.

109 *Researchers at the University of Illinois*: For a good review of photosynthetic optimization, see: Miko U. F. Kirschbaum, "Does Enhanced Photosynthesis Enhance

Growth? Lessons Learned from CO_2 Enrichment Studies," American Society of Plant Biologists, 2011. Available: www.plantphysiol.org/content/155/1/117.full.

109 *the 70 percent of us who will soon live in cities*: "Human Population: Urbanization," Population Reference Bureau, www.prb.org/Educators/TeachersGuides/HumanPopulation/Urbanization.aspx.

Protein

109 *optimal health means 10 percent to 20 percent*: "Dietary Protein Recommendations for Adequate Intake and Optimal Health: A Tool Kit for Healthcare Professionals," Egg Nutrition Center, 2011.

109 *Cattle, for starters, are energy*: Mark Bittman, "Rethinking the Meat-Guzzler," *New York Times*, January 27, 2008.

110 *As people rise out of poverty*: Christopher Delgado, "Rising Consumption of Meat and Milk in Developing Countries Has Created a New Food Revolution," *Nutrition: The Journal of the American Society for Nutritional Sciences* 133, November 2003, 3907S–3910S.

110 *Aquaculture is nothing new*: Barry A. Costa-Pierce, *Ecological Aquaculture: The Evolution of a Blue Revolution* (Wiley-Blackwell, 2003), pp. 9–19.

110 *global aquaculture yields increased*: Charles Mann, "The Bluewater Revolution," *Wired*, May 2004.

110 *the journal* Nature *reported that 90 percent of all large fish*: Ransom A. Myers and Boris Worm, ibid.

110 *oceanographer Sylvia Earle*: www.nationalgeographic.com/adventure/environment/what-it-takes-07/Sylvia-earle.html.

111 *the National Oceanic and Atmospheric Administration (NOAA) believes that fish farming*: www.noaanews.noaa.gov/stories2011/20110711_aquaculture.html.

111 *Others are more cautious*: The World Wildlife Fund does a good job on the issues: http://wwf.panda.org/about_our_earth/blue_planet/problems/aquaculture.

111 *shrimp industry is starting to clean up*: For example, see Jill Schwartz, "Tsunami Region's Shrimp Industry: Building It Back Better," www.worldwildlife.org/what/globalmarkets/aquaculture/featuredpublication-tsunami.html.

111 *Improved vegetable proteins and rendered animal by-products*: Brian Halweil, *Fish Farming for the Future* (Worldwatch Institute, 2008).

111 *Asian rice farmers use fish to fight*: "Integrated Aquaculture: Rice Paddy Success," *Sustainable Harvest International Newsletter*, spring 2010.

111 *In Africa, farmers are installing fish ponds*: For a look at UNESCO-backed efforts: www.ihe.nl/Fingerponds/Publications.

111 *Will Allen, the MacArthur Genius Award–winning force*: www.growingpower.org/aquaponics.htm.

111 *"If we value the ocean"*: www.nationalgeographic.com/adventure/environment/what-it-takes-07/Sylvia-earle.html.

Cultured Meat

112 *In 1932 Winston Churchill*: Abigail Paris, "In Vitro Meat, a More Humane Treat," *Policy Innovations*, May 22, 2008.

112 *process was pioneered by NASA*: "Lab Meat," PBS, January 10, 2006.

112 *goldfish cells were being used to create edible muscle*: Abigail Paris, "In Vitro Meat, a More Humane Treat," *Policy Innovations*, May 22, 2008. See: http://www.policyinnovations.org/ideas/briefings/data/000054.

112 *People for the Ethical Treatment of Animals (PETA) created a $1 million*: John Schwartz, "PETA's Latest Tactic: $1 Million for Fake Meat," *New York Times*, April 21, 2008.

112 *"Cattle ranching is always going to be an environmental disaster"*: For some fast facts, see: www.news.cornell.edu/releases/aug97/livestock.hrs.html. For an amazing overview: Eric Schlosser, *Fast Food Nation: The Dark Side of the All-American Meal* (HarperPerennial, 2002).

113 *(70 percent of emerging diseases come from livestock)*: Mario Herrero, with Susan MacMillan, Nancy Johnson, Polly Erickson, Alan Duncan, Delia Grace, and Philip K. Thornton, "Improving Food Production from Livestock," *State of the World 2011: Innovations That Nourish the Planet*, the Worldwatch Institute.

113 *30 percent of the world's surface that is currently used for livestock*: *Livestock's Long Shadow—Environmental Issues and Options*, Food and Agriculture Organization of the United Nations, November 29, 2006.

113 *PETA president Ingrid Newkirk*: Michael Specter, "Annals of Science: Test-Tube Burgers," *New Yorker*, May 23, 2011.

Between Now and Then

113 *the GE industry is dominated by three seeds*: personal interview with UC Davis plant pathologist Pamela Ronald, 2010.

113 *Golden rice*: Michael Pollan, "The Way We Live Now: The Great Yellow Hype," *New York Times Magazine*, March 4, 2001. And for a counter opinion: Pamela Ronald and James E. McWilliams, "Genetically Engineered Distortions," *New York Times*, May 14, 2010.

114 *Known as agroecology*: C. Francis et al., "Agroecology: The Ecology of Food Systems," *Journal of Sustainable Agriculture* 22, no. 3 (2003).

114 *A recent UN survey found that agroecology projects*: Olivier De Schutter, *Agroecology and the Right to Food*, report presented at the sixteenth session of the United Nations Human Rights Council [A/HRC/16/49], March 8, 2011.

114 *push-pull system*: Zeyaur Khan, David Amudavi, and John Pickett, "Push-Pull Technology Transforms Small Farms in Kenya," *PAN North America Magazine*, spring 2008.

114 *as UC Davis plant pathologist Pamela Ronald*: www.economist.com/debate/days/view/606.

A Tough Row to Hoe

115 *it's called primary productivity*: Richard Manning, "The Oil We Eat: Following the Food Chain Back to Iraq," *Harper's*, February 2004.

PART FOUR: THE FORCES OF ABUNDANCE

CHAPTER TEN: THE DIY INNOVATOR

Stewart Brand

119 *opening pages of* The Electric Kool-Aid Acid Test: Tom Wolfe, *The Electric Kool-Aid Acid Test* (Bantam, 1999), p. 4.

119 *Brand was reading a copy of Barbara Ward's*: Andrew Kirk, *Counterculture Green: The Whole Earth Catalog and American Environmentalism* (University Press of Kansas, 2007), p. 1.

119 *ethic has a long history*: www.emersoncentral.com/selfreliance.htm.

119 *Arts and Crafts renaissance of the early twentieth century*: Oscar Lovell Triggs, *Chapters in the History of the Arts and Crafts Movement* (Cornell University Library, 2009).

119 *the late 1960s marked the largest communal uprising in American history*: personal interview with Andrew Kirk, 2009.

120 *"I was in the thrall of Buckminster Fuller," Brand recalls*: "Counterculture to Cyberculture: The Legacy of the *Whole Earth Catalog*," a great panel discussion held at Stanford, viewable here: www.youtube.com/watch?v=B5kQYWLtW3Y.

120 *Out of all of this was born the* Whole Earth Catalog: Much of the information in this section was culled from interviews done while researching: Steven Kotler, "The Whole Earth Effect," *Plenty* magazine, May 2009.

120 *SRI was both at the cutting edge*: John Markoff, *What the Dormouse Said: How the 60s Counterculture Shaped the Personal Computer Industry* (Penguin, 2005), pp. 152–57.

120 *"American culture's acceptance of the personal computer"*: Kotler, ibid.

121 *the movement's adoption of two more* WEC *principles*: Most people believe that the first time Brand told anyone "information wants to be free" was at the first Hackers Conference in 1984.

Homebrew History

121 *DIY innovator named Fred Moore . . . This was the birth of the Homebrew Computer Club*: Markoff, ibid.

122 *"The* WEC *not only gave you permission to invent your life," Kevin Kelly once said*: Kotler, ibid.

The Power of Small Groups (Part I)

123 *our relationship with the final frontier began in the spring of 1952*: Al Blackburn, "Mach Match," *Air & Space*, June 1, 1999.

123 *X-series of experimental aircraft*: For NASA's short history of the X-Plane program: http://history.nasa.gov/x1/appendixa1.html.

123 *X-15 was an extreme machine*: http://history.nasa.gov/x15/cover.html.

123 *Rutan, on the other hand, is prolific*: W. J. Hennigan, "Aerospace Legend Burt Rutan Ready for Landing," *Los Angeles Times*, April 1, 2011; personal interview and data provided by Burt Rutan.

124 *In his mind, the problem was one of volume*: personal interview with Burt Rutan, 2010.

124 *"When Buzz [Aldrin] first walked on the Moon"*: Steven Kotler, "Space Commodity," *LA Weekly*, June 24, 2004.

124 *SpaceShipOne, outperformed the government's X-15*: Alan Boyle, "SpaceShipOne Wins $10 Million X Prize," msnbc.com, October 5, 2004.

125 *"The success of SpaceShipOne altered the perceptions of what a small group of developers can do"*: personal interview with Gregg Maryniak, 2010.

The Maker Movement

125 *Chris Anderson did the same thing for unmanned air vehicles (UAV)*: Much of this section is based on personal interviews with Chris Anderson and a presentation he made at SU in August 2011. But for a great look at his work, check out: "DIY Drones: An Open Source Hardware and Software Approach to Making 'Minimum UAVs,'" O'Reilly Where 2.0 Conference, available here: http://blip.tv/oreilly-where-20-conference/chris-anderson-diy-drones-an-open-source-hardware-and-software-approach-to-making-minimum-uavs-973054.

125 *The cheapest military-grade UAV on the market*: www.globalsecurity.org/intell/systems/raven.htm.

126 *By the 1950s, tinkering had become a middle-class virtue*: personal interview with Dale Daugherty, 2011. But for a look at the rise of the Maker Movement, also see Rob Walker, "Handmade 2.0," *New York Times*, December 16, 2007.

126 *resurfacing as the bedrock ethos of punk-rock culture*: Teal Triggs, "Scissors and Glue: Punk Fanzines and the Creation of the DIY Aesthetic," *Journal of Design History* 19, no. 1 (2006), pp. 69–83.

126 *Matternet, a Singularity University (SU) 109+ company*: www.matternet.net.

DIY Bio

127 *a biologist named Drew Endy*: Jon Mooallem, "Do-It-Yourself Genetic Engineering," *New York Times,* February 10, 2010.

128 *founded the International Genetically Engineered Machine (iGEM) competition*: http://igem.org.

128 *These standardized parts, known technically as BioBricks*: Alok Jha, "From the Cells Up," *Guardian,* March 10, 2005.

129 *take a look at "Splice It Yourself"*: Rob Carlson, "Splice It Yourself," *Wired,* May 2005.

The Social Entrepreneur

129 *The term itself was coined in 1980 by Ashoka founder Bill . . . Drayton*: Caroline Hsu, "Entrepreneur for Change," *US News & World Report*, October, 21, 2005.

130 *Take Kiva*: Sonia Narang, "Web-Based Microfinancing," *New York Times,* December 10, 2006.

130 *"Your money is safer in the hands of the world's poor than in your 401(k)"*: Adam Fisher, "Best Websites 2009," *Time*, August 24, 2009.

130 *By 2007, this third sector employed around forty million people, with 200 million volunteers*: Charles Leadbeater, "Mainstreaming of the Mavericks," *Observer*, March 25, 2007.

130 *by 2009, according to B Lab*: Stacy Perman, John Tozzi, Amy S. Choi, Amy Barrett,

Jeremy Quittner, and Nick Leiber, "America's Most Promising Social Entrepreneurs," *Bloomberg Businessweek*, September 2004.

130 *J. P. Morgan and the Rockefeller Foundation analyzed*: Nick O'Donohoe, Christina Leinjonhufvud, Yasemin Saltuk, Anthony Bugg-Levine, and Margot Brandenburg, "Impact Investments: An Emerging Asset Class," J. P. Morgan Global Research, November 29, 2010.

130 *KickStart, started in July 1991 by Martin Fisher and Nick Moon*: www.techawards.org/laureates/feature/kickstart.

130 *An even bigger example is Enterprise*: Ellen McGirt, "Edward Norton's $9,000,000,000 Housing Project," *Fast Company*, December 1, 2008.

CHAPTER ELEVEN: THE TECHNOPHILANTHROPISTS

The Robber Barons

132 *the X PRIZE Foundation is holding its annual Visioneering meeting*: Each year the X PRIZE board of trustees and Vision Circle members (www.xprize.org) gather to debate and discuss the world's grand challenges and to design prize concepts to address them. The Foundation calls this process "visioneering."

133 *This sphere of caring expanded during the Renaissance*: Matthew Bishop and Michael Green, *Philanthrocapitalism: How the Rich Can Save the World* (Bloomsbury, 2008), pp. 20–27.

133 *the titans of industrialization known collectively as the robber barons*: Maury Klein, "The Robber Baron's Bum Rap," *City Journal*, winter 1995.

133 *these Gilded Age magnates who invented*: Ibid.

133 BusinessWeek *wrote: "John D. Rockefeller"*: "The Robin Hood Robber Baron," *BusinessWeek*, November 27, 2008.

133 *Great-great grandson Justin Rockefeller, an entrepreneur and political activist, disagrees*: personal interview with Justin Rockefeller, 2010.

133 *In 1910 Rockefeller took $50 million*: Ron Chernow, *Titan: The Life of John D. Rockefeller, Sr.* (Warren Books, 1998), pp. 563–66.

134 *When Warren Buffett wanted to inspire philanthropy*: Robert A. Guth and Geoffrey A. Fowler, "16 Tycoons Agree to Give Away Fortunes," *Wall Street Journal*, December 9, 2010.

134 *his major contribution was to construct 2,500 public libraries*: www.pbs.org/wgbh/amex/carnegie/sfeature/p_library.html.

The New Breed

134 *attempted to identify every millionaire*: Klein, ibid.

134 *Even Carnegie was prone to the tendency*: A. A. Van Slyck, "Spaces of Literacy: Carnegie Libraries and an English-Speaking World." Paper presented at the annual meeting of the American Studies Association, March 13, 2011.

134 *Osman Ali Khan, known as Asaf Jah VII*: "Hyderabad: Silver Jubilee Durbar," *Time*, February 22, 1973.

135 *"Today's technophilanthropists are a different breed"*: personal interview with Jeff Skoll, 2011.

135 *When Skoll cashed out of eBay*: Michael S. Malone, "The Indie Movie Mogul," *Wired*, February 2006.

135 *The Skoll Foundation attempts*: personal interview with Jeff Skoll, but see: www.skoll foundation.org.

135 *in an article for the Huffington Post*: www.huffingtonpost.com/2011/06/02/jeff-skoll -foundation-climate-change_n_869457.html.

136 *Rockefeller-backed Acumen Fund*: "2008 Social Capitalist Awards," *Fast Company*, 2008. Available: www.fastcompany.com/social/2008/index.html.

136 *eBay founder Pierre Omidyar's Omidyar Network*: Jim Hopkins, "EBay Founder Takes Lead in Social Entrepreneurship," *USA Today*, November 3, 2005.

136 *can use their donations to create a profitable solution*: Bishop, ibid., p. 6.

136 *"impact investing"*: Paul Sullivan, "With Impact Investing, a Focus on More Than Returns," *New York Times*, April 23, 2010.

136 *the research firm the Monitor Group*: "Investing for Social & Environmental Impact: A Design for Catalyzing an Emerging Industry," www.monitorinstitute.com/impact investing.

136 *Another of those secrets is a hands-on approach*: personal interview with Paul Shoemaker, 2011.

137 *Paul Schervish of the Boston College Center on Wealth and Philanthropy calls* hyper-agents: Paul G. Schervish, Albert Keith Whitaker, *Wealth and the Will of God* (Indiana University Press, 2010), p. 8.

137 *As Matthew Bishop explains*: Bishop, ibid., p. 12.

How Many and How Much?

138 *Naveen Jain grew up*: personal interviews with Naveen Jain, 2011. See: www.naveen jain.com.

138 *2010 Credit Suisse Global Wealth Report*: *Global Wealth Report*, Credit Suisse Research Institute, October 2010.

138 *"The Internet's rich are giving it away"*: Sam Howe Verhovek, "The Internet's Rich Are Giving It Away, Their Way," *New York Times*, February 11, 2000.

138 *By 2004, charitable giving in America*: "Charitable Giving in US Nears Record Set at End of Tech Boom," *USA Today*, June 19, 2006.

139 *By 2007, CNBC had taken to calling our era*: Christina Cheddar Berk, "Rich and Richer: A New Golden Age of Philanthropy," CNBC, May 2, 2007. See: www.cnbc .com/id/18333214/Rich_Richer_Golden_Age_of_Philanthropy.

139 *Foundation Giving reported a record-setting*: Stephanie Strom, "Foundations' Giving Is Said to Have Set Record in '06," *New York Times*, April 3, 2007.

139 *2 percent in 2008, 3.6 percent in 2009*: Tom Watson, "Philanthropy's Double Dip: Giving Numbers Tumble for Second Straight Year," *On Philanthropy*, June 10, 2010.

139 *the year Bill Gates put $10 billion*: Alexander Higgins, "Gates Foundation Pledges $10 Billion to Vaccine Research," *Washington Post*, January 30, 2010.

139 *Gates and Warren Buffett*: http://givingpledge.org.

139 *the Ibrahim Prize for Achievement in African Leadership*: "Prize Offered to Africa's Leaders," BBC, October 26, 2006.

139 *PayPal cofounder Elon Musk*: personal interview with Elon Musk, but also see Tad Friend, "Letter from California: Plugged In," *New Yorker*, August 24, 2009.

CHAPTER TWELVE: THE RISING BILLION

The World's Biggest Market

140 *Stuart Hart met Coimbatore Krishnarao Prahalad*: personal interview with Stuart Hart, 2010.

140 *ideas about "core competencies" and "cocreation" sparked a revolution*: C. K. Prahalad and Gary Hamel, "The Core Competencies," *Harvard Business Review*, May-June 1990. And: C. K. Prahalad and Venkat Ramaswamy "Co-Opting Customer Competence," *Harvard Business Review*, January 2000. But for a really good C.K. bio, try: Schumpter, "The Guru at the Bottom of the Pyramid," *Economist*, April 24, 2010.

140 *Prahalad had a reputation for unorthodoxy*: www.businessweek.com/magazine/content/06_04/b3968089.htm.

140 *he wrote his now-seminal "Beyond Greening: Strategies for a Sustainable World"*: Stuart Hart, "Beyond Greening: Strategies for a Sustainable World," *Harvard Business Review*, January 1, 1997. This article won the McKinsey Award for Best Article in the *HBS* in 1997.

141 *" 'The Fortune at the Bottom of the Pyramid' "*: C. K. Prahalad and S. L. Hart, "The Fortune at the Bottom of the Pyramid," *Strategy+Business* 26 (2002), pp. 54–67.

141 *the majority of BoP consumers lived*: For a summary of the argument: C. K. Prahalad, The Fortune at the Bottom of the Pyramid," *Fast Company*, April 13, 2011. See: www.fastcompany.com/1746818/fortune-at-the-bottom-of-the-pyramid-ck-prahalad.

141 *Arvind Mills, for example, the world's fifth-largest denim*: Prahalad and Hartibid.

142 *He opened with a strong statement of purpose*: C. K. Prahalad, *The Fortune at the Bottom of the Pyramid* (Wharton School Publishing, 2005), p. 25.

142 *the telecom Grameenphone*: "Power to the People," *Economist*, March 9, 2006.

142 *adding ten phones per one hundred people*: Nicholas Sullivan, *You Can Hear Me Now: How Microloans and Cell Phones Are Connecting the World's Poor to the Global Economy* (Jossey-Bass, 2007), p. xxxiv.

142 *Nicholas Sullivan*: Ibid.

142 *essentially one of commodification*: *Wired* editor in chief Chris Anderson makes this argument here: http://longtail.typepad.com/the_long_tail/2005/03/long_tail_vs_bo.html.

143 *a hygiene-based marketing campaign for BoP markets in India*: Prahalad, ibid., pp. 207–39.

143 *which kills 660,000 people*: Mindy Murch, Kate Reeder, C. K. Prahalad, "Selling Health: Hindustan Lever Limited and the Soap Market," Department of Corporate Strategy and International Business, University of Michigan, December 12, 2003, p. 2.

143 *1995 book* Capitalism at a Crossroads: Stuart Hart, *Capitalism at a Crossroads: Aligning Business, Earth, and Humanity* (Wharton School Publishing, 2007).

143 *Honda began selling very stripped-down and inexpensive motorized bicycles*: Ibid., p. 121.

143 *he created the Nano*: "The New People's Car," *Economist*, March 26, 2009. Also see: www.businessweek.com/innovate/content/mar2009/id20090318_012120.htm.

143 *the* Financial Times *reported*: David Pilling, "India Hits Bottleneck on Way to Prosperity," *Financial Times*, September 24, 2008.

143 *jump-started an innovation trend*: "A Global Love Affair," *Economist*, November 13, 2008.

144 *This new generation growing up with freedom*: personal interview with Ratan Tata, 2011.

Quadir's Bet

144 *Iqbal Quadir was working as a venture capitalist*: Much of the information in this section is from a personal interview with Iqbal Quadir, 2010. But he tells the founding story of Grameenphone in his Ted Talk: www.ted.com/talks/iqbal_quadir_says_mobiles_fight_poverty.html.

145 *Grameenphone transformed life in Bangladesh*: For the full story, see Sullivan, ibid.

145 *fifteen million new cell phone users were being added*: Ibid., pp. xvii–xx.

145 *2.7 billion people in the developing world without access to financial services*: The World Bank provides a pretty good overview here: web.worldbank.org/WBSITE/EXTERNAL/NEWS/0,,contentMDK:20433592~menuPK:34480~pagePK:64257043~piPK:437376~theSitePK:4607,00.html.

145 *Tanzania, for example, less than 5 percent of the population . . . Ethiopia, there's one bank for every 100,000 people*: "Africa's Mobile Banking Revolution," BBC, August 12, 2009. Available: http://news.bbc.co.uk/2/hi/8194241.stm.

145 *In Uganda (circa 2005), there were 100 ATM machines . . . Opening an account in Cameroon*: Efam Dovi, "Boosting Domestic Savings in Africa," *African Renewal* 22, no. 3 (October 2008), p. 12.

145 *M-banking allows people*: Sullivan, ibid., pp. 125–44.

145 *M-PESA, launched in Kenya in 2007*: See www.thinkm-pesa.com/2011/07/m-pesa-mobile-money-for-unbanked.htm. And for a slightly bigger picture: Alex Perry and Nick Wadhams, "Kenya's Banking Revolution," *Time*, January 21, 2011.

145 *A market that did not exist as of 2007*: "Mobile Payment Market to Almost Triple Value by 2012, Reaching $670 Billion," Juniper Research Limited, July 5, 2011.

146 *incomes of Kenyan households using M-PESA*: "The Power of Mobile Money," *Economist*, September 24, 2009.

146 *information available on everything*: Richard Lester, Paul Ritvo et al., "Effects of a Mobile Phone Short Message Service on Antiretroviral Treatment Adherence in Kenya," *Lancet* 376, no. 9755 (November 23, 2010), pp. 1938–1945.

146 *fishermen can check in advance*: Kevin Sullivan, "For India's Traditional Fisherman, Cellphones Deliver a Sea Change," *Washington Post*, October 15, 2006.

146 *turns an iPhone into a stethoscope*: Amelia Hill, "iPhone Set to Replace the Stethoscope," *Guardian*, August 30, 2010.

146 6,000 *health care apps*: Francesca Lunzer Kritz, "A Guide to Healthcare Apps for Your Smart Phone," *Los Angeles Times*, July 12, 2010.

The Resource Curse

146 the "resource curse": Paul Collier, *The Bottom Billion: Why the Poorest Countries Are Failing and What Can Be Done About It* (Oxford University Press, 2007), pp. 39–44.

146 *economist William Easterly has frequently pointed out*: William Easterly, *White Man's Burden: Why the West's Efforts to Aid Have Done So Much Ill and So Little Good* (Penguin, 2006), p. 11.

147 *Oxford University economist Paul Collier writes in* The Bottom Billion: Collier, ibid., p. 40.

147 *no easy way to break the resource curse*: George Soros, "Transparency Essential to Lifting the 'Resource Curse,'" *Taipei Times*, March 22, 2002.

147 *freelancers the world over are*: Kermit Pattison, "Enlisting a Global Work Force of Freelancers," *New York Times*, June 24, 2009.

147 *makes it harder for individuals or groups to corner resources*: Charles Kenny, "What Resource Curse," *Foreign Policy*, December 6, 2010.

148 *former Harvard business professor Jeffrey Rayport . . . writes*: Jeffrey F. Rayport, "Seven Social Transformations Unleashed by Mobile Devices," *Technology Review*, November 30, 2010.

The World Is My Coffee Shop

148 *the impact of coffeehouses on the Enlightenment culture*: Steven Johnson, *Where Good Ideas Come From: The Natural History of Innovation* (Riverhead Books, 2010), or see his TED talk: www.ted.com/talks/steven_johnson_where_good_ideas_come_from.html.

148 *Malcolm Gladwell explains it this way*: Malcolm Gladwell, "Java Man," *New Yorker*, July 30, 2001. Available on Gladwell's website: www.gladwell.com/2001/2001_07_30_a_java.htm.

148 *In his book* London Coffee Houses, *Bryant Lillywhite explains it . . .* : Bryant Lillywhite, *London Coffee Houses: A Reference Book of Coffee Houses of the Seventeenth, Eighteenth and Nineteenth Centuries* (George Allen and Unwin, 1963).

149 *the more complicated, multilingual, multicultural, wildly diverse the city*: Brand, *The Whole Earth Discipline*, ibid., pp. 25–73.

149 *physicist Geoffrey West found that when a city's population doubles*: Helen Coster, "Physicist Geoffrey West on Solving the Urban Puzzle," *Forbes*, April 11, 2011.

149 *by 2020, nearly 3 billion people*: Carolyn Duffy Marsan, "Analysis: The Internet in 2020," *Network World US*, January 10, 2009.

150 *"Indeed," writes Stuart Hart, "new technologies"*: Stuart Hart and Ted London, *Next Generation Business Strategies for the Base of the Pyramid* (FT Press, 2010), p. 80.

Dematerialization and Demonetization

150 *Bill Joy, cofounder of Sun Microsystems turned venture capitalist*: personal interview with Bill Joy, 2011.

151 *the Android and Apple App stores boasted 250,000 and 425,000 applications, respectively*: As of July 2011, http://en.wikipedia.org/wiki/Android_Market; https://www.mylookout.com/mobile-threat-report.

151 *Stanford economist Paul Romer*: Charles I. Jones and Paul M. Romer, "The New Kaldor Facts: Ideas, Institutions, Population, and Human Capital," National Bureau of Economic Research Working Paper 15094, p. 6. Romer's work is amazing.

152 *this trend, as Stuart Hart explains, will only continue*: *Capitalism at a Crossroads*, ibid., p. 33.

152 *Craigslist, which demonetized advertising*: Chris Anderson, "Free! Why $0.00 Is the Future of Business," *Wired*, February 2, 2008.

PART FIVE: PEAK OF THE PYRAMID

CHAPTER THIRTEEN: ENERGY

Energy Poverty

155 *when humanity first tamed fire*: Steven R. James, "Hominid Use of Fire in the Lower and Middle Pleistocene: A Review of the Evidence." *Current Anthropology* 30 (1989), pp. 1–26. And Nire Alperson-Afil, "Continual Fire-Making by Hominins at Gesher Benot Ya'aqov, Israel," *Quaternary Science Reviews* 27 (2008), 1733–39.

155 *United Nations estimates that one and a half billion people live without electricity*: The Secretary-General's Advisory Group on Energy and Climate Change, *Energy for a Sustainable Future: Report and Recommendations*, United Nations, April 28, 2010, p. 7.

155 *three billion still rely on primitive fuels*: Ibid.

155 *sub-Saharan Africa, the numbers are even higher*: The UNDP/WHO 2009 report, *The Energy Access Situation in Developing Countries, A Review Focusing on the Least Developed Countries and Sub-Saharan Africa*, can be downloaded from www.undp.org/energy.

155 *the United Nations Development Programme warned*: www.undp.org/energy/ and www.unmillenniumproject.org/documents/MP_Energy_Low_Res.pdf.

155 *85 percent of her nation is still ravaged by energy poverty*: Elizabeth Rosenthal, "African Huts Far from the Grid Glow with Renewable Power," *New York Times*, December 24, 2010.

157 *Emem Andrews, a former senior program manager*: personal interview with Emem Andrews, 2010.

157 *the Trans-Mediterranean Renewable Energy Cooperation*: German Aerospace Center, Institute for Technical Thermodynamics, "Trans-Mediterranean Interconnection for Concentrating Solar Power," Federal Ministry for the Environment, Nature Conservation and Nuclear Safety, November 2007.

157 *German Aerospace Center estimates*: Ibid.

157 *David Wheeler*: Vijaya Ramachandran, Alan Gelb, and Manju Kedia Shah, *Africa's Private Sector: What's Wrong with the Business Environment and What to Do About It,* Center for Global Development, 2009. Also see: www.cgdev.org/content/article/detail/1421353.

157 *there is a significant price paid for hauling and safeguarding kerosene*: Wim Naude and Marianna Matthee, "The Significance of Transport Costs in Africa," *UN Policy Brief,* vol. 05/2007, August 2007.

157 *with existing solar options at 20 cents*: personal interview with Bill Joy, 2011.

A Bright Future

158 *Andrew Beebe got out just in time*: "Andrew Beebe: Lesson Learned: Grow Slowly, Conserve Cash, Treat Employees Well," *Bloomberg Businessweek*, July 5, 2005. Available: www.businessweek.com/magazine/content/04_27/b3890407.htm.

159 *Over the past thirty years, the data show that for every cumulative doubling of global PV production . . . now known as Swanson's law*: personal interview with Andrew Beebe, 2011.

159 *the mission of 1366 Technologies, a solar start-up*: Christine Lagorio, "Innovation: Let There Be Light," *Inc.*, October 1, 2010, and www.1366tech.com.

159 *the number of clean-tech patents hit a record high of 379*: "Clean Energy Patent Growth Index," Heslin Rothenberg Farley & Mesiti, June 2010, p. 2.

159 *Scientists at IBM recently announced that they've found a way to replace expensive, rare Earth elements*: Teodor K. Todorov, Kathleen B. Reuter, David B. Mitzi, "High-Efficiency Solar Cell with Earth-Abundant Liquid-Processed Absorber," *Advanced Materials* 22, no. 20 (May 25, 2010), pp. E156-E159.

159 *MIT, meanwhile, using carbon nanotubes to concentrate solar energy*: Jae-Hee Han, Geraldine L. C. Paulus, Ryuichito Maruyama, Daniel A. Heller, Woo-Jae Jim, et al., *Nature Materials* 9 (September 12, 2010), pp. 833–39.

159 *Maryland-based New Energy Technologies*: www.gizmag.com/new-energy-technologies-solar-window/17777.

160 *University of Michigan, physicist Stephen Rand*: Mark Brown, "Light's Magnetic Field Could Make Solar Power Without Solar Cells," *Wired UK*, April 15, 2011.

160 *"thirty percent California tax credit"*: www.gosolarcalifornia.org/consumers/taxcredits.php.

160 *SunShot Initiative*: www.eere.energy.gov/solar/sunshot. Also, Matthew Wald, environmental blogger for the *New York Times*, is excellent in general, see: http://green.blogs.nytimes.com/2011/02/04/from-sputnik-to-sunshot.

160 *wind power is also approaching grid parity*: Grist.com published a thorough review article here: www.grist.org/article/2011–02–07-report-wind-power-now-competitive-with-coal-in-some-regions.

161 *Vestas, one of the world's largest wind energy firms*: Vestas, *Annual Report 2010*, February 2011. Available here: www.vestas.com/Default.aspx?ID=10332&action=3&NewsID=2563.

Synthetic Life to the Rescue

161 *In 2010 Emil Jacobs, ExxonMobil's vice president*: Jad Mouawad, "Exxon to Invest Millions to Make Fuel from Algae," *New York Times*, July 13, 2009.

161 *the older generation of biofuels, primarily corn-based ethanol, was a disaster*: See, for two examples: Joseph Fargione, et al., "Land Clearing and the Biofuel Carbon Debt," *Science* 319, no. 5867 (February 7, 2008), pp. 1235–38; and Timothy Searchinger, et al., "Use of US Cropland for Biofuels Increases Greenhouse Gases," *Science* 310, no. 5867 (February 7, 2008), pp. 1238–40. For a general review (although the author does not seem to understand there's a huge difference between algae-based biofuels and traditional ones), Michael Grunwald, "The Clean Energy Scam," *Time*, March 27, 2008.

161 *US Department of Energy says that algae can produce thirty times more energy*: "A Promising Oil Alternative: Algae Energy," *Washington Post*, January 6, 2008.

161 *tested at several major power plants as a carbon dioxide absorber*: Ibid.

162 *biology's bad boy, Craig Venter*: personal interviews with Craig Venter, 2010 and 2011. Also see: Mouawad, ibid., and "Craig's Twist: Algae Inch Ahead in Race to Produce the Next Generation of Biofuels," *Economist*, July 15, 2009.

162 *Paul Roessler*: personal interview with Paul Roessler, 2011.

163 *today's average of twenty-five miles per gallon and twelve thousand miles driven per year*:

"Emission Facts: Greenhouse Gas Emissions from a Typical Passenger Vehicle," US Environmental Protection Agency, February 2005, EPA420-F-05–004, www.epa.gov/otaq/climate/420f05004.htm.

163 *LS9 has partnered with Chevron*: Michael Kanellos, "Chevron Invests in LS9; Microbe Diesel by 2011?," *Greentech Media,* September 24, 2009.

163 *Amyris Biotechnologies has done the same with Shell*: www.amyris.com/en/newsroom/128-amyris-enters-into-off-take-agreement-with-shell. Also see: Paul Vaosen, "Biofuels Future That US Covets Takes Shape—in Brazil," *New York Times*, June 1, 2011.

163 *The Boeing Company and Air New Zealand*: Candice Lombardi, "Air New Zealand Tests Biofuel Boeing," CNET, January 2, 2009.

163 *Virgin Airlines is already using a partial biofuels*: "Airline in First Biofuel Flight," BBC, February 24, 2008.

163 *Solazyme delivered 1,500 gallons of algae-based biofuels*: Candace Lombardi, "US Navy Buys 20,000 Gallons of Algae Fuel," CNET, September 15, 2010.

163 *DOE is funding three different biofuel institutes*: See: www1.eere.energy.gov/biomass/news_detail.html?news_id=17698; http://greeneconomypost.com/department-of-energy-funding-biofuels-2469.htm; http://techcrunch.com/2011/06/13/doe-biofuels-funding-anti-valley-bias.

163 *Clean Edge, which tracks the growth of renewable energy markets*: Ron Pernick, Clint Wider, et al., *Clean Energy Trends 2011*, Clean Edge, 2011.

163 *Secretary Chu says, production has to be increased a millionfold*: Fiona Harvey, "Second Generation Biofuels—Still Five Years Away?," *Energy Source*, May 29, 2009.

163 *the Joint Center for Artificial Photosynthesis*: http://solarfuelshub.org.

164 *Dr. Harry Atwater, director of the Caltech Center for Sustainable Energy Research*: personal interview with Harry Atwater, 2011.

The Holy Grail of Storage

164 *solar and wind can provide reliable 7x24 baseload power*: The baseload debate is nothing if not loud. For an overview: Lena Hansen and Amory B. Lovins, "Keeping the Lights On While Transforming Electric Utilities," Rocky Mountain Institute, see: www.rmi.org/rmi/Transforming+Electric+Utilities. Also see: "The Coming Baseload Crisis," Thomas Blakeslee, Clearlight Foundation, www.renewableenergyworld.com/rea/news/article/2008/04/the-coming-baseload-power-crisis-52157.

164 *Buckminster Fuller proposed a global energy grid*: This was first proposed by Fuller in 1969, but the idea also shows up in Buckminster Fuller, *Critical Path* (St. Martin's Griffin, 1982), p. 206.

164 *lithium-ion batteries are woefully inadequate*: personal interview with Donald Sadoway, 2011.

165 *Kleiner Perkins Caufield & Byers*: www.kpcb.com.

165 *Bill Joy, formerly of Sun Microsystems and now KPCB's lead green energy partner*: personal interview with Bill Joy, 2011. Also see: Martin LaMonica, "Bill Joy Chases Green-Tech Breakthroughs," CNET, April 6, 2011.

165 *Primus Power . . . energy storage system in Modesto, California*: Eric Wesoff, "Primus Gets $11M from KP and Others for Energy Storage," *Greentech Media*, May 31, 2011.

165 *Aquion Energy*: Monica LaMonica, "Aquion Energy Takes Plunge into Bulk Grid Storage," CNET, July 22, 2011. Also see: www.aquionenergy.com.

165 *MIT professor Donald Sadoway*: personal interview with Donald Sadoway, 2011, but also see: Eric Wseoff. "MIT's Star Prof. Don Sadoway on Innovations in Energy Storage," Greentechmedia.com, March 20, 2011.

165 *Liquid Metal Battery*: Ibid.

Nathan Myhrvold and the Fourth Generation

167 *Nathan Myhrvold likes a good challenge*: personal interview with Nathan Myhrvold, but also see: Malcolm Gladwell, "Annals of Innovation: In the Air," *New Yorker*, May 12, 2008.

167 *with a sum that, as* Fortune *once said*: Nicholas Varchaver, "Who's Afraid of Nathan Myhrvold," *Fortune*, June 26, 2006.

167 *Civilization currently runs on sixteen terawatts of power*: Saul Griffith, "Climate Change Recalculated," talk at the Long Now Foundation, January 16, 2009.

167 *450 parts per million*: Ibid. Also, the "Avoiding Dangerous Climate Change Symposium" (held in Exeter, February 2005) established this number; see a summary of its report here: www.stabilisation2005.com.

167 *we humans dump nearly 26 billion tons of* CO_2 *into the atmosphere*: "Carbon Budget 2009," Global Carbon Project, November 21, 2010. Also see Bill Gates's TED talk: "Energy: Innovating to Zero," available: www.ted.com/talks/bill_gates.html.

167 *increasing global energy production to meet the needs of the rising billion*: "Energy and Climate Change: Facts and Trends to 2050," World Business Council on Sustainable Development, available here: www.wbcsd.org/DocRoot/xxSdHDlXwf1J2J3ql0I6/Basic-Facts-Trends-2050.pdf.

167 *there are plenty who believe that solar will scale and storage will materialize*: See Amory Lovins et al., *Ending the Oil Endgame,* Rocky Mountain Institute, 2005. Or: www.ted.com/talks/amory_lovins_on_winning_the_oil_endgame.html.

167 *Both the George W. Bush administration*: Thor Valdmanis, "Nuclear Power Slides Back onto the Agenda," *USA Today*, September 26, 2004.

167 *the current Obama administration*: Ben Geman, "White House Restates Nuclear Power Support, Committed to 'Learning' from Japanese Crisis," *Hill*, March 13, 2011.

167 *as do serious greens*: *Whole Earth Discipline,* ibid., pp. 75–116.

167 *Tom Blees, author of* Prescription for the Planet: personal interview with Tom Blees, 2009.

168 *Generation I reactors were built in the 1950s and 1960s . . . Generation III is considerably cheaper and safer*: Gwyneth Cravens, *Power to Save the World: The Truth About Nuclear Energy* (Vintage, 2007), pp. 178–80.

168 *fast reactors, which burn at higher temperatures*: "A Technology Roadmap for Generation IV Nuclear Energy Systems," US DOE Nuclear Research Advisory Committee and the Generation IV International Forum, December 2002.

168 *the liquid fluoride thorium reactor*: A great place to start is Richard Martin, "Uranium Is So Last Century—Enter Thorium, the New Green Nuke," *Wired,* December 21, 2009. For the full deep dive: http://energyfromthorium.com.

168 *generation IV technologies are "passively safe"*: Peter Coy, "The Prospect for Safe Nuclear Power," *Bloomberg Businessweek*, March 24, 2011. Also see: http://ecohearth.com/eco-zine/green-issues/391-meltdown-or-mother-lode-the-new-truth-about-nuclear-power.html.

168 *retired Argonne National Laboratory nuclear physicist George Stanford*: personal interview with George Stanford, 2009.

168 *so-called backyard nukes*: Brand, ibid.; Kevin Bullis, "Small Nuclear," *Technology Review*, November 10, 2005.

168 *Nathan Myhrvold's Company TerraPower*: personal interview with Nathan Myhrvold. Also see: Peter Behr, "Futuristic US Power Reactor May Be Developed Overseas," *New York Times*, June 23, 2011. And Robert Guth, "A Window into the Nuclear Future," *Wall Street Journal*, February 28, 2011.

Perfect Power

169 *an intelligent network of power lines, switches, and sensors*: For a general description, try: http://energy.gov/oe/technology-development/smart-grid; In 2009 the Obama administration developed smart grid standards, see: Henry Pulizzi, "Obama Administration Unveils New Set of Smart-Grid Standards," wsj.com, May 18, 2009. Also see: Peter Behr, "Smart Grid Costs Are Massive, but Benefits Will Be Larger, Industry Study Says," *New York Times*, May 25, 2011.

170 *Bob Metcalfe*: personal interview with Bob Metcalfe. For a slightly older profile, see: Scott Kirsner, "The Legend of Bob Metcalfe," *Wired*, November 1998. For his Inventor Hall of Fame bio: http://invent.org/Hall_Of_Fame/353.html. For Metcalfe on the smart grid: Elizabeth Corcoran, "Metcalfe's Power Law," *Forbes*, August 12, 2009.

171 *Cisco . . . has made a huge commitment to build the smart grid*: David Bogoslaw, "Smart Grid's $200 Billion Investment Lures Cisco, ABB," *Bloomberg Businessweek*, September 23, 2010.

171 *Laura Ipsen*: personal interview with Laura Ipsen, 2011.

So What Does Energy Abundance Really Mean?

172 *Travis Bradford, chief operating officer of the Carbon War Room*: Eric Wesoff, "A Lifetime in the Solar Industry: Travis Bradford," Greentechmedia.com, March 30, 2010.

CHAPTER FOURTEEN: EDUCATION

The Hole-in-the-Wall

174 *the Indian physicist Sugata Mitra got interested in education*: For an overview of his work, see Mitra's fairly astounding TED talk: www.ted.com/index.php/talks/sugata_mitra_shows_how_kids_teach_themselves.html. It was Mitra's work that served as the inspiration for the film *Slumdog Millionaire*. See Lucy Tobin, "Slumdog Professor," *Guardian*, March 2, 2009. And if you want to look over Mitra's general data, try: www.hole-in-the-wall.com/Findings.html.

175 *kids, working in small, unsupervised groups, and without any formal training*: For a look at the research: Sugata Mitra, Ritu Dangwal, Shiffon Chatterjee, Swati Jha, Ravinder S. Bisht, and Preeti Kapur (2005), "Acquisition of Computer Literacy on Shared

Public Computers: Children and the 'Hole in the Wall,'" *Australasian Journal of Educational Technology*, 2008, vol. 24, no. 3, pp. 339–54.

176 *Matt Ridley wrote in the* Wall Street Journal: Matt Ridley, "Turning Kids from India's Slums into Autodidacts," *Wall Street Journal,* December 4, 2010.

176 *a new model of primary school education he calls "minimally invasive education"*: www.hole-in-the-wall.com/MIE.html.

One Tablet Per Child

177 *Papert delivered a now-famous paper, "Teaching Children Thinking"*: This originally appeared in the report: *World Conference on Computer Education,* IFIPS, Amsterdam, 1970, but can be found at: www.citejournal.org/articles/v5i3seminal3.pdf.

177 *an architect named Nicholas Negroponte*: Most of this section was based on a personal interview with Nicholas Negroponte, 2011, but for an amazing tour through his world, check Stewart Brand's *The Media Lab: Inventing the Future at MIT* (Penguin, 1998).

177 *the 23 percent of the world's children*: See "The Global Expansion of Primary Education," Charles Kenny, available here: http://charleskenny.blogs.com/weblog/files/the_global_expansion.pdf. Also see: UNESCO's 2011 Global Monitoring Report: www.unesco.org/new/en/education/themes/leading-the-international-agenda/efa report.

178 *One Laptop Per Child*: Negroponte outlines the vision at TED: www.ted.com/talks/nicholas_negroponte_on_one_laptop_per_child.html.

178 *the computer's fabled $100 price tag*: Douglas McGray, "The Laptop Crusade," *Wired,* August 2006.

178 *only two-thirds of American public school students finish*: Tony Wagner, *The Global Achievement Gap: Why Even Our Best Schools Don't Teach the New Survival Skills Our Children Need—and What We Can Do About It* (Basic Books, 2010), p. 114.

178 *writes Tony Wagner*: Ibid.

178 *North American version*: David Pogue, "Laptop with a Mission Widens Its Audience," *New York Times,* October 4, 2007.

Another Brick in the Wall

179 *Standardization was the rule*: Ken Robinson, *Out of Our Minds: Learning to Be Creative* (Capstone, March 2011), pp. 57–58.

179 *as Sir Ken Robinson put it*: Ibid.

179 *killing creativity and squelching talent*: personal interview with Ken Robinson, 2011.

180 *Harvard's Tony Wagner isn't so sure*: Tony Wagner, ibid., p. 92.

180 *Mackinac Center for Public Policy estimates that remediation costs*: Jay P. Greene, "The Cost of Remedial Education," Mackinac Center for Public Policy, August 31, 2000.

180 *the Heritage Foundation observed*: "Education Notebook: The Cost of American Education," Heritage Foundation, September 15, 2006. Available: www.heritage.org/research/education-notebook/education-notebook-the-cost-of-american-education.

181 *the National Governors Association interviewed*: Wagner, ibid., p. 23.

181 *executives from four hundred major corporations were asked*: Wagner, ibid., p. 20.

181 *"twenty-first-century learning"*: The best place to learn about this is the Partnership

for 21st Century Skills, www.p21.org, or James Bellanca, Ron Brandt, *21st Century Skills: Rethinking How Students Learn* (Solution Tree, 2010).

181 *Ellen Kumata*: Wagner, ibid., p. 20.

182 *Schools in America are falling apart*: The 2009 Report Card for America's Infrastructure by the American Society of Civil Engineers gives our public school infrastructure a D grade. See report here (p. 125): www.infrastructurereportcard.org/sites/default/files/RC2009_full_report.pdf.

James Gee Meets Pajama Sam

182 *Dr. James Gee sat down to play*: Most of this section is based on personal interviews with James Gee, 2011, but also see: James Gee, *What Video Games Have to Teach Us About Learning and Literacy* (Palgrave Macmillan, 2003).

183 *academic research includes the phrase:* "The Legend of Zelda: The Windwalker": See James Gee, "The Legend of Zelda and Philosophy," Open Court, August 31, 2008.

183 *games outperform textbooks*: There's a lot to choose from here, but for starters: J. P. Akpan and T. Andre, "Using a Computer Simulation Before Dissection to Help Students Learn Anatomy," *Journal of Computers in Mathematics and Science Teaching* 19, no. 3 (2000), pp. 297–313; M. P. J. Habgood, S. E. Ainsworth, and S. Benford, "Endogenous Fantasy and Learning in Digital Games, *Simulation & Gaming* 36, no. 4 (2005), pp. 483–98; James Gee, "Why Are Video Games Good for Learning?," Available: www.academiccolab.org/resources/documents/MacArthur.pdf.

183 *surgeons and pilots trained on video games*: Robert T. Hays et al., "Flight Simulator Training Effectiveness: A Meta-Analysis," *Military Psychology* 4 (1992). Also see: Verena Dobnik, "Surgeons May Err Less by Playing Video Games," Associated Press, April 7, 2004.

183 *develop planning skills and strategic thinking*: "Video Games Stimulate Learning," BBC.com, March 18, 2002.

183 *Interactive games are great teachers*: Federation of American Scientists, "'Shoot-'em-up' Video Game Increases Teenagers' Science Knowledge," December 8, 2009.

183 *"educators compare game play to the scientific method"*: Hama Yusuf, "Video Games Start to Shape Classroom Curriculum," *Christian Science Monitor,* September 18, 2008.

183 *Jeremiah McCall, a history teacher*: See: http://gamingthepast.net/theory-practice/mccall-simulation-games-as-historical-interpretations.

184 *Lee Sheldon, meanwhile, a professor at the University of Indiana*: Liz Taylor, "Employers: Look to Gaming to Motivate Staff," *ITNews*, March 18, 2010. Also, the website Gaming the Classroom has compiled a full list of links about Sheldon's work here: http://gamingtheclassroom.wordpress.com/2010/03/23/mentions-of-lees-game-design-class/.

184 *Jesse Schell*: Gives an amazing presentation of this topic here: www.g4tv.com/videos/44277/dice-2010-design-outside-the-box-presentation.

184 *new schools like Quest2Learn*: Yusuf, ibid.

184 Popular Science *explains it this way*: Jeremy Hsu, "New York Launches Public School Curriculum Based on Playing Games," *Popular Science*, September 16, 2009.

184 *President Obama said*: www.gamepolitics.com/2011/03/09/president-obama-make-educational-software-compelling-video-games.

The Wrath of Khan

184 *Salman Khan was a successful*: See Khan's TED talk: www.ted.com/talks/salman_khan_let_s_use_video_to_reinvent_education.html.

185 *the Khan Academy . . . became an underground Internet sensation*: Most of this section is based on a personal interview with Shantanu Sinha, Khan Academy president and COO, 2011, but also see: Clive Thompson, "How Khan Academy Is Changing the Rules of Education," *Wired*, July 15, 2011.

185 *Khan Academy has recently partnered with the Los Altos School District*: NPR presented a great *All Things Considered* about the partnership, in June 2011: www.khanacademy.org/video/npr-story-on-ka-los-altos-pilots—june-2011?playlist=Khan%20Academy-Related%20Talks%20and%20Interviews.

186 *John Martinez, a thirteen-year-old from Los Altos*: Anya Kamenetz, "The 100 Most Creative People in Business 2011; Sal Khan: Kahn Academy," *Fast Company*, September 15, 2011.

This Time It's Personal

186 *we also need to change the way progress*: personal interview with James Gee, 2011.

187 *peer-to-peer tutoring networks*: Cathy N. Davidson and David Theor Goldberg, "The Future of Learning Institutions in a Digital Age," *The John D. and Catherine T. MacArthur Foundation Reports on Digital Media and Learning* (MIT Press, 2009).

187 *Apangea Learning's math tutor*: www.apangea.com/results/successStories/successStory_BillArnold_TX.htm.

187 *author Neal Stephenson*: Neal Stephenson, *The Diamond Age* (Spectra, 1996).

188 *Neil Jacobstein*: personal interview with Neil Jacobstein, 2011.

188 *relationship between health and education*: David M. Cutler and Adriana Lleras-Muney, "Education and Health," *National Poverty Center*, Policy Brief No. 9, March 2007.

188 *a well-educated population and a stable, free society*: For a good overview, see Harvard economist Edward L. Glaeser blogging for the *New York Times* here: http://economix.blogs.nytimes.com/2009/11/03/want-a-stronger-democracy-invest-in-education.

188 *two-thirds of them are girls*: See: www.unicef.org/media/media_11986.html. Also see: Nicholas Kristof and Sheryl WuDunn, "The Women's Crusade," *New York Times*, August 17, 2009.

CHAPTER FIFTEEN: HEALTH CARE

Life Span

189 *average life expectancy*: Data are everywhere. For a brief overview: "Mortality," *Encyclopedia Britannica. Encyclopedia Britannica Online*. Encyclopedia Britannica, 2011. Web. September 15, 2011, www.britannica.com/EBchecked/topic/393100/mortality. And for a breakdown of current life expectancies, see the CIA World Factbook: www.cia.gov/library/publications/the-world-factbook/rankorder/2102rank.html.

189 *"Natural selection favors the genes"*: Marvin Minsky, "Will Robots Inherit the Earth?," *Scientific American*, October 1994.

189 *as our living conditions improved*: "Health: A Millennium of Health Improvement," BBC, December 27, 1997.

189 *Socrates a seventy-year-old anomaly*: This is according to Plato, see John Burnett, *Plato: Phaedo*, 1911, p. 12.

190 *During the early sixteen hundreds in England*: W. J. Rorabaugh, Donald T. Critchlow, and Paula C. Baker, *America's Promise: A Concise History of the United States*, Rowman & Littlefield, 2004, p. 47.

190 *industrial revolution that started us*: Clark Nardinelli, "Industrial Revolution and the Standard of Living," *Concise Encyclopedia of Economics* (Liberty Fund), 2008.

190 *the early twentieth century*: Laura B. Shrestha, "Life Expectancy in the United States," *CRS Report for Congress*, August 16, 2006. Also see: www.pbs.org/fmc/timeline/dmortality.htm.

190 *centenarians and supercentenarians*: For starters, here's the validated list of supercentenarians (those over 110 years old): www.grg.org/Adams/E.HTM. Also see "Supercentenarians Around the World," *Christian Science Monitor*, available: www.csmonitor.com/World/2010/0810/Supercentenarians-around-the-world/Italy.

190 *verified age record*: *The Guinness Book of Records*, Guinness World Records, 1999 edition, p. 102.

The Limits of Being Human

191 *And with ten trillion cells in our body*: D. C. Savage, "Microbial Ecology of the Gastrointestinal Tract," *Annual Review of Microbiology* 31 (1977), pp. 107–33.

191 *RAND Corporation report*: Anna-Marie Vilamvska and Annalijn Conklin, "Improving Patient Safety: Addressing Patient Harm Arising from Medical Errors," *Policy Insight* 3, no. 2 (April 2009).

192 *Fifty-seven countries currently don't have enough health care workers*: "More Than a Quarter of the World's Countries Struggling to Provide Basic Health Care Due to Health Worker Shortfalls," World Health Organization, Second Global Forum of Human Resources for Health, Bangkok, Thailand, January 25–29, 2011.

192 *Africa has 2.3 health care workers*: Saraladevi Naicker, Jacob Plange-Rhule, Roger C. Tutt, and John B. Eastwood, "Shortage of Healthcare Workers in Developing Countries—Africa," *Ethnicity & Disease* 19 (spring 2009), p. 1.

192 *the Americas, which have 24.8*: Ibid., p. 2.

192 *Association of American Medical Colleges*: Suzanne Sataline and Shirley Wang, "Medical Schools Can't Keep Up," *Wall Street Journal*, April 12, 2010.

Watson Goes to Medical School

192 *"IBM Watson Vanquishes Human* Jeopardy! *Foes"*: Joab Jackson, "IBM Vanquishes Human *Jeopardy!* Foes," *PCWorld*, February 16, 2011.

192 *Deep Blue had beaten world chess champion*: Bruce Weber, "Swift and Slashing, Computer Topples Kasparov," *New York Times*, May 12, 1997.

192 *Watson had access to 200 million pages*: Bill Hewitt, "Big Data: Big Costs, Big Risks, and Big Opportunity," *Forbes*, May 27, 2011.

193 *to send Watson to medical school*: Collin Berglund, "Watson Artificial Intelligence Being Directed Toward Medicine at UMD," Capital News Service, April 21,

2011. Also see: www.huffingtonpost.com/2011/05/21/ibm-watson-supercomputer-_n_865157.html.

193 *Dr. Herbert Chase*: Jim Fitzgerald, "IBM Watson Delving into Medicine," *USA Today*, May 21, 2011.

193 *Dr. Eliot Siegel*: See: www.youtube.com/watch?v=NByCczOfN4k.

Zero-Cost Diagnostics

194 *Carlos Camara*: Katherine Bourzac, "X-rays Made with Scotch Tape," *Technology Review*, October 10, 2008.

194 *cover of* Nature: Carlos G. Camara, Juan V. Escobar, Jonathan R. Hird, and Seth J. Putterman, "Correlation Between Nanosecond X-ray Flashes and Stick-Slip Friction in Peeling Tape," *Nature* 455 (October 23, 2008), pp. 1089–92.

194 *an episode of* Bones: season 6, episode 16.

194 *Dale Fox to found Tribogenics*: personal interview with Dale Fox, 2010.

195 *George Whitesides*: personal interview with George Whitesides, 2011. Also see: www.ted.com/talks/george_whitesides_toward_a_science_of_simplicity.html.

195 *the edge of Whitesides's paper*: Whitesides, TED, ibid.

195 *Dr. Anita Goel at her company, Nanobiosym*: personal interview with Anita Goel. Also see: www.nanobiosym.com and www.technologyreview.com/tr35/profile.aspx?trid=97.

196 *mChip, a technology out of Columbia University*: Abbie Smith, "'Lab in a Chip' Card to Revolutionize Blood Tests," Healthcareglobal.com, August 1, 2011. See a live demonstration: http://singularityhub.com/2011/08/10/new-lab-on-a-chip-is-an-hiv-test-that-fits-in-your-pocket-video.

196 *the Qualcomm Tricordor X PRIZE*: Paul Jacobs, CEO of Qualcomm, underwrote the cost of developing the Tricorder X PRIZE design. As of this book's publication, X PRIZE and Qualcomm are still in discussions regarding the funding and launch of this competition. The prize is designed to accelerate the technology required to bring about health care abundance.

Paging Dr. da Vinci to the Operating Room

196 *age-related cataracts*: See www.who.int/blindness/causes/priority/en/index1.html.

197 *ORBIS International*: www.orbis.org.

197 *Da Vinci*: personal interview with Catherine Mohr, 2011. Also see: www.intuitivesurgical.com.

197 *at the behest of cardiac surgeons*: Steve Sternberg, "Robot Reinvents Bypass Surgery," *USA Today*, April 30, 2008. Also see: http://spectrum.ieee.org/biomedical/devices/doc-at-a-distance.

197 *prostatectomies and gastric bypasses*: Barnaby J. Feder, "A Medical Robot Makes Headway," *New York Times*, February 12, 2008.

197 *MAKO surgical robot*: Katherine Bourzac, "Robotic Guidance for Knee Surgery," *Technology Review*, March 27, 2008.

Robo Nurse

198 *When the trend peaks in 2030*: James R. Knickman and Emily K. Snell, "The 2030 Problem: Caring for Aging Baby Boomers," *Health and Human Services* 34, no. 4 (August 2002), pp. 849–84.

198 *centenarian population is doubling*: Matthew Sedensky, "Latest US Census Reveals Doubling of Centenarian Population," *Spectator*, April 27, 2011, and www.prcdc.org/300million/The_Aging_of_America.

198 *growth rate of those over eighty*: And all this aging is seriously changing our world; see Steven Heller, "Let the 80s Roll," *Theatlantic.com*, September 1, 2011: www.theatlantic.com/life/archive/2011/09/let-the-80s-roll-in-the-design-world-octogenarians-rule/244452.

198 *In 2050 we'll have 311 million*: www.un.org/esa/population/publications/wpp2002/WPP2002-HIGHLIGHTSrev1.PDF.

198 *National Center for Health Statistics*: See: www.cdc.gov/nchs/nnhs.htm, but also Sandra Block, "Eldercare Shifting Away from Nursing Homes," *USA Today*, February 1, 2008.

198 *Dr. Dan Barry*: personal interview with Dan Barry, 2010, but for his NASA bio: www.jsc.nasa.gov/Bios/htmlbios/barry.html; and for his *Survivor* bio*: www.cbs.com/shows/survivor/cast.*

198 *robots can be applied to the future of health care*: For a great talk by Barry on the future of robotics: http://singularityhub.com/2009/11/18/dan-barry-the-future-of-robotics-singularity-university-video.

199 *"with sexual dysfunction or need"*: In his 2007 book *Love and Sex with Robots*, British chess player and artificial intelligence expert David Levy argues that robots will become significant sexual partners for humans, and the trend is moving in that direction. Jack Scholfield covered the story for the *Guardian* here: www.guardian.co.uk/technology/2009/sep/16/sex-robots-david-levy-loebner.

199 *"I expect the initial robots will cost on the order of a thousand dollars"*: According to Dr. Barry, $1,000 in five years will be the price for a basic home/office telepresence robot with no arms but with reasonable indoor autonomous navigation abilities. Adding arms and making them into robo nurses will increase the price to around $5,000, as it is hard to make robots safe and reliable enough to work in the home environment. They could cost less without liability concerns.

199 *massive scale of production for Microsoft's Xbox Kinect*: personal interview with Dr. Dan Barry, 2011, www.informationweek.com/news/windows/microsoft_news/229300784. Microsoft sold more than ten million Kinect sensor units since launching in November 2010, making it one of the hottest-selling gadgets in tech industry history.

199 *all the other components are on similar price-performance reduction curves*: personal interview with Dr. Dan Barry, 2011.

The Mighty Stem Cell

199 *surgeon Robert Hariri*: personal interview with Robert Hariri, 2011. But for a good Hariri talk on the subject: www.youtube.com/watch?v=eF3IaYyz8js.

200 *Hariri*: The 1984 cult movie classic *The Adventures of Buckaroo Banzai Across the 8th Dimension!* portrayed the efforts of the multitalented Dr. Buckaroo Banzai (Peter

Weller), a physicist, neurosurgeon, pilot, and rock musician, to save the world. I don't know if brain surgeon Robert Hariri can play any musical instruments, but I do know that this military aviator (and Rocket Racing League vice chairman) has developed one of the hottest stem cell companies, which does have the potential to save much of the world from pain and suffering.

200 *stem cells*: For a good overview about stem cells and the future of medicine: Sarah Boseley, "Medical Marvels," *Guardian*, January 29, 2009. Also, the National Institutes of Health (NIH) has a fantastic stem cell database: http://stemcells.nih.gov/info.

200 *Dr. Daniel Kraft*: personal interview with Daniel Kraft, 2010 and 2011. Also see Kraft at TED: www.ted.com/talks/daniel_kraft_invents_a_better_way_to_harvest_bone_marrow.html.

201 *Anthony Atala of Wake Forest University Medical Center*: For starters, it's hard to beat watching Atala print a kidney onstage at TED: www.ted.com/talks/anthony_atala_printing_a_human_kidney.html. Also see: Megan Johnson, "Anthony Atala: Grinding Out New Organs One at a Time," *US News and World Report*, January 30, 2009.

201 *80 percent of patients on the transplant list*: http://optn.transplant.hrsa.gov/data. In fall 2011 the number was 89,807 individuals waiting for kidneys out of a total of 112,264 on the waiting list.

201 *sixteen thousand kidney transplants in*: See http://newsinhealth.nih.gov/issue/mar2011/Feature1.

201 *"induced pluripotent stem cells"*: Kazutoshi Takahashi, et al., "Induction of Pluripotent Stem Cells from Adult Human Fibroblast by Defined Factors," *Cell* (2007). Also see: http://news.sciencemag.org/sciencenow/2007/11/20–01.html and http://www.sciencedaily.com/releases/2011/07/110720115252.htm.

Predictive, Personalized, Preventive, and Participatory

201 *P4 medicine*: For an introduction: Emily Singer, "A Vision for Personalized Medicine," *Technology Review*, March 9, 2010. Also see: www.systemsbiology.org/Intro_to_Systems_Biology/Predictive_Preventive_Personalized_and_Participatory.

202 *$100 million genome*: Emily Singer, "The $100 Genome," *Technology Review*, April 17, 2008.

202 *trillion-dollar sequencing market*: Richard Troyer and Jamie Kiggen, "New Technologies Spur the Race to Affordable Genome Sequencing," *Bernestein Journal*, Fall 2007.

202 *Genetic profiles will be part of standard patient care*: www.mayoclinic.com/health/personalized-medicine/CA00078.

202 *global epidemic: obesity*: Benjamin Caballero, "The Global Epidemic of Obesity: An Overview," *Epidemiologic Reviews* 29, no. 1 (May 13, 2007), pp. 1–15. Also see: www.who.int/nutrition/topics/obesity/en.

202 *genetic culprit here is the fat insulin receptor gene*: www.scientificamerican.com/article.cfm?id=reprogramming-biology.

202 *Harvard researchers used RNAi*: "Ray Kurzweil, Reprogramming Biology," *Scientific American* 295, no. 38 (2006), pp. 706–38.

202 *23andMe and Navigenics*: Amy Harmon, "My Genome, Myself: Seeking Clues in DNA," *New York Times*, November 17, 2007.

203 *Sensors have plummeted in cost, size, and power consumption*: Another 10^9+ SU company, Senstore, is focused on amplifying this trend by empowering the DIY community to build a new generation of low-power, low-cost sensors for health care. See: www.senstore.com.

203 *Thomas Goetz*: personal interview with Thomas Goetz, 2010. Also see: Thomas Goetz, *The Decision Tree: Taking Control of Your Health in the Era of Personalized Medicine* (Rodale, 2010).

203 *tracking everything from sleep cycles, to calories burned, to real-time electrocardiogram signals*: For starters, here's a blog about how to use your iPhone to monitor sleep cycles: http://blog.snoozester.com/2011/06/08/sleep-cycle-turn-your-iphone-into-a-sleep-tracking-device. Also see: Amanda Schaffer, "In Which I Bug Myself," Slate.com, November 7, 2007. Available: www.slate.com/id/2177551. And for heart monitor apps: www.iphoneness.com/iphone-apps/best-heart-rate-monitors-for-iphone.

An Age of Health Care Abundance

203 *entering a period of explosive transformation*: See Daniel Kraft at TED: www.ted.com/talks/daniel_kraft_medicine_s_future.html.

204 *mobile-phone-enabled education programs*: Vital Wave Consulting, "mHealth for Development: The Opportunity of Mobile Technology in the Developing World," United Nations Foundation, Vodafone Foundation, February 2009.

204 *South Africa, uses text messages to broadcast an HIV-awareness bulletin*: Stephanie Busari, "Texts Used to Tackle South Africa HIV Crisis," CNN, December 9, 2008.

204 *Johnson & Johnson's Text4Baby*: Brian Dolan, "White House CTO Officially Launches Text4Baby," Mobihealthnews.com, February 4, 2010.

204 *Bill Gates and his war on malaria*: See his TED talk on the subject: www.ted.com/talks/bill_gates_unplugged.html, Also see: www.gatesfoundation.org/topics/Pages/malaria.aspx.

CHAPTER SIXTEEN: FREEDOM

Power to the People

205 *Nobel laureate Amartya Sen*: For a pretty good summary of Sen's argument, see Harvard economist Richard Cooper's article in *Foreign Affairs* (January–February 2000), available here: www.foreignaffairs.com/articles/55653/richard-n-cooper/the-road-from-serfdom-amartya-sen-argues-that-growth-is-not-enough; or see Sen's *Development as Freedom*.

206 *website Ushahidi*: Megha Baree, "Citizen Voices," *Forbes*, November 20, 2008.

206 *sexual minorities in Namibia*: Denis Nzioka, "Security Initiative for Kenyan LGBTI Launched," Gaykenya.com, March 28, 2011.

206 *potential victims of military abuse*: www.newtactics.org/en/blog/new-tactics/geo-mapping-human-rights#comment-3114.

206 *World Is Witness document stories*: http://blogs.ushmm.org/worldiswitness.

206 *WikiLeaks blow the whistle*: http://wikileaks.org.

206 *Mexican citizens self-police their elections*: www.cuidemoselvoto.org.

206 *$130,000 Enough Is Enough Nigeria*: "'Enough Is Enough Nigeria' Receives Grant from Omidyar Network to Promote Transparency Around the Presidential Elections," PR Newswire, February 22, 2011.

207 *Google executive chairman Eric Schmidt*: personal interview with Eric Schmidt, 2011.

207 *Great Firewall of China*: Oliver August, "The Great Firewall: China's Misguided—and Futile—Attempt to Control What Happens Online," *Wired*, October 23, 2007.

207 *Ben Scott, Secretary of State Hillary Clinton's policy advisor*: As reported by Rosebell Kagumire, a guest blogger for the *Christian Science Monitor*: www.csmonitor.com/World/Africa/Africa-Monitor/2011/0613/Africa-and-the-Internet-a-21st-century-human-rights-issue.

One Million Voices

207 *Jared Cohen decided*: personal interview with Jared Cohen, 2011.

207 Children of Jihad: Jared Cohen, *Children of Jihad: A Young American's Travels Among the Youth of the Middle East* (Gotham, 2007), p. 3.

208 *Two-thirds of Iran*: Ibid. Or see: Caroline Berson, "The Iranian Baby Boom," Slate.com, June 12, 2009.

208 *"the FARC"*: The *New York Times* has a fairly good overview page here: http://topics.nytimes.com/top/reference/timestopics/organizations/r/revolutionary_armed_forces_of_colombia/index.html; the Center for International Policy has another: www.ciponline.org/colombia/infocombat.htm.

208 *FARC controlled 40 percent*: Harvey W. Kushner, *The Encyclopedia of Terrorism* (Sage, 2003), p. 252.

208 *Hostage taking had become so common*: Mark Potter, "Colombian Kidnapping Nightmare," the *Daily Nightly* on msnbc.com, March 28, 2008. Available: http://dailynightly.msnbc.msn.com/_news/2008/03/28/4372333-colombian-kidnapping-nightmare.

209 *a Colombian computer engineer named Oscar Morales*: For Cohen and Morales: Rick Schmitt, "Diplomacy 2.0," *Stanford* magazine, May–June 2010; for a broad overview: Martia Camila Pacrez, "Facebook Brings Protest to Colombia," *New York Times*, Febuary 8, 2008.

209 *mobilized some 12 million people*: For a great overview of the exponential growth of One Million Voices and a great video of Morales telling the story: www.movements.org/case-study/entry/oscar-morales-and-one-million-voices-against-farc.

210 *"twenty-first-century statecraft"*: For a broad overview of the trend: Jesse Lichtenstein, "Digital Diplomacy," *New York Times Magazine*, July 16, 2010. For what the State Department has to say: www.state.gov/statecraft/index.htm.

210 *Secretary Clinton*: www.state.gov/statecraft/index.htm.

210 *shutdown of the Twitter site*: Rick Schmitt, ibid.

Bits Not Bombs

211 *Internet has proved to be a fantastic recruiting tool*: Bob Drogin and Tina Susman, "Internet Making It Easier to Become a Terrorist," *LA Times*, March 11, 2010. Also see this *60 Minutes* report: www.cbsnews.com/stories/2007/03/02/60minutes/main2531546.shtml.

211 *terrorists who sailed from Karachi to Mumbai*: Rhys Blakely, "Google Earth Accused of Aiding Terrorists," London *Sunday Times*, December 9, 2009. Also see: Emily

Wax, "Mumbai Attackers Made Sophisticated Use of Technology," *Washington Post*, December 3, 2008.

211 *Kenya, hateful text messages*: Tim Querengesseri, "Cellphones Spread Kenyans' Messages of Hate," *Globe and Mail*, February 29, 2008.

211 *Cohen left the State Department*: Christina Larson, "State Department Innovator Goes to Google," *Foreign Policy*, September 7, 2010.

211 *"The Digital Disruption"*: Eric Schmidt and Jared Cohen, "The Digital Disruption," *Foreign Policy*, November–December 2010.

212 *the Arab Spring*: *Technology Review* did an overview of the use of technology in the Arab Spring: www.technologyreview.com/ontopic/arabspring; FORATV did a short interview with Jared Cohen about technology and the Arab Spring: www.dailymotion.com/video/xjgxg9_jared-cohen-technology-s-role-in-arab-spring-protests_news. And for a general overview of the Arab Spring: Jack Gladstone, "Understanding the Revolutions of 2011," *Foreign Affairs*, May–June 2011.

212 *one activist summed this up nicely in a tweet*: Philip N. Howard, "The Arab Spring's Cascading Effects," *Miller McCune*, February 23, 2011.

212 *government shut down the Internet . . . Daniel B. Baer*: Mary Beth Sheridan, "Autocratic Regimes Fight Web Savvy Opponents with Their Own Tools," *Washington Post*, May 22, 2011.

212 *Evgeny Morozov*: Evgeny Morozov, *The Dark Side of Internet Freedom: The Net Delusion* (Public Affairs, 2011), pp. 97–98.

213 *Schmidt and Cohen*: "The Digital Disruption," ibid.

PART SIX: STEERING FASTER

CHAPTER SEVENTEEN: DRIVING INNOVATION AND BREAKTHROUGHS

Fear, Curiosity, Greed, and Significance

217 *There are four major motivators*: These four motivators and their relative importance are the personal opinion of the authors. Interestingly, significance—the quest for meaning—turns out to be a much stronger motivator than most suspect. For the long version of that argument, see *Drive* by Daniel Pink; for the short version, see: www.youtube.com/watch?v=u6XAPnuFjJc.

217 *Kennedy's Apollo program*: Monika Gisler and Didier Sornette, "Exuberant Innovation: The Apollo Program," *Springer Science* and *Business Media*, November 25, 2008, available here: www.rieti.go.jp/jp/events/09030501/pdf/5-4_E_Sornette_Paper5_o.pdf.

218 *defense budget*: http://comptroller.defense.gov/defbudget/fy2011/fy2011_budget_request_overview_book.pdf.

218 *to the science budget*: Dan Vergano, "Proposed Budget Cuts Target Science and Research," *USA Today*, March 1, 2011.

The New Spirit of St. Louis

218 *Raymond Orteig grew up a shepherd*: See: www.charleslindbergh.com/plane/orteig.asp.

218 *John Alcock and Arthur Whitten Brown*: See www.century-of-flight.net/Aviation%20history/daredevils/Atlantic%202.htm.

218 *he laid out his plan in a short letter*: Ibid.

219 *Charles W. Clavier and Jacob Islamoff*: *Salt Lake City Tribune* 113, no. 161 (September 22, 1926).

219 *Commander Noel Davis and Lieutenant Stanton H. Wooster*: Charles A. Lindbergh, Reeve Lindbergh, *The Spirit of St. Louis* (Scribner, 2003), p. 119.

219 *French aviators Charles Nungesser and François Coli*: "History of Flight: Checking In on the Missing Persons File," *Air & Space Magazine*, September 1, 2010.

219 *Charles A. Lindbergh*: http://www.charleslindbergh.com.

219 *The Orteig Prize captured the world's attention*: www.charleslindbergh.com/plane/orteig.asp.

219 *Gregg Maryniak*: personal interview with Gregg Maryniak, 2010.

220 *In 1993 it was also Maryniak who gave me a copy of Lindbergh's*: I often give Gregg credit for helping inspire the creation of the X PRIZE, but our friendship goes much deeper. We have known each other since the early 1980s, when he was the executive director of the Space Studies Institute and an advisor to my first organization, Students for the Exploration and Development of Space (SEDS). Maryniak, a trial lawyer by training, is the only counselor I know who can teach both orbital mechanics and lecture on the future of energy. Once the X PRIZE Foundation was started, Gregg joined full time, moving his family from Princeton, New Jersey, to Saint Louis to become my partner and the executive director of the X PRIZE. Much of the foundation's success is owed to him.

220 The Spirit of St. Louis: Lindbergh and Lindbergh, ibid.

220 *Nine teams cumulatively spent $400,000 to try*: Charles A. Lindbergh, *The Spirit of St. Louis* (Scribner, 2003).

220 *I called it the X PRIZE*: After I read *The Spirit of St. Louis* and had the initial idea for the X PRIZE, a number of key people were critical in advising me and helping to found the competition. Some of the first people I turned to for advice and help included Gregg Maryniak, James Burke (who educated me about Paul McCready's efforts on the Kremer Prize), and Bill Gaubatz, who was running the DC-X program at McDonnell Douglas. Two individuals who deserve significant credit as early founders and coconspirators include Dr. Byron K. Lichtenberg, a fellow MIT alumnus and a two-time Space Shuttle payload specialist, and Colette M. Bevis, who had been instrumental in the first-ever serious space tourism business by the Seattle-based Society Expeditions.

220 *I met our ultimate purse benefactors: Anousheh, Hamid, and Amir Ansari*: I first read about Anousheh Ansari in the 2001 issue of *Fortune* magazine's "40 Under 40." In the article, much to my amazement, Anousheh expressed a desire to fly on a suborbital flight into space. I made it my mission to track her down. I found her and her husband, Hamid, vacationing in Hawaii, and was their first meeting when they returned home to Dallas. Byron Lichtenberg and I jointly presented the opportunity, and they quickly offered up the sponsorship. We changed the name of the competition to the Ansari X PRIZE in their honor. Anousheh, born in 1966 in Mashhad, Iran, has cowritten an excellent memoir, *My Dream of Stars* (along with Homer Hickam), spanning the decades from her childhood through her private flight to the International Space Station. Along with Hamid and her brother in-law Amir, Anousheh has been a serial entrepreneur, starting four different telecommunications-related companies. Their third company, Telecom Technologies, developed a software IP-telephony

product that was sold to Sonus Networks in 2000. This sale afforded them the capital to sponsor the X PRIZE. Since then, the three of them have founded Prodea Systems. Anousheh and Amir (who is also a huge space fan and the family's CTO) both sit on the X PRIZE board of trustees.

The Power of Incentive Competitions

221 *Parliament wanted some help crossing the Atlantic*: Dava Sobel, *Longitude: The True Story of a Lone Genius Who Solved the Greatest Scientific Problem of His Time* (Walker and Company, 1995).

221 *Napoléon I offered a 12,000-franc*: Steve Lohr, "Change the World, and Win Fabulous Prizes," *New York Times*, May 21, 2011.

221 *recent McKinsey & Company report*: "And the Winner Is: Capturing the Promise of Philanthropic Prizes," available: www.mckinsey.com/app_media/reports/sso/and_the_winner_is.pdf.

221 *a half dozen companies were formed*: The Ansari X PRIZE stimulated twenty-six teams from seven countries to register. Many of these teams remain active as space companies today. In addition to the registered X PRIZE teams, a significant number of private space companies formed following the publicity, regulatory changes, and capital interests resulting from the Ansari X PRIZE. A listing of all the competing teams is available on the competition wiki, http://en.wikipedia.org/wiki/Ansari_X_Prize.

221 *nearly $1 billion has been invested*: As a direct result of the Ansari X PRIZE, the Virgin Group invested over $100 million in Virgin Galactic. Following that, the Aabar Investments group (of Abu Dhabi) took a 32 percent stake for $280 million (with plans to invest another $100 million). See: www.spacenews.com/venture_space/abu-dhabi-company-invest-virgin-galactic.html. At the same time, the government of New Mexico has invested over $200 million in building out a spaceport. See: http://online.wsj.com/article/SB10001424053111903352704576540690208736946.html. Additional private space companies such as Blue Origin and SpaceX have invested hundreds of millions in private launch capabilities. Finally, there are numerous smaller companies with millions to tens of millions invested, including: Zero Gravity Corporation, Space Adventures, Armadillo Aerospace, Rocket Racing League, XCOR, and Masten, to name a few.

221 *hundreds of millions of dollars' worth of tickets*: This figure includes both hundreds of suborbital tickets sold by Virgin Galactic at $200,000 per person and hundreds of seats sold by Space Adventures for suborbital tickets at $105,000 per person. The figure also includes tickets sold by Space Adventures for orbital flights to the International Space Station. Since 2001, Space Adventures, which I cofounded and for which I serve as vice chairman, has sold eight tickets in published prices ranging from $20 million (to Dennis Tito in 2001) to $35 million (to Gui LaLiberte in 2009). See: www.huffingtonpost.com/2009/09/30/guy-laliberte-billionaire_n_303980.html.

221 *the failure of the BP Deepwater Horizon oil platform created a disaster*: The *New York Times* maintains a file of all its coverage of the spill here: http://topics.nytimes.com/top/reference/timestopics/subjects/o/oil_spills/gulf_of_mexico_2010/index.html; *Mother Jones* has done the same: http://motherjones.com/category/primary-tags/bp.

222 *"flash prize"*: Normally, an X PRIZE take six to nine months to design, fund, and launch. The idea of doing this in a compressed time frame in response to a disaster

such as the Deepwater Horizon explosion was proposed by X PRIZE's newest trustee, James Cameron.

222 *technology used to clean up the BP spill*: Henry Fountain, "Advances in Oil Spill Cleanup Lag Since Valdez," *New York Times*, June 24, 2010. Also see: Eric Nalder, "Decades After Exxon Valdez, Cleanup Technology Still Same," *Houston Chronicle*, May 17, 2010.

222 *Philanthropist Wendy Schmidt*: personal interview with Wendy Schmidt, 2011.

222 *their ability to cast a wide net*: For a look at just ten of the teams entered in the Oil Cleanup Prize, see: Morgan Clendaniel, "The 10 Contenders for X Prize's Latest Challenge: Removing Oil from Water," *Fast Company*, May 26, 2011.

222 *John Harrison*: Sobel, ibid.

The Power of Small Groups (Part II)

223 *anthropologist Margaret Mead*: Widely attributed; for example, *And I Quote: The Definitive Collection of Quotes, Sayings, and Jokes for the Contemporary Speechmaker* (St. Martin's, 1992), edited by Ashton Applewhite, Tripp Evans, and Andrew Frothingham.

223 *Northrop Grumman Lunar Lander X CHALLENGE*: Alan Boyle, "Lunar Lander Contest Cleared for Liftoff," MSNBC.com, May 5, 2006.

223 *Not since the Defense Department's DC-X program*: Jeff Foust, "The Legacy of DC-X," *Space Review*, August 25, 2008.

223 *two teams that ultimately split this purse*: www.nasa.gov/home/hqnews/2009/nov/HQ_09–258-Lunar_Lander.html.

223 *John Carmack*: personal interview with John Carmack, 2010. Also see: www.armadillo aerospace.com/n.x/Armadillo/Home. Also see: Loeonard Davis, "Armadillo Rocket Takes $350,000 Prize," MSNBC.com, October 26, 2008, www.msnbc.msn.com/id/27368176/ns/technology_and_science-space/t/armadillo-rocket-takes-prize/#.TnONNK44ubE.

224 *with the Progressive Insurance Company, the X PRIZE*: See: www.wired.com/autopia/2010/01/auto-x-prize-cruises-into-michigan-for-2010-competition and http://www.wired.com/wiredscience/2008/03/x-prize-rolls-o.

224 *president and vice chairman, Robert K. Weiss*: Robert (Bob) Weiss joined the X PRIZE in 1996 as vice chairman of the foundation and then in 2008 became its full-time president, running all activities and finances. Bob, who is principally responsible for the foundation's growth and success since his full-time engagement, spent the first twenty-five years of his career as a very successful TV and film producer. See: www.imdb.com/name/nm0919154. Having produced twenty motion pictures, he is best known for films such as *The Blues Brothers, Kentucky Fried Movie, The Naked Gun* (sequels), *A Night at the Roxbury, Tommy Boy, The Ladies Man, and Scary Movie 3 and 4.* He also produced some classic TV series, including *Police Squad!, Sliders,* and *Weird Science.*

224 *"Google Lunar X PRIZE"*: www.googlelunarxprize.org. The Google Lunar X PRIZE (or GLXP) was launched in September 2008, with $30 million offered up by Google as the sole sponsor, to any team able to build and launch a robot to the surface of the Moon. The prize was green-lit principally by Sergey Brin and Eric Schmidt, given that Larry Page was a board member of the X PRIZE. Because of the importance of

this technology to NASA, in 2010 the agency announced a complementary program offering up to $30 million in contracts to teams fulfilling the principal objectives of GLXP. See: www.space.com/9343-nasa-spend-30-million-private-moon-data.html. The competition is now being run by Alexandra Hall, the previous CEO of Airship Ventures and the Chabot Space and Science Center.

224 *"ten-million-dollar Archon Genomic X PRIZE"*: See: http://genomics.xprize.org. The Archon Genomics X PRIZE presented by Medco is a $10 million purse funded by philanthropists Stewart and Marilyn Blusson, and support by diagnostics giant Medco, Inc. The competition asks teams to sequence 100 human genomes of health centenarians in under 10 days, for less than a cost of $1,000, with an accuracy better than one error per million basepairs. This is a price-time performance increase of more than 365 millionfold over the work done by Craig Venter in 2001. This X PRIZE is now active and has not yet been claimed.

The Power of Constraints

224 *Dan and Chip Heath*: Dan and Chip Heath, "Get Back in the Box," *Fast Company*, December 1, 2007.

225 *many companies started selling whole genome-sequencing*: Peter Aldhous, "Genome Sequencing Falls to $5,000," *New Scientist*, February 6, 2009.

Fixed-Price Solutions

226 *We've launched six competitions, awarded four of them*: The following have been launched and awarded: Ansari X PRIZE, Progressive Automotive X PRIZE, Northrop Grumman Lunar Lander X CHALLENGE, and Wendy Schmidt Oil Cleanup X CHALLENGE. The following have been launched but not awarded: Archon Genomics X PRIZE and Google Lunar X PRIZE. At the time of this book's publication, three X PRIZEs are in development, any one of which might be launched in early 2012: Qualcomm Tricorder X PRIZE, Autonomous Auto X PRIZE, and Tristate Carbon Capture X PRIZE.

226 *AIDS costs the US government over $20 billion a year*: The federal budget request for fiscal year (FY) 2011 included a total of $20.4 billion for domestic HIV and AIDS, a 4 percent increase from the FY 2010 funding, which totaled $19.6 billion. See: www.avert.org/america.htm#contentTable7.

226 *(to paraphrase computer scientist Alan Kay)*: at a 1971 PARC meeting.

CHAPTER EIGHTEEN: RISK AND FAILURE

The Evolution of a Great Idea

227 *Sir Arthur C. Clarke*: personal interviews with Arthur C. Clarke, 1982, 1987, and 1989. I first met Clarke in Vienna at the United Nations Conference on the Peaceful Uses of Outer Space. Clarke became a friend and advisor to my first organization, SEDS, and later the chancellor to the International Space University (ISU), the university I cofounded with Todd B. Hawley and Robert D. Richards (www.Isunet.edu). These interviews took place during my two visits to Sri Lanka and our many visits in New York and DC with regard to his chancellor role at ISU (www.youtube.com/watch?v=d_VRxkuzIbI). I am very proud to be the winner of the Arthur C. Clarke Award for Innovation (www.clarkefoundation.org/news/031008.php).

227 *Tony Spear was given the job of landing*: personal interview with Tony Spear, 2011. For his official NASA bio: http://marsprogram.jpl.nasa.gov/MPF/bios/team/spear1.html. Tony also worked for me as the program manager of a company called BlastOff!, for which I was CEO between 1999 and 2001. BlastOff! was an Idealab company focused on doing a first private mission to the Moon very similar to what would later become GLXP.

227 *Viking, a complex and expensive mission*: www.nasa.gov/mission_pages/viking.

227 *total development cost of only $150 million*: Mars Pathfinder director's logs, ACE logs, and command request forms collection, 1996–98, JPL 264 available: http://pub-lib.jpl.nasa.gov/docushare/dsweb/Get/Document-1031/JPL264,%20Mars%20Pathfinder%20Director's%20Logs,%20ACE%20Logs,%20and%20Command%20Request%20Forms%20Collection,%20%201996–1998.pdf.

228 *air bags to cushion the initial impact*: for JPL's description of the air bag innovation process: http://mars.jpl.nasa.gov/MPF/mpf/edl/edl1.html. And for NASA's take: www.nasa.gov/centers/glenn/about/history/marspbag.html.

228 *"the administrator took a bold tack"*: "One Marvelous Martian Week," CNN, July 11, 1997.

229 *"Tony Spear was a legendary project manager"*: See: http://mars.jpl.nasa.gov/msp98/news/news68.html.

229 *Burt Rutan puts it, "Revolutionary ideas come from nonsense"*: personal interviews with Rutan 2002–08.

The Upside of Failure

229 *Professor Baba Shiv*: Baba Shiv, "Why Failure Drives Innovation." *Stanford GBS News*, March 2011.

230 *Edison responded, "I have not failed"*: Attributed. But see James Dyson, "No Innovator's Dilemma Here: In Praise of Failure," *Wired*, April 8, 2011.

230 *take the Newton*: Bryan Gardiner, "Learning from Failure: Apple's Most Notorious Flops," *Wired*, January 24, 2008.

230 *as the iPhone*: http://en.wikipedia.org/wiki/IPhone.

230 *Arianna Huffington*: personal interview with Arianna Huffington, 2011, but also see: Arianna Huffington, *On Becoming Fearless* (Little, Brown, 2006).

230 *Sri Ramakrishna*: Joseph Campbell, *A Joseph Campbell Companion: Reflections on the Art of Living* (Harper Perennial, 1995), p. 202.

Born Above the Line of Supercredibility

231 *a group of visionary Saint Louisians*: The first person to suggest bringing the foundation to the arched city was Doug King, who had just taken the role as president of the St. Louis Science Center. Through King, I was introduced to two key civic leaders: Alfred Kerth and Dick Fleming. Kerth, president of Civic Progress and SVP of Fleishman Hillard, deserves much of the credit for our success in fund-raising. He conceived of the New Spirit of St. Louis (NSSL) organization and helped launch the X PRIZE above the line of supercredibility on May 18, 1996. Also critical was

Fleming, who helped introduce Gregg Maryniak, our first executive director, and me, to many of the financial patrons of St. Louis. Families such as McDonnell, Taylor, Danforth, Busch, Maritz, and Holton contributed generously. Marc Arnold, an early New Spirit of St. Louis member, and Ralph Korte, our first member of NSSL, also contributed. All members of NSSL contributed $25,000 each. Some, like author Tom Clancy, contributed as much as $100,000. All funds went to supporting the foundation and its educational mission.

231 *On stage with me were Erik and Morgan Lindbergh*: I first met Erik and Morgan Lindbergh through their aunt Reeve Lindbergh. Erik would go on to become a trustee of the foundation, and in 2002, on the seventy-fifth anniversary of his grandfather's flight, he re-created the now famous San Diego to St. Louis to New York to Paris flight as a fund-raiser to support the foundation.

231 *and twenty veteran NASA astronauts*: Credit for bringing these astronauts together goes to one of the early X PRIZE founders, Dr. Byron K. Lichtenberg, who was also cofounder of the Association of Space Explorers (ASE). ASE was a supporting organization of X PRIZE, and Andy Turnage and Rusty Schweickart (also an ASE cofounder) and Lichtenberg assembled the group, which included Buzz Aldrin and many Mercury, Gemini, Apollo, and Space Shuttle astronauts.

231 *Patti Grace Smith, the associate administrator for spaceflight*: As head of the Office for Commercial Spaceflight, Smith helped create and pass the legislation required for private commercial spaceflight.

Think Different

232 *Apple introduced*: Text only: http://americandigest.org/mt-archives/004924.php; video: http://www.youtube.com/watch?v=4oAB83Z1ydE.

233 *Henry Ford agreed*: Henry Ford, *My Life and Work: An Autobiography of Henry Ford* (Create Space, 2011), p. 66.

233 *engineers who got us to the Moon*: Joe P. Hasler, "Is America's Space Administration Over the Hill? Next-Gen NASA," *Popular Mechanics*, May 26, 2009.

Getting Comfortable with Failure

234 *Intuit*: "How Failure Breeds Success," *Bloomberg Businessweek*, July 10, 2006. Available: www.businessweek.com/magazine/content/06_28/b3992001.htm.

234 *Ratan Tata*: "Out of India," *Economist,* March 3, 2011.

234 *5x5x5 Rapid Innovation Method*: Michael Schrage, "Exploring and Exploiting Experimentation for Enterprise Innovation: A 5X5X5 Approach," *European Financial Review,* April 15, 2011.

CHAPTER NINETEEN: WHICH WAY NEXT?

The Adjacent Possible

236 *theoretical biologist Stuart Kauffman*: Ursula Goodenough, "Emergence into the Adjacent Possible," NPR, January 2, 2010. Also see: http://edge.org/memberbio/stuart_a_kauffman.

236 *author Steven Johnson*: Steven Johnson, "The Genius of Tinkerer," *Wall Street Journal*, September 25, 2010.
237 What Technology Wants: Kelly, ibid., p. 350–51.

The Pursuit of Happiness

237 *Kahneman set aside the question of cognitive biases.* See: D. Kahneman and A. Deaton, *Proceedings of the National Academy of Sciences.* USA advance online publication doi: 10.1073/pnas.1011492107 (2010); Kahneman himself talks about this work in the Q&A after his "The Riddle of Experience Versus Memory" TED talk: www.ted.com/talks/daniel_kahneman_the_riddle_of_experience_vs_memory.html. Also see: David Leonhardt, "Maybe Money Does Buy Happiness After All," *New York Times*, April 16, 2008.
238 *typical American spending breakdown*: The US Department of Labor has a breakdown available here: http://www.creditloan.com/infographics/how-the-average-consumer-spends-their-paycheck.
238 *well-being and money diverge is roughly $10,000*: See Barry Schwartz, "The Paradox of Choice: Why More Is Less," a talk at Google: April 27, 2006. See: http://video.google.com/videoplay?docid=6127548813950043200.
239 *Proverbs*: Proverbs 29:18, King James Bible.

APPENDIX: DANGERS OF THE EXPONENTALS

Why the Future Doesn't Need Us

293 *Bill Joy*: Bill Joy, "Why the Future Doesn't Need Us," *Wired*, April 2000.
294 *Eric Drexler*: Eric Drexler, *Engines of Creation* (Anchor, 1987), p. 172.

Bioterrorism

294 *Andrew Hessel*: personal interviews with Andrew Hessel, 2010 and 2011.
295 *Lord Martin Rees*: John Tierney, "Can Humanity Survive? Want to Bet on It?," *New York Times*, January 30, 2007.
295 *Dr. Larry Brilliant*: Larry Brilliant, "The Age of Pandemics," *Wall Street Journal*, May 2, 2009.
295 *organized crime was the main result*: Mark Thornton, "Alcohol Prohibition Was a Failure," Cato Institute Policy Analysis No. 157, June 17, 1991.
295 *John D. Rockefeller Jr.*: Letter on Prohibition. See Daniel Okrent, *Great Fortune: The Epic of Rockefeller Center* (Viking, 2003), pp. 246–47.
296 *Rob Carlson*: Rob Carlson, "Synthetic Biology 101," see: http://osdir.com/ml/diybio/2010-05/msg00214.html.
297 *UCLA*: Jovana Lara, "UCLA Unveils New Laboratory to Fight Bioterrorism," KABC Los Angeles, May 20, 2011.
297 *Larry Brilliant imagines a scenario*: personal interviews with Larry Brilliant, 2010.
297 *automobile kills about forty thousand Americans*: See: www-fars.nhtsa.dot.gov/Main/index.aspx.
297 *Stan Lee*: Stan Lee, *Amazing Fantasy*, no. 15, August 1962.

Cyber Crime

298 *Marc Goodman*: personal interview with Marc Goodman, 2011.

299 *Connecticut Democrat Richard Blumenthal*: Nick Bilton, "Senator Introduces Online Security Bill," *New York Times*, September 8, 2011.

Robotics, AI, and the Unemployment Line

300 *1862, 90 percent of our workforce*: "Timeline of Farming in the US," PBS: *The American Experience*, see: www.pbs.org/wgbh/amex/trouble/timeline.

300 *1930s, the number was 21*: There are different percentages out there. The more conservative number, 21 percent, comes from: www.agclassroom.org/gan/timeline/1930.htm. In "US Subsidies Help Big Business, but Crush Farmers from Developing Countries," *The Final Call*, November 8, 2002, writers claim 25 percent.

300 *less than 2 percent*: National Institute of Food and Agriculture. See: www.csrees.usda.gov/qlinks/extension.html.

300 *Second Life creator Philip Rosedale*: personal interview with Philip Rosedale, 2011.

301 *Vivek Wadhwa*: personal interview with Vivek Wadhwa, 2011.

301 *Neil Jacobstein*: personal interview with Neil Jacobstein, 2011.

301 *Douglas Rushkoff*: Douglas Rushkoff, "Are Jobs Obsolete?," CNN.com, September 7, 2011.

302 *commonly used definitions of economics . . . complexity economics*: For a great discussion of the entire problem, see Eric D. Beinhocker, *Origin of Wealth: Evolution, Complexity, and the Radical Remaking of Economics* (Harvard Business Press, 2007).

303 *Marvin Minsky*: Marvin Minsky, "Will Robots Inherit the Earth?," *Scientific American*, October 1994.

303 *designing clothing for Second Life avatars*: Rosedale, ibid.

Unstoppable

303 *Bill Joy's suggestion*: Joy, ibid.

304 *Susan Fisher*: Gareth Cook, "US Stem Cell Research Lagging," *Boston Globe*, May 23, 2004.

304 *Matt Ridley*: *The Rational Optimist*, ibid., p. 358.

Acknowledgments

The authors received a lot of great help from a lot of wonderful people along the way. For starters, our wives, Kristen Hladecek Diamandis and Joy Nicholson, without whose love and support this book could never have been written (we're also indebted to Kristen for designing this beautiful book jacket). Our agent, John Brockman, and our editor, Hilary Redmon, were both warriors for this project. We'd also like to thank everyone at Free Press, whose hard work helped bring this vision to fruition. Of course, a deep and special note of appreciation goes to Ray Kurzweil for his inspiration and (in Peter's case) partnership in the creation of Singularity University. We are grateful to the dozens of innovators, philanthropists and thinkers who gave freely of their time to be interviewed for this book.

Incredible feedback along the way was provided by a host of great minds: Carl Bass, Salim Ismail, Dan Barry, Gregg Maryniak, Naveen Jain, Doug Mellinger, Andrew Hessel, Marc Goodman, Kathryn Myronuk, Bob Hariri, Rafe Furst, Tim Ferriss, Chris Anderson, and Neil Jacobstein (we'd also like to thank Neil for suggesting the book's title). Kathryn Myronuk, SU's Knowledge Sommelier, did a great job gathering and editing the data in the reference section. Claire Lin, our creative marketing instigator, coordinated and implemented a world-class marketing campaign with enthusiasm and grace. Connie Fox handled two driven individuals and two impossible schedules and made it all look easy. We are grateful to Mark Fortier for his PR leadership, Joe Diaz for his social media prowess, Jesse Dylan for his cinematic kung fu, and Vj Anma for his assistance on taking the pulse of the public. Thank you to everybody at Singularity University—students, faculty, alumni, and staff—and the X PRIZE Foundation team—for their ideas, enthusiasm, and support. Lastly, the authors want to thank Dezso Molnar, who brought us together over a decade ago.

Index

NOTE: Bold page numbers refer to charts.

About the Authors

PETER H. DIAMANDIS is the chairman and CEO of the X PRIZE Foundation, cofounder and executive chairman of Singularity University, and the founder of more than a dozen space and high-tech companies, including Zero Gravity Corporation, Space Adventures, and the Rocket Racing League. He is also founder of International Space University.

Dr. Diamandis attended MIT, where he received degrees in molecular biology and aerospace engineering, and Harvard Medical School, where he received his medical degree.

He is the winner of dozens of awards, including the Robert H. Heinlein Award, the Neil Armstrong Award, the Arthur C. Clarke Innovation Award, the Economist No Boundaries Innovation Award, and the Charles A. Lindbergh Award. Diamandis's personal motto is "The best way to predict the future is to create it yourself!"

STEVEN KOTLER is a bestselling author, award-winning journalist, cofounder of the Rancho de Chihuahua dog sanctuary (www.ranchodechihuahua.org), and cofounder and director of research of the Flow Genome Project (www.flowgenomeproject.com). His books include the nonfiction works *Abundance, A Small Furry Prayer*, *West of Jesus*, and the novel *The Angle Quickest for Flight*. His articles have appeared in more than sixty publications, including *The New York Times Magazine*, *Wired*, *GQ*, *Outside*, *Popular Science*, and *Discover*. He also writes "The Playing Field," a blog about the science of sport and culture for PsychologyToday.com.

Mr. Kotler attended the Johns Hopkins University, where he received a degree in creative writing, and the University of Wisconsin, Madison, where he received degrees in English and creative writing.